Wide Bandgap Semiconductors

As Moore's law approaches its physical limit, research results beyond Moore's law based on novel materials, structures and devices is providing a new development direction for the semiconductor industry. GaN, SiC and other wide bandgap semiconductor materials are the third generation of semiconductor materials, following silicon, germanium, GaAs and InP. They are the preferred materials for manufacturing solid-state light sources, power electronic devices, and microwave RF devices due to their superior frequency, efficiency, voltage resistance, temperature resistance, and radiation resistance. The study of new materials can meet national strategic requirements for energy conservation and emission reduction, intelligent manufacturing and information security among others. It has become the forefront of global semiconductor technology research and a new industrial focus, and has a significant impact on industrial growth.

The book series "Wide Bandgap Semiconductors" is dedicated to clarifying the fundamental frontiers, key scientific and technological issues of wide bandgap semiconductors. It provides a comprehensive and prospective discussion of the characteristics of new generation semiconductor materials, fundamental physical properties, device fabrication, international R&D trends, key scientific and technological problems encountered, major R&D results achieved, device applications, and development prospects. This series represents the latest findings in fundamental semiconductor theory research and has significant application value.

The book series "Wide Bandgap Semiconductors" brings together the research experience and achievements of more than twenty of China's leading experts in this field. It is a creative exploration of the "synergy between fundamental and applied disciplines". This series fills a gap in the publication of monographs in this field both domestically and internationally, and can serve as a resource for scientific and technical workers engaged in the research and development of wide bandgap semiconductor electronic materials and devices, as well as for enterprise engineers and technicians seeking to strengthen their fundamental knowledge and investigate the problems encountered in practical scientific research.

Ke Xu · Jianfeng Wang · Guoqiang Ren ·
Zongliang Liu

GaN Single Crystal Growth and Application

Springer

Ke Xu
Suzhou Institute of Nano-Tech and Nano-bionics, Chinese Academy of Sciences
Suzhou, Jiangsu, China

Guoqiang Ren
Suzhou Institute of Nano-Tech and Nano-bionics, Chinese Academy of Sciences
Suzhou, Jiangsu, China

Jianfeng Wang
Suzhou Nanowin Science and Technology Co. Ltd.
Suzhou, Jiangsu, China

Zongliang Liu
Jiangsu Institute of Advanced Semiconductors, Co. Ltd.
Suzhou, Jiangsu, China

ISSN 2948-2607 ISSN 2948-2615 (electronic)
Wide Bandgap Semiconductors
ISBN 978-981-96-7571-5 ISBN 978-981-96-7572-2 (eBook)
https://doi.org/10.1007/978-981-96-7572-2

Jointly published with Xidian University Press
The print edition is not for sale in China (Mainland). Customers from China (Mainland) please order the print book from: Xidian University Press.

Translation from the Chinese Simplified language edition: “氮化镓单晶材料生长与应用” by Ke Xu et al., © Xidian University Press 2022. Published by Xidian University Press. All Rights Reserved.

This Springer imprint is published by the registered company Springer Nature Singapore Pte Ltd.
The registered company address is: 152 Beach Road, #21-01/04 Gateway East, Singapore 189721, Singapore

Preface

As one of the typical representative materials of wide band gap semiconductors, gallium nitride material has been developing rapidly in past 40 years, and has shown a very important role in the application fields of novel displays, power electronics, and microwave communications. The wide band gap semiconductor technologies represented by gallium nitride are promoting the evolution of many industry aspects, and are gradually becoming important support for the intelligent society. At present, even there are many publications about nitride semiconductors, most of the related works in this research field are written by foreign researchers, and they emphasize on the introduction of basic theories, and there are still laking systematic reference books for ordinary researchers to quickly grasp the knowledge in the field of gallium nitride single-crystal materials. In order to concisely show the cutting-edge achievements and research progress in this field, this book focuses on summarizing the key technologies and basic laws of the growth and application of gallium nitride single-crystal materials, and the specific process technology is combined with the reader to deepen the understanding of the actual operation. With the improvement of material growth equipment manufacturing process technology, in the past ten years, gallium nitride single crystal material growth technology has been rapid developed, has achieved a dislocation density as low as 10^2 cm^{-2}, substrate wafer size has reached 6 inches (1 inch = 2.54 cm). The doping technology is also developed to provide a variety of conductive, semi-insulating, high-resistance single-crystal materials. At the same time, with the improvement of the quality of single crystal, GaN single crystal materials for the development of new type of devices to provide the possibility and innovative ideas. The study of high-quality GaN single-crystal materials and devices is an evolving discipline, and its related scientific and technical level will be gradually improved and mature, readers should learn and refer to the knowledge of this book dialectically from the perspective of development.

The book is divided into eight chapters.

Chapters 1 and 2 discuss the core physical fundamentals involved in the preparation of gallium nitride single-crystal materials, including single-crystal structural properties, physical and chemical properties, phase diagrams, and growth drivers, which are the core fundamentals of single-crystal material preparation. The data used are from the open literature and do not cover all the fundamental properties of GaN single-crystal materials. We do not intentionally pursue the comprehensiveness, given the author's many years of experience in materials growth, this book only focuses on the key properties and basic laws to analyze and discuss, beginners can be combined with the relevant literature to further expand the relevant knowledge.

Chapters 3–5 respectively on the current mainstream methods of GaN single crystal preparation, namely, hydride vapor phase epitaxial growth method, ammonothermal method and Na-flux method of the basic growth principles, research progress, growth technology core issues are discussed, the content of this part is more suitable for readers familiar with the crystal growth process and have some practical experience to provide reference. Hydride vapor phase epitaxial growth method is currently the only GaN single-crystal material growth to achieve market-scale preparation of growth technology, product-level dislocation density has been as low as 10^5 cm^{-2}, R&D-level dislocation density has been as low as 10^4 cm^{-2}, the maximum wafer size of up to 6 inches, but there are still many problems in the control of stress, production costs and other aspects. ammonothermal method as an emerging preparation technology route, the rapid development in recent years, especially with the Polish Institute of High Pressure Physics, Japan, Mitsubishi Chemical as the representative of the R&D team, has been achieved dislocation density as low as 10^2 cm^{-2}, the wafer size of up to 4 inches of high-quality single-crystal material preparation, but the stability of the mineralizing agent, impurity control, size enlargement, etc. are also facing great challenges. Na-flux method, also as an emerging growth technology, because of its relatively low pressure of ammonothermal method preparation, and easy to realize the preparation of large size, in recent years has been the focus of attention and rapid development, has also realized the dislocation density as low as 10^2 cm^{-2}, the wafer size of up to 6 inches, but in the control of impurities, stress control and other aspects of the problem also exists. These problems faced by the growth technology urgently need to be solved by researchers together. It is also hoped that the researchers can be accommodating, grasp the latest developments in the field and actively cooperate in order to accelerate the development and growth of GaN single-crystal material preparation technology in China. Chapter 6 is a discussion of the basic methods and techniques of homogeneous epitaxial growth based on GaN single-crystal substrates. Epitaxial technology based on gallium nitride single crystal substrate as an important component of nitride growth technology and one of the technical routes, is also becoming one of the mainstream trends in the development of nitride semiconductor technology. This chapter focuses on the epitaxial growth interface control, substrate surface treatment, defect control and other aspects of the discussion. Epitaxial research based on high-quality GaN single-crystal substrates is still in the early stage of research, and there are still many problems that need to be solved, and more interested researchers are needed to devote themselves to it.

Chapters 7 and 8 discuss the progress of GaN single-crystal materials for optoelectronic devices and power electronics applications, including typical applications such as high luminous efficiency LEDs, Micro-LEDs, Schottky barrier diodes, field effect transistors, and high electron mobility transistors, respectively. Although important progress has been made, including lower turn-on resistance, breakdown voltage approaching the theory, and record-breaking mobility, there are still many problems due to process equipment and process methods. However, it is certain that high-quality GaN single-crystal materials will provide an important guarantee to promote the development of high-performance devices.

The authors uphold an open and learning attitude and are committed to discussing the future development of gallium nitride material growth and device applications with their counterparts in the field, hoping that they can present the current status and prospects of the cutting-edge development of gallium nitride material growth and applications to the people in the related fields. At the same time, it is hoped that it can bring readers more thinking and make them generate positive innovation points, which is also one of the original purposes of compiling this book.

This book is written by Dr. Ke Xu, with Chaps. 1 and 2 written by Dr. Xujun Su, Chap. 3 by Dr. Jianfeng Wang, and Chap. 4 by Dr. Guoqiang Ren and Dr. Tengkun Li. Chap. 5 was prepared by Dr. Zongliang Liu and Dr. Zhiwei Si, Chap. 6 was prepared by Dr. Yu Xu. Chapter 7 was prepared by Dr. Miao Wang, and Chap. 8 was prepared by Dr. Yumin Zhang. In addition to the above compilers, De-Min Cai, Mingyue Wang, Xiao-Jian Hu, Lin Shi, and Jicai Zhang participated in the research of gallium nitride materials, and we would like to express our thanks! The research related to this book has been supported by the National Natural Science Foundation of China, the 973 and 863 projects of the Ministry of Science and Technology of China, the Key R&D Program of the Ministry of Science and Technology of China, the Key Projects of the Chinese Academy of Sciences, and the Science and Technology Projects of Jiangsu Province.

In the process of compiling this book, we have received the support and encouragement of many colleagues, especially the director of the editorial committee of this series of books, academician Professor Yue Hao, has given valuable advice on the direction of the compilation, and we would like to express our heartfelt thanks to all of you! At the same time, the publisher's Ting Chen editor also put forward a lot of positive suggestions for revision, and to express my sincere thanks!

In view of the limited level of the authors, the shortcomings in this book are inevitable, and we sincerely hope that the readers will criticize and correct us!

Suzhou, China
December 2024

Ke Xu

About This Book

The growth and application of GaN single-crystal materials are developing rapidly, this book is based on the research work of the author's team over the years, and has been expanded and supplemented to present the latest research results in this field. Specifically, the book is divided into eight chapters detailing the basic principles, technical methods, development status and application trends of the growth of gallium nitride single-crystal materials. The book is systematic, with a complete explanation of specialized knowledge and abundant experimental data. Although the content of this book does not cover all the content of the growth of GaN single crystal materials, but has a typical, concise, forward-looking. This book can be engaged in the growth of GaN single-crystal materials related to research and engineering personnel to read reference, but also can be used as a higher education related professional fields of graduate students and senior undergraduate reference materials.

Contents

Chapter 1
Basic Characteristics of GaN Monocrystals

1.1 GaN Crystal Structure

1.1.1 Wurtzite Structure

Nitride semiconductor materials mainly refer to thin films and monocrystalline materials including gallium nitride (GaN), aluminum nitride (AlN), indium nitride (InN), and their alloys such as InGaN, AlGaN, etc. There are mainly two types of crystal structures for nitride semiconductor materials: wurtzite structure and zinc-blende structure. Only the wurtzite structure is stable under normal temperature and pressure. The wurtzite structure is a hexagonal crystal system, defined by point group 6 mm, space group $P6_3$ mc (186). It consists of two interpenetrating hexagonal close-packed sub-lattices that are shifted along the c-axis direction by 3c/8, as shown in Fig. 1.1. Nitride semiconductor materials belong to AB-type covalent bond (mixed with a certain component of ionic bond) crystals, where nitrogen atoms make hexagonal close packing, and metal atoms fill in the tetrahedral voids formed by N atoms, with a coordination number of 4. From the perspective of crystal stacking, atoms in the hexagonal GaN crystal are arranged in the ...AaBbAaBb... stacking manner along the [0001] direction. The location in hexagonal crystal system is given by a four-index crystal axis system, where the basis vectors are a_1, a_2, a_3, c, as shown in Fig. 1.2. The angle between any two of a_1, a_2, a_3 is 120°, and the c-axis is perpendicular to a_1, a_2, and a_3. The 4-index [u v t w] and (hkil) are more convenient for representing crystal directions and crystal faces, where $u + v = - t$ and $h + k = -i$. Taking gallium nitride as an example, its equilibrium lattice parameters are $a = 0.3189$ nm and $c = 0.5185$ nm, and the unit cell includes four atoms: N1 (0, 0, 3/8); N2 (1/3, 2/3, 7/8); Ga1 (0, 0, 0) and Ga2 (1/3, 2/3, 1/2).

K. Xu et al., *GaN Single Crystal Growth and Application*, Wide Bandgap Semiconductors, https://doi.org/10.1007/978-981-96-7572-2_1

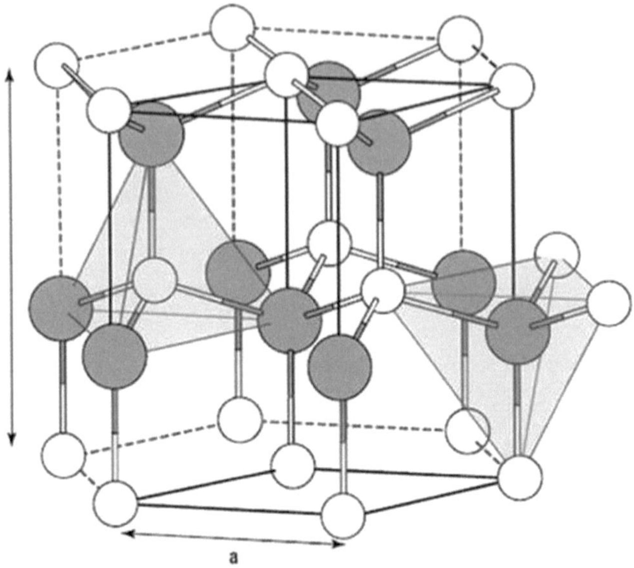

Fig. 1.1 Schematic diagram of GaN wurtzite structure

Fig. 1.2 Polarity of Wurtzite GaN crystals [1]

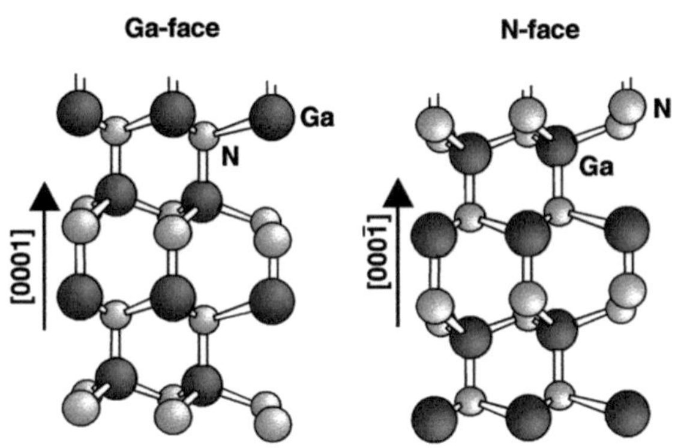

1.1.2 *Polarity*

Wurtzite GaN belongs to non-centrosymmetric crystal, and the crystal has a polar axis, which is the c-axis for wurtzite GaN. Due to the non-coincidence of the positive and negative charge centers in the unit cell, a dipole moment is formed, showing the polarity. This polarization phenomenon that exists without the action of an external electric field is called spontaneous polarization. At the same time, under the action of external stress, the deformation of the wurtzite GaN lattice causes the separation of the positive and negative charge centers, showing the piezoelectric polarization effect. The direction of polarization in the hexagonal GaN crystal is along the $-c$ direction, forming a polarization electric field along the $+c$ direction in the crystal of ~ MV/cm. When the normal direction of the crystal surface is along the + c direction, the crystal is of Ga polarity, otherwise, it is of N polarity, as shown in Fig. 1.2.

The polarization effect can be described by the polarization intensity P. The spatial variation of the polarization intensity induces polarization bound charges σ_{pol}. When a crystal is subjected to in-plane extensive or compressive stress, the direction of resultant piezoelectric polarization is parallel or opposite to the direction of spontaneous polarization, respectively. For self-supporting gallium nitride crystals, theoretically, negative polarization charges $-\sigma_{pol}$ are formed on the Ga-polar surface, while positive polarization charges $+\sigma_{pol}$ are formed on the N-polar surface. Since the polarization bound charges are equal in value and opposite in electricity, an internal electric field is formed between the Ga-polar and N-polar surfaces, and its direction is opposite to the spontaneous polarization field. In fact, many factors like crystal surface atom reconstruction, foreign adsorbed charges, free carriers, etc., have a shielding effect on bound polarization charges, so heterostructures are usually used in experiments to study the polarization effects.

The polarity of hexagonal GaN crystals leads to significant differences in the thermodynamics and kinetics of material growth, physical and chemical properties, such as surface morphology, Schottky barrier height, heterojunction interface band step, band tilt, and reactions with acids and bases, etc.

1.1.3 Band Structure

For wurtzite GaN of the hexagonal crystal system, its first Brillouin zone in the wave vector space is a regular hexagonal prism, as shown in Fig. 1.3a. k_x, k_y, and k_z are the three basis axes of the first Brillouin zone in the Cartesian coordinate system, corresponding to the $[10\underline{1}0]$, $[\underline{1}2\underline{1}0]$, and [0001] directions given by the four-index system of real-space hexagonal coordinates. The symbols in the figure represent a part of high symmetry points and symmetry axes. The coordinates of these points and axes in the wave vector space are shown in Tables 1.1 and 1.2, respectively. Figure 1.3b shows the band structure of wurtzite GaN. The symbols in the figure have the same meaning as the symbols in the first Brillouin zone. Some features are indicated by Fig. 1.3. Firstly, the energetic minimum of the conduction band and the maximum of the valence band of wurtzite GaN are both found at the Γ point, which represents the Brillouin zone center $k = 0$. This feature is known as direct band gap, and the band gap width at room temperature is 3.39 eV. Secondly, there are two additional energy minima in the GaN conduction band, located at the A valley and the M-Γ valley, respectively. The A valley is along the k_z direction, of which the minimum is eV higher than the valence band maximum. The M-Γ valley is in the k_x direction, and its minimum is eV higher than the valence band maximum. Thirdly,

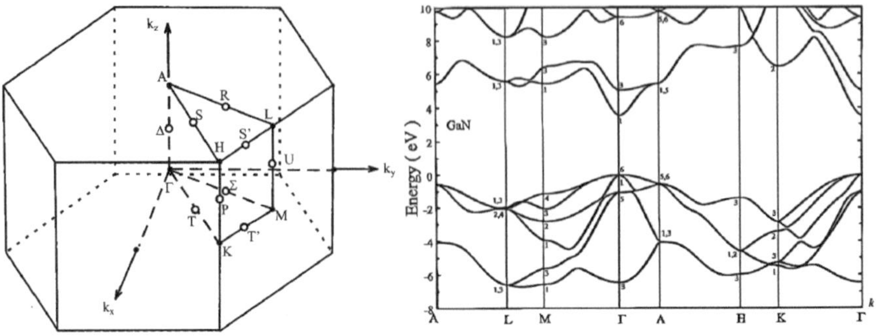

Fig. 1.3 First Brillouin zone and band structure of gallium nitride [6]

the GaN valence band splits into three bands at Γ point: heavy hole band, light hole band, and spin–orbit coupling split band [2–4], as shown in Fig. 1.4. Finally, the GaN valence band is also split by the crystal field [5].

Table 1.1 High symmetry points of the first Brillouin zone

Symmetric point symbol	Coordinate	Note
Γ	$\sqrt{2\pi/a}(0, 0, 0, 0)$	The center of Brillouin
A	$\sqrt{2}\pi/a(0, 0, \sqrt{3}/4)$	The point where the [0001] axis intersects the boundary of the first Brillouin zone
M	$\sqrt{2}\pi/a(1/\sqrt{6}, 1/\sqrt{2}, 0)$	The point where the $[10\bar{1}0]$ axis intersects the boundary of the first Brillouin zone
K	$\sqrt{2\pi/a}(0, \sqrt{2/3}, 0)$	The point where the $[11\bar{2}0]$ axis intersects the boundary of the first Brillouin zone
L	$\sqrt{2}\frac{\pi}{a}\left(\frac{1}{\sqrt{6}}, \frac{1}{\sqrt{2}}, \frac{\sqrt{3}}{4}\right)$	The point where the $[10\bar{1}1]$ axis intersects the boundary of the first Brillouin zone
H	$\sqrt{2}\pi/a(0, \sqrt{2}/3, \sqrt{3}/4)$	The point where the $[11\bar{2}3]$ axis intersects the boundary of the first Brillouin zone

Table 1.2 High symmetry axes of the first Brillouin zone

Symbol	Crystallographic axis	symbol	Crystallographic axis
Δ	Γ–A or [0001]	U	M–L
Σ	Γ–M or $[10\bar{1}0]$	P	K–H
T	Γ–K or $[11\bar{2}0]$	T'	M–K
S	A–H	S'	L–H
R	A–L		

Fig. 1.4 Band structure of gallium nitride [7]

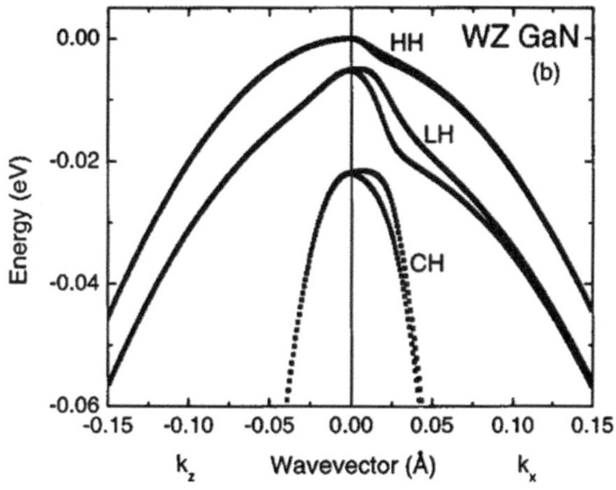

1.2 Crystal Defects in Gallium Nitride

1.2.1 Point Defects

Point defects substantially influence, or even primarily determine the physical and chemical properties of semiconductor materials. Optical and electrical properties of materials are affected by point defects, leading to the reduced luminescence efficiency and device leakage, etc. Common point defects include (1) intrinsic point defects: vacancies, interstitials, and antisites; (2) impurity defects: unintentional doping, intentional doping; (3) complex defects: the complex defects between intrinsic point defects and impurity defects are more complicated. The study of point defects in GaN stands a long history, and there have been many detailed review articles yet [8, 9]. Here, only recent research progresses of the point defects in GaN will be introduced.

1. Intrinsic Point Defects

Intrinsic point defects are a type of thermodynamic equilibrium defect. The formation energies of different point defects are closely related to the Fermi energy level and growth conditions. At the same time, even the point defects belonging to the same type can have different energy level positions due to their different valence states, as shown in Fig. 1.5. Tuomisto et al. [10] used positron annihilation techniques to study point defects in HVPE-GaN grown with different polarities. The results showed that the concentration of Ga vacancies in N-polarity grown GaN was at least two orders of magnitude higher than that in Ga-polarity GaN, that is, the concentration of V_{Ga} was $\leq$ 1E15 cm^{-3}, while the counterpart in N-polarity grown materials reached 1E17–1E18 cm^{-3}. Chichibu et al. [11, 12] applied time-resolved photoluminescence (TR-PL) and positron annihilation (PAM) to study the non-radiative recombination behavior of homoepitaxial GaN materials. The results indicated the presence of complex defect $V_{Ga}(V_N)_n$ (n = 2 or 3) composed of a single Ga vacancy (V_{Ga})

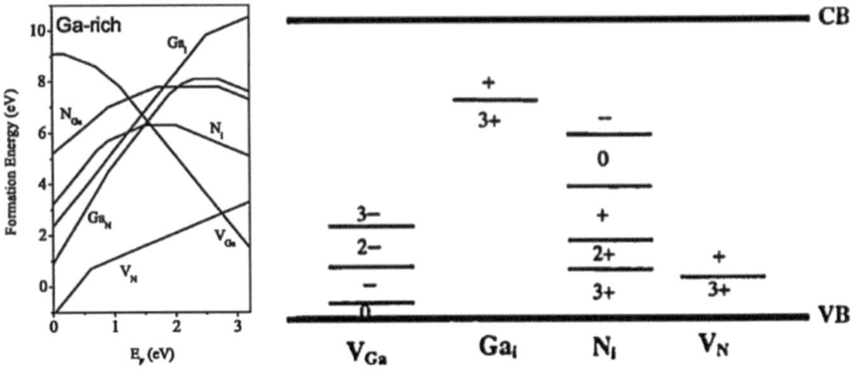

Fig. 1.5 Formation energies of intrinsic point defects [13]. **a** The relationship between the formation energy of intrinsic point defect and the Fermi energy level; **b** Energy levels of intrinsic point defects

and multiple N vacancies (V_N) in p-type GaN. Meantime, in n-type GaN, $V_{Ga}V_N$ was the main intrinsic non-radiative recombination center (NRCs). In ion-implanted Mg-doped GaN films, large vacancy complexes were observed, such as $(V_{Ga})_3(V_N)_3$. These vacancy complex defects were the main non-radiative recombination centers.

2. Unintentional doping

Oxygen and silicon impurities are two common unintentional dopants, which cause n-type conduction as shallow donor impurities. M. A. Reshchikov et al. [14] studied the thermal quenching behavior of the yellow band ~ 2.2 eV in unintentional doped, Si-, C-, Fe-doped GaN grown by MBE and MOCVD. The results showed that the YL bands in different GaN samples were caused by the same defect, namely the YL1 center. The YL1 center was deduced to be C_N, or C_NO_N defects.

3. Intentional doping—conductive type

(1) N-type doping—Si and Ge impurities

High n-type doping is beneficial for the fabrication of high-power electronics and optoelectronic devices, and is also beneficial for current dispersion, and ohmic contact, etc. n-type GaN is usually achieved by Si doping, but there is an upper limit for Si doping. When the doping concentration exceeds 10^{19} cm^{-3}, the surface quality and crystal quality of the material were rapidly degraded, and resultantly the stress accumulation leads to crack failure. Ge doping has the potential to achieve high doping ~ 10^{20} cm^{-3} [15]. Recent research showed that at room temperature, a hole concentration of up to 6.7×10^{20} cm^{-3} in p-type GaN can be achieved through Ge doping, as shown in Fig. 1.6 [16].

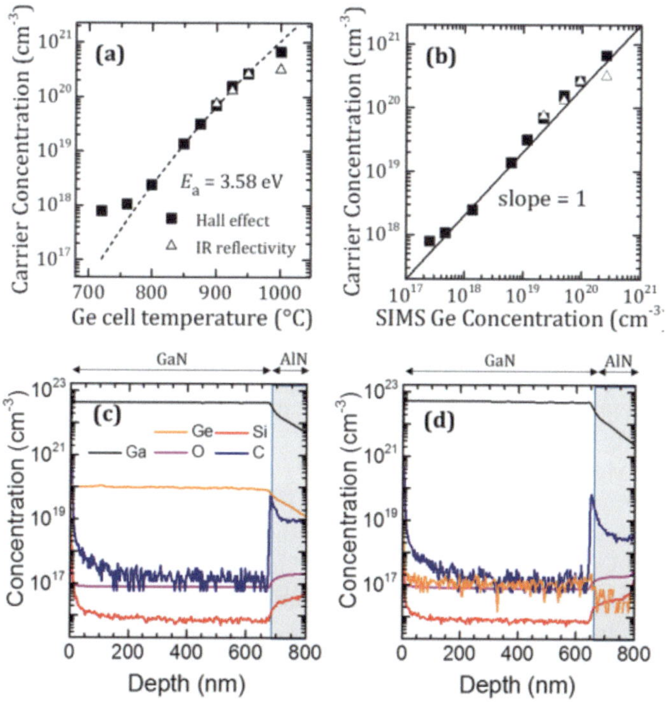

Fig. 1.6 Relationship between Ge doping concentration and hole concentration

(2) P-type Mg impurities

Mg can be doped into GaN through in-situ growth or high-energy ion implantation. Mg doping is the only effective way to achieve P-type GaN, but there are still problems such as low doping efficiency, existence of doping upper limit, and low activation efficiency remained for Mg doping. Mg impurities in the GaN lattice are presented mainly in two positions, substitutional and interstitial, i.e., Mg_{Ga} and Mg_i. The relaxation state and symmetry of the local lattice around Mg impurity will cause completely different electronic structures and luminescent properties. The electronic structure of Mg doping and its influence on the luminescent properties (PL) of GaN materials are still controversial. Substitutional doping not only leads to the breaking of lattice symmetry and the formation of Mg-H complexes, but also results in other intrinsic point defects. In addition, there are unintentional doping O_N and the substitutional impurities Si_{Ga} that is of shallow donor state. Furthermore, the interaction between various defects leads to a more complex luminescence mechanism. Over the past decades, several key theoretical models have been proposed for the physical nature of substitutional Mg_{Ga} acceptor. Lany and Zunger [17] proposed a dual nature model for Mg_{Ga} acceptor, speculating that Mg_{Ga} acceptor has two states: a ground state of deep energy level with localized holes, and a shallow effective mass state in a metastable state, as shown in Fig. 1.7a. For the deep energy level state, the hole is localized in the P orbit of the adjacent N atoms in the (0001) plane. According to the Jahn–Teller symmetry breaking theory, the distance from the N atom where the hole is located to the Mg site is 2.23 angstroms, which is much larger than the other three bonds with about 2.02 angstroms length. For the shallow effective mass state in the metastable state, all N-Mg bond lengths are the same, about 2.05 angstroms. The thermodynamic transition (0/−) energy levels of the localized state and non-localized state are similar, 0.18 and 0.15 eV, respectively. However, the optical transitions through these states are fundamentally different, 2.93 and 3.35 eV, respectively. John L. Lyons and the co-workers [18] found that Mg_{Ga} induced a shallow acceptor level 260 meV above the valence band maximum, resulting in a wide blue light emission peak, and the Mg_{Ga}-H complex was revealed to be associated with the UVL emission peak, as shown in Fig. 1.7b. Mg exists in two states in GaN, shallow acceptor and deep acceptor, and Mg_{Ga} has two different 0/− valence states. For $Mg_{Ga}{}^0$, the hole is strongly localized, that is, the hole is bound at the neighboring N position coaxial with Mg_{Ga}, the local lattice symmetry is broken, and the Mg_{Ga}–N bond is 15% longer than the Ga-N bond, as shown in Fig. 1.7b. $Mg_{Ga}{}^-$ does not cause local lattice symmetry breaking, it is just local relaxation. The transition from the bottom of the conduction band or shallow donor level to $Mg_{Ga}{}^0$ results in a blue light peak. The gap H_i^+ interacts with $Mg_{Ga}{}^-$ to form a stable Mg_{Ga} –H complex, and the optical transition luminescence related to the complex is the cause of UVL. In addition, the Mg_{Ga} –H complex has a stable + 1 valence state when the Fermi level is close to the top of the valence band, and the (+/0) transition has a significant impact on the electrical properties of GaN. If electron irradiation or high-temperature annealing is performed, the Mg_{Ga} –H complex still exists in the material, which not only causes passivation of the holes but even captures free holes to reduce the hole concentration, which may be an important reason for the existence of a concentration limit for the hole concentration. Recently, Sun et al. [19] demonstrated that the neutral acceptor in $Mg_{Ga}{}^0$ exhibits three different defect states. Two of them are small polarons bound by local defects, and the third is an anisotropic delocalized state—in the [11–20] direction it is similar to effective mass, while in other directions it is localized. The transition levels of all three states (0/−) are very similar, at 0.21–0.23 eV, one of the localized states is the hole ground state. Giacomo Miceli et al. [20] further studied that the lowest energy defect state on the surface is a small polaron along the [11–00] direction, where the Mg–N bond length increases to 2.20 A, while the other three Mg-N bond lengths are 2.01 A, resulting in self-limiting of the hole. The second state is along the [0001] direction, 12 meV higher than the first state, and the Mg-N bond extends in this direction to 2.27 A. Finally, the anisotropic delocalized state, 31 meV higher than the first state, the Mg-N bond length along the [11–20] direction is 2.06 A, slightly larger than the other two directions (2.03 A), as shown in Fig. 1.7c. M. A. Reshchikov et al. [21] showed that the hole combined with $Mg_{Ga}{}^0$ is weakly localized, resulting in a UVL emission peak, and the experiment found that the BL peak appears in heavily doped samples.

When achieving P-type doping through high-energy Mg ion implantation, the repair of lattice damage caused by Mg implantation has always been a hot and difficult point of research. Ashutosh Kumar et al. [23] found that Mg implantation can lead to

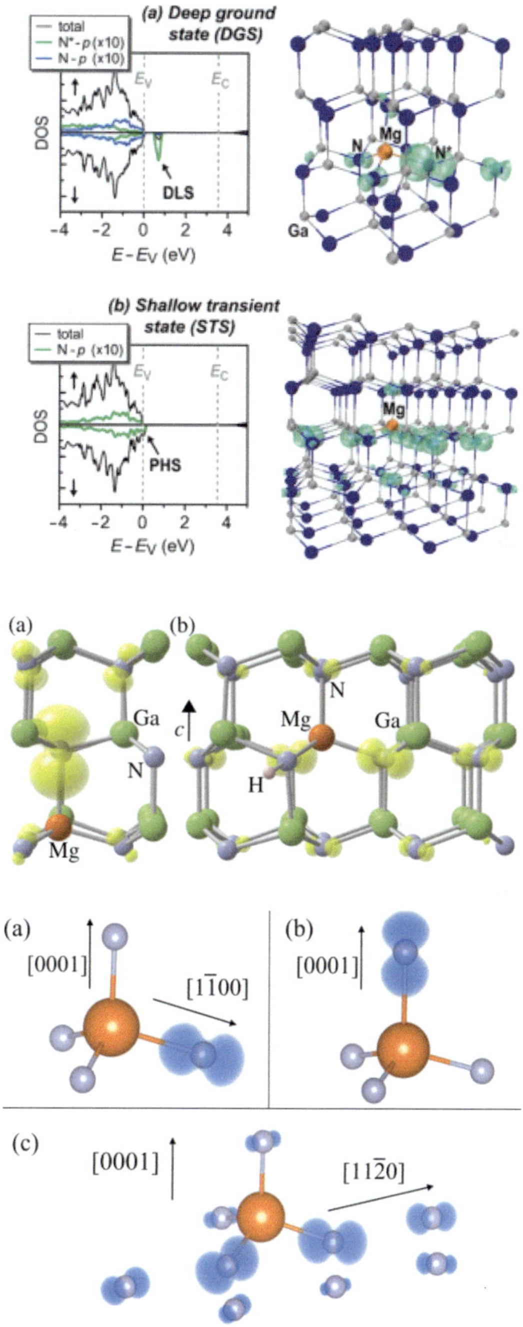

Fig. 1.7 Mg impurity defect structure and defect state. **a** Mg_{Ga} energy state [17], **b** Mg_{Ga} localized state [18], **c** Mg_{Ga}^{0} of three defect states [22]

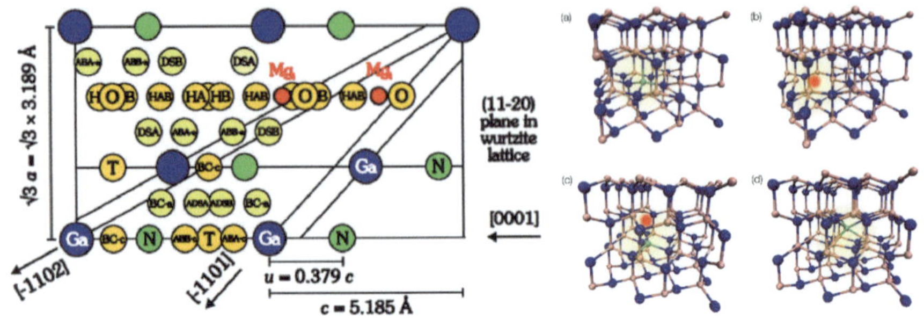

Fig. 1.8 Position of Mg interstitial impurities in GaN. **a** Mg impurity interstitial position [18], **b** proportion in interstitial position [20]

aggregation, stacking faults and other defects. Another problem faced by Mg doping GaN is the doping limit. It was found that when the Mg doping concentration reaches a certain concentration ~ 10^{19} ~ 10^{20} cm^{-3}, the hole concentration of the material will not increase with the further increase of Mg impurities. Wahl et al. [24] theoretically found that this may be related to the existence of Mg in two positions, substitutional Mg_{Ga} and interstitial Mg_{in}, as shown in Fig. 1.8a. Among them, the ratio of Mg concentration in the interstitial position to the substitutional position is closely related to the ion implantation temperature, GaN doping type and implantation process. Giacomo Miceli et al. [20] further found that under p-type conditions, the formation energy of Mg interstitial is significantly lower than that of substitutional, when Mg replaces Ga, it acts as an acceptor, and when Mg is in the interstitial position, it acts as a donor, as shown in Fig. 1.8b. This position-determined self-compensation mechanism may limit the doping limit of Mg.

(3) Semi-insulating Fe doping

Zhang et al. [25] used low-temperature PL and time-resolved PL to study the luminescent characteristics of Fe-doped HVPE-GaN, and the results showed the existence of Fe_{Ga}-V_N complex defects, as shown in Fig. 1.9. Puzyrev et al. [26] used first-principles to calculate the defects related to Fe impurities. The structure shows that the electronic energy levels of Fe-O_N, Fe-C_N, Fe_{Ga}-V_N are E_C-1.5 eV, E_C-0.25 eV and Ec-0.5 eV, but the concentrations of the first two defects are limited by the Fe doping concentration, which are about 1E16~1E17 and 1E17~1E19 cm^{-3} respectively. Among them, the formation energy of Fe_{Ga}-V_N is the lowest, and the theoretical concentration limit is 1E18 cm^{-3}.

(4) C doping

Obtaining semi-insulating GaN through C doping is the mainstream method for developing GaN-based electronic devices. However, as an element of Group IV, C impurities in GaN have amphoteric characteristics. They can replace either N atoms or Ga atoms, or form complexes with other impurities and defects, making the doping mechanism of C in GaN very complex. Determining the lattice position

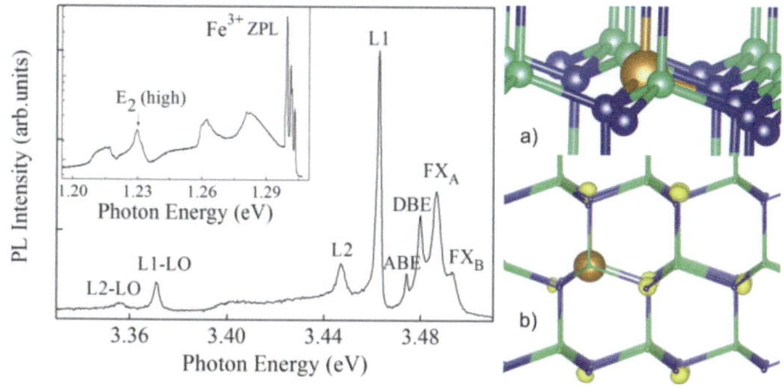

Fig. 1.9 Position of Fe in GaN lattice and defect complex. **a** Energy level of Fe^{3+} in GaN [25]. **b** Schematic diagram of the lattice structure of the Fe_{Ga}-V_N complex defect [27]

of C impurities in GaN is crucial for solving the above problems. Recently, Wu et al. [28] found that C atoms occupy the N position in gallium nitride and carry a -1 charge state, i.e., $C_N{}^{-1}$. For the $C_N{}^{-1}$ defect, after local lattice relaxation, the C–Ga bond along the C axis is 1.9025 Å (less than the N–Ga bond 1.9198 Å), and the other three C–Ga bonds are all 1.8955 Å (less than the N–Ga bond 1.9201 Å), showing the local point symmetry of C_{3v}, as shown in Fig. 1.10a. Xu et al. [29] found that different doping methods can significantly affect the lattice position of carbon (C). Intrinsic doping caused by TMGa source gas replaces the Ga position, annealing treatment can achieve the conversion of C atoms from Ga position to N position, and C doping with propane as the precursor, C atoms are distributed on almost all N sites. K. Irmscher et al. [30] used mid-infrared and ultraviolet absorption spectra to study the lattice position and defect state of C in HVPE-GAN, and found that the defect type of Tri-C exists. Narita et al. [28] used the variable temperature Hall effect to study the behavior of C impurities in MOCVD-grown P-GaN, and found that C_N impurities in the p-GaN layer act as ionized donors, thereby compensating for the carriers. Deep-level transient spectroscopy (DLTS) found two defect states, trap energy levels Hd and Ha at the top of the valence band 0.88 and 0.29 eV, Hd defect energy level originated from the change of C_N defect (0/−1) valence state and Ha defect energy level originated from the change of C_N defect (+1/0) valence state. In summary, C_N has two different charge states, compensating for one electron and one hole on the n-type and p-type gallium nitride layers, respectively, as shown in Fig. 1.10b.

1.2.2 Dislocation

Depending on the different Burgers vectors b, there are three common types of threading dislocations, in GaN, namely a, c, and a + c types, corresponding to edge, screw, and mixed types. The atomic structure of the dislocation core determines

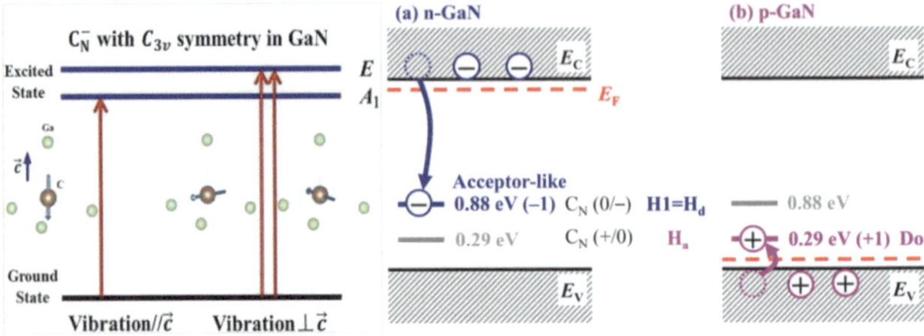

Fig. 1.10 Position and optical properties of C_N. **a** Schematic diagram of the phonon energy level and dipole transition of $C_N{}^{-1}$ [28]. **b** Charge state of C^N in N-type and P-type GAN

the electronic structure, further affecting the local luminescent properties and electrical properties of the material. In recent years, the development of atomic-scale characterization technology, especially aberration-corrected transmission electron microscopy technology, has made it possible to study the atomic structure of the dislocation core at the sub-angstrom scale.

1. Atomic structure of dislocation

Recent theoretical and experimental works have shown that for a-type dislocations in gallium nitride, the most stable core structure includes adjacent 5- and 7-atom rings on the (0001) plane. Growth conditions and doping do not affect the core structure of the dislocation, as shown in Fig. 1.11. Rhode et al. [31] used atomically resolved STEM to find that pure edge-type a has a 5/7 ring atomic structure. Rhode et al. [32] further found that Si doping does not affect the core structure of the dislocation, but Si tends to segregate or cluster at the edge dislocation. The 5/7 ring structure is different from the complete 8-atom core structure in earlier theories [33] and experiments [34].

For screw dislocations, theoretical calculations show that there are two structures: hollow and solid. Elsner et al. [33] used first-principles calculations to demonstrate that the open-core structure of c screw dislocations in GaN is more stable

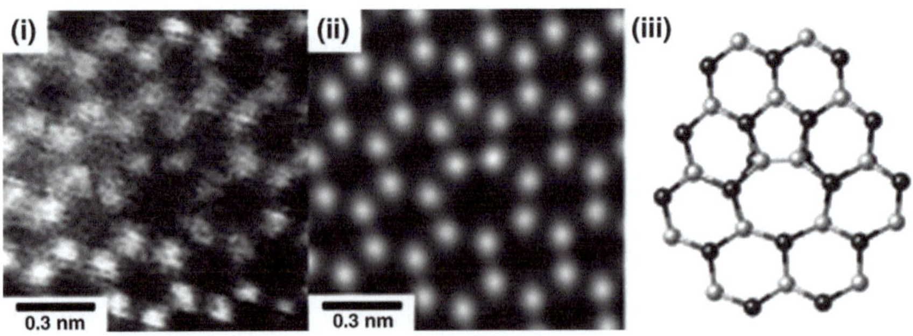

Fig. 1.11 The atomic structure of edge-type penetrating dislocation: HAADF-STEM image, simulated image, and atomic structure model [31]

than the solid structure, with a dislocation line energy of 4.55 eV/Å, as shown in Fig. 1.12a. Because the Ga (N) atoms in the dislocation core adopt a similar sp2 (p3) hybridization, this dislocation core is electrically inactive. Subsequently, Northrup [35] also used first-principles calculations to predict that under Ga-rich conditions, the Ga-filled solid structure model (Ga:N ratio 6:0) is more stable than the open-core model, and under N-rich conditions, the structure model with half of the N atoms removed (Ga:N ratio 6:3) is more stable, as shown in Fig. 1.12b, c. These two models produce electronic states dispersed within the bandgap, making this type of dislocation a highly effective non-radiative recombination center and current leakage channel. Therefore, regardless of the growth conditions of the sample, screw dislocations should be electrically active. Xin et al. [36] experimentally confirmed that pure screw dislocations do indeed have a solid core structure, but did not investigate the stoichiometry of the core. Shigeyoshi Usami et al. [37] experimentally confirmed that 1c pure screw dislocations are associated with the reverse leakage of vertical p-n diodes. These four theoretical models (hollow, solid, Ga-filled, and N half-filled) share a common belief that the dislocation line should be located at the center of the projected hexagon to the base plane, collectively referred to as the "A" structure model. Matsubara et al. [38] used first-principles calculations to propose another screw dislocation structure, believing that the dislocation line is located in the middle of the bond connecting two adjacent hexagons. This structure is collectively referred to as the "B" model, as shown in Fig. 1.12d, e. At the same time, the formation energies of the above four structure types were compared under Ga-rich and N-rich conditions. The results show that this structure originates from the "B" type screw dislocation under N-rich conditions, exhibiting an open-core structure, without introducing deep states in its bandgap, i.e., the screw dislocation in this structure is electrically inactive.

For mixed dislocations, Horton et al. [40] used classical atomic theory models and atomically resolved scanning transmission electron microscopy to show that there are multiple different core structures. Density functional theory calculations indicate that all these structures will introduce localized states in the bandgap, thereby impacting device leakage and efficiency. Deng et al. [31] found that in unintentionally doped GaN, mixed dislocations a + c not only possess a 5/6 atomic ring structure, but also undergo decomposition. Mg doping prevents this decomposition and maintains the 5/6 atomic ring structure. Arslan et al. [41] also observed that a + c type mixed dislocations decompose into two Frank partial dislocations a + c/2 and stacking faults located at $\{11\underline{2}0\}$, as shown in Fig. 1.13a, b. Rhode et al. [32] further found that Si doping does not affect the core structure of the dislocation. Xiong et al. [42] theoretically confirmed that the double 5/6 atomic ring core structure is stable, while the 5/7 atomic ring core structure tends to decompose due to the anisotropy in local atomic stress, as shown in Fig. 1.13c–e.

In summary, it remains unclear whether the electrical characteristics of dislocations in gallium nitride are intrinsic properties of the dislocations, or caused by point defect agglomeration around the dislocations. Different growth methods, growth conditions, doping types, etc., can result in core defect states of the dislocation.

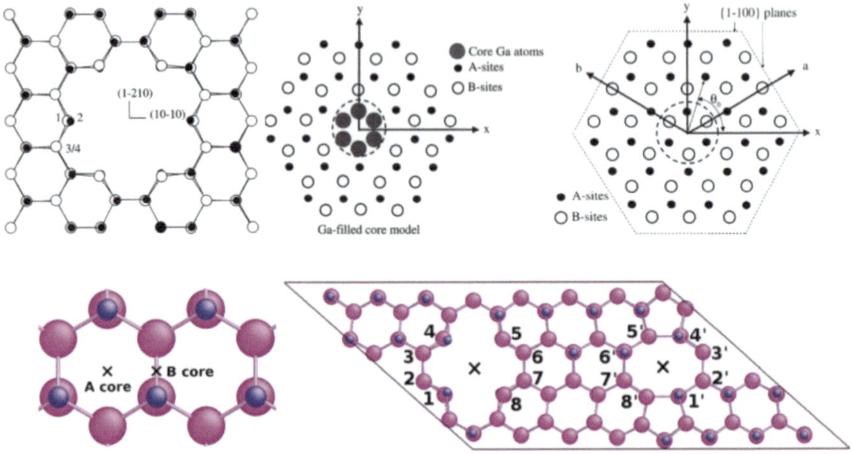

Fig. 1.12 Atomic model of screw-type penetrating dislocation structure. **a** Hollow structure [33]. **b** Under Ga-rich conditions, Ga-filled structure model [35]. **c** Under N-rich conditions, N atom half-filled model [39]. **d** Schematic diagram of the dislocation line at positions A and B. **e** Two hollow structures based on position B [38]

2. Dislocation and Leakage

(1) Dislocation and leakage phenomenon

As a type of line defect, dislocations can cause device leakage and reliability issues. The electrical properties of dislocations depend not only on the structure of their cores, but also on the agglomeration of impurities at the core. Usami et al. [43] applied atomic probe research and found that in Mg-doped GaN, Mg impurities accumulate and diffuse along the dislocation. Rhode et al. theoretically calculated that Si impurities in intentionally doped GaN also tend to agglomerate at edge dislocations [32], while Mg doping also tends to agglomerate at a + c dislocation cores [31]. Similarly, unintentionally doped O will also agglomerate at the dislocation core [44]. Kim et al. [45] applied Conductive Atomic Force Microscopy (C-AFM) to investigate the leakage channels in MOCVD-grown n-GaN, and identified two types of leakage paths: open core dislocations (nanotubes) and pure screw dislocations. Usami et al. [37] used a combination of photoluminescence microscopy, TEM, CL, etc., to study the leakage characteristics of vertical structure p-n diodes. The structure shows that 1c screw dislocations correspond one-to-one with the reverse leakage points of vertical p-n diodes, as shown in Fig. 1.14a. Galiano et al. [46] used Scanning Probe Deep Level Transient Spectroscopy (SP-DLTS) and Electron Channeling Contrast Imaging (ECCI) to investigate the spatial distribution of trap states (located 0.57 eV below the conduction band bottom) in NH_3-MBE grown n-GaN. The results show that the area with a high concentration of E_C-0.57 eV trap states is spatially related to pure edge dislocations in GaN, while it has no correlation with mixed and spiral dislocations, as shown in Fig. 1.14b.

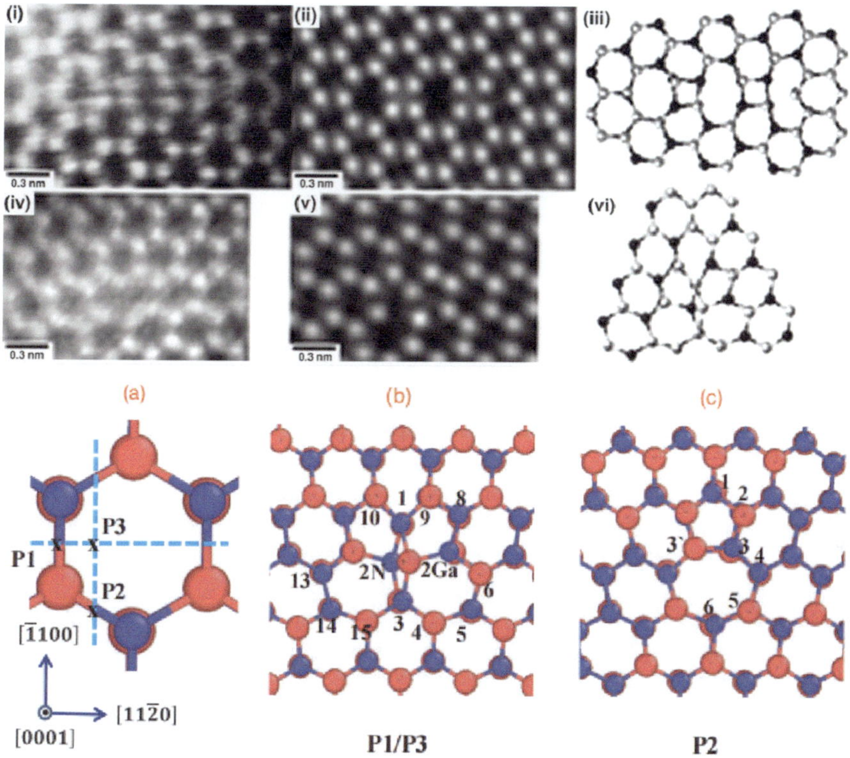

Fig. 1.13 HAADF images, simulated images, and atomic structure models of mixed a + c threading dislocation: **a** decomposed 7/4/8/4/9 structure and **b** stable 5/6 ring structure [31], **c** three possible locations introduced by the mixed dislocation line and **d** double 5/6 ring structure and **e** 5/7 ring structure schematic [42]

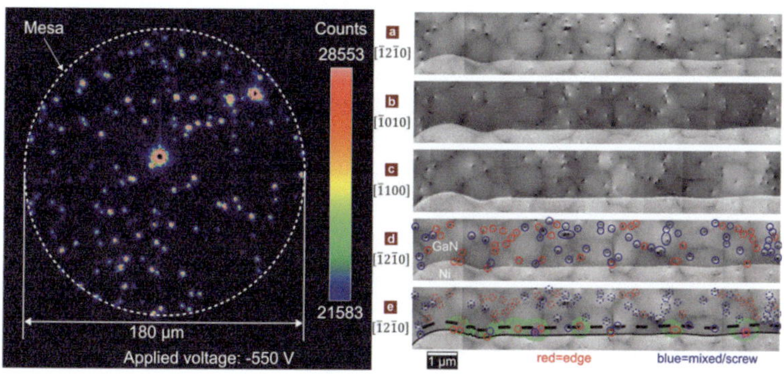

Fig. 1.14 Dislocation and leakage. **a** Light reflection micrograph under − 550 V reverse bias of vertical pn diode [37], **b** ECCI image of dislocations [46]

(2) Dislocation leakage mechanism

The position of the defect state energy level, as well as its relationship with the Fermi energy level and quasi-Fermi energy level, determines the behavior of the capturing or releasing electrons by trap states. Meanwhile, the defect state density and the doping concentration of the material determine the filling state of the defect state. The interaction between dislocations and impurities leads to dislocation-related defect states that can act as deep donors or deep acceptors. It is generally believed that the electronic properties of dislocations in semiconductors can be described by the electron energy band (partially filled) within the forbidden band. Assuming that a partially filled dislocation defect band is located at the middle of the forbidden band, when the occupancy limit E_{dis} (similar to the Fermi energy level of the dislocation band) coincides with the Fermi energy level E_f, the state of the dislocation is electrically neutral. For n-type semiconductors, the empty dislocation energy levels are filled by electrons until the occupancy limit Edis reaches E_f, additional charges are distributed along the dislocation, generating an electrostatic potential Vdis with a maximum value of V_B, causing local bending of the energy band near the dislocation. For p-type semiconductors, positive excess charges accumulate on the dislocation core, causing the band structure to bend in the opposite direction, as shown in Fig. 1.15.

Cherns et al. [48] used electron holography to study the electrostatic potential V_{dis} near a single edge dislocation in n-type GaN (doping concentration [Si] ~ 6E17 cm^{-3}) prepared by the MOCVD method. The study found that the edge dislocation core in n-GaN carries a negative charge, but for p-type GaN [49] (doping concentration [Mg] ~ 1E20 cm^{-3}), the edge dislocation indeed carries a positive charge, further proposing that the Fermi energy level is pinned at the dislocation energy level, which is located 2.5 eV below the conduction band. Cai et al. [50] further studied different types of dislocations in n-GaN, finding that screw, edge, and mixed dislocations all carry negative charges, with charge densities of 1, 0.3, and 0.6 e/c, respectively. Müller et al. [47] further found that in a cylindrical area with a radius of 5 nm around the dislocation core, the negative charge density is between 5E19 and 1E20 cm^{-3}, suggesting that the spatial extension of the charge around the dislocation core is related to point defects near the dislocation core rather than the intrinsic core state of the dislocation. Robertson et al. [51] used finite element simulation to study the dislocation defect state, and the study showed that threading dislocations in GaN

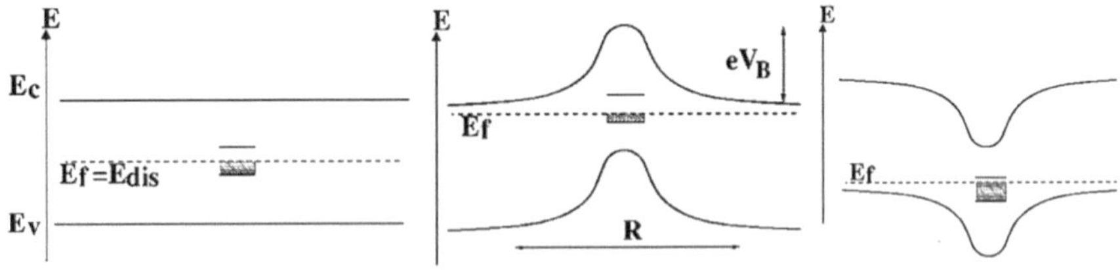

Fig. 1.15 Relationship between dislocation-related defect energy level band and Fermi energy level [47]

introduce a deep acceptor trap state in the forbidden band. The line density is about one electron trap state per unit c-direction lattice length [52, 53].

In addition to the nature intrinsic deep energy level of dislocations, Rackauskas et al. [54] proposed a deep donor energy level model for the impurity band distributed along the dislocation, which may originate from the aggregation of vacancy defects during the growth process, as shown in Fig. 1.16a. During the study of the leakage mechanism of GaN-on-GaN vertical p-n diodes using current and capacitance transient methods, a peak was observed in the current transient that could not be explained by the conventional capture model, As a result, the impurity band leakage mechanism was proposed. Bian et al. [55] studied the leakage mechanism of quasi-vertical GaN Schottky barrier diodes (SBDs). Based on the VRH (variable-range-hopping) mechanism (Fig. 1.16c) [56], a VRH mechanism along the dislocation line was proposed, as shown in Fig. 1.16b. Dislocations can also lead to the formation of V-pit defects in the LED quantum well, which become weak links in device breakdown [57].

2. Dislocation and non-radiative recombination center

Threading dislocations are typically considered as non-radiative recombination centers, causing a reduction in luminescence intensity around them. This decrease

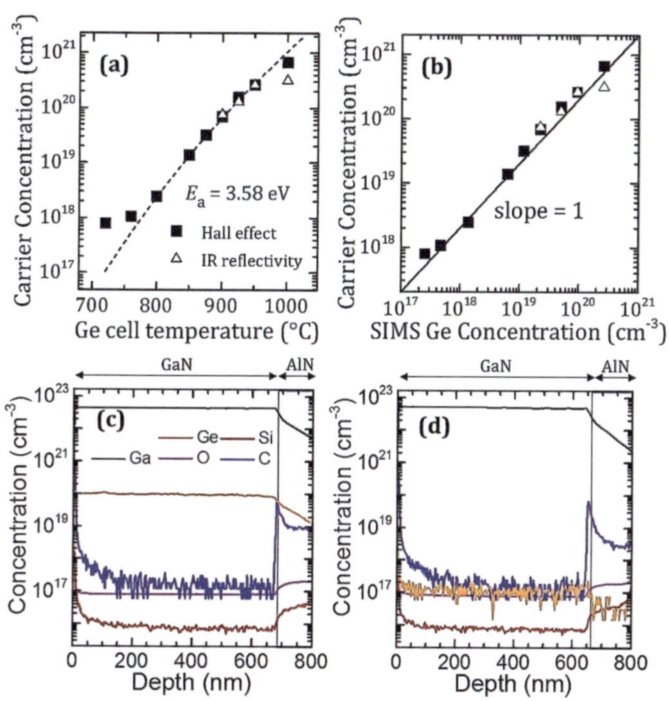

Fig. 1.16 Leakage mechanism of dislocation. **a** Dislocation-related impurity band model [54]. **b** Dislocation-assisted VRH mechanism [55]. **c** Deep energy level assisted VRH and tunneling mechanism [56]

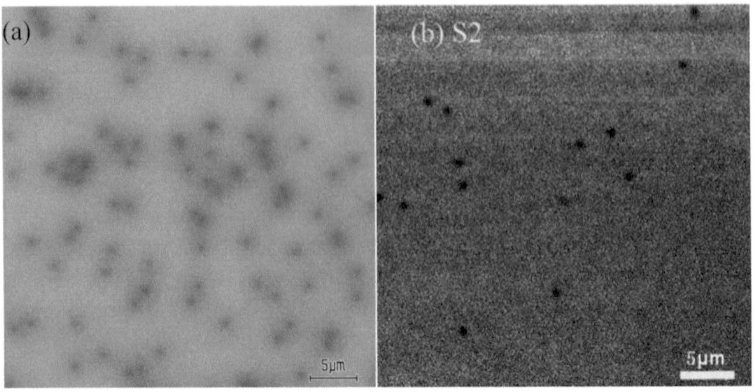

Fig. 1.17 EBIC and CL images of dislocations. **a** EBIC image of HVPE-GaN dislocations [58]. **b** CL image of dislocations [59]

is mainly attributed to the diffusion length of carriers or excitons, resulting in a dark contrast in cathodoluminescence (CL) and electron beam induced current (EBIC) images, as shown in Fig. 1.17.

Therefore, the extension of dark contrast (black spots) around dislocations in CL and EBIC luminescence images is often used to determine the exciton diffusion length. Pauc et al. [60] employed cathodoluminescence (CL) to analyze the CL intensity distribution near a single dislocation core, deriving the diffusion length of excitons and minority carriers. Recently, Kaganer et al. investigated how strain relaxation of the surface exposure of dislocations affects the exciton on lifetime. They proposed that strain relaxation leads to an uneven distribution of piezoelectric polarization charges, creating a bulk electric field around the exposed dislocation and causing excitons to ionize in this field, thereby reducing their lifetime. Theoretical calculations indicate that, even in the absence of exciton diffusion, dislocations in undoped gallium nitride can result in black spots with a diameter of up to 100 nm in CL images. Further theoretical and experimental studies on the drift, diffusion, and recombination of excitons in the edge dislocation area exposed on the GaN{0001} surface reveal that the main reason for the reduction in CL intensity caused by dislocations is primarily due to the exciton decomposition resulting from the piezoelectric field around the edge dislocation, rather exciton diffusion. It is proposed that analyzing the distribution of CL luminescence peak energy around dislocations provides a sensitive method for determining the diffusion length of excitons or carriers in GaN, rather than the intensity distribution [61]. Yakimov et al. [58] used electron beam induced current (EBIC) to investigate the recombination characteristics of dislocations in n-GaN with varying doping concentrations and dislocation densities. The results show that the observed contrast in EBIC images corresponds to charged dislocations surrounded by a space charge region, which is supported by the observed variation of the dislocation recombination strength with doped concentration.

3. Dislocation and Residual Stress

The curvature of GaN single crystals essentially results from the bending of the material lattice parameters. The inconsistency in lattice parameters caused by growth or the presence of defects leads to lattice bending. Current research has found that the crystal curvature is related to the bending of threading dislocations, but there are two explanations for the mechanism of curvature caused by dislocation bending. Yamane et al. [62] quantitatively studied the lattice constant of HVPE-GaN single crystals as the film thickness changed. The results showed that the curvature radius and lattice constant of the upper surface were basically consistent with those of the lower surface. Furthermore, it was found that the bending of threading dislocations introduced a component within (0001) plane, proposing that the dislocation bending is primarily responsible for the observed lattice curvature phenomenon, as shown in Fig. 1.18a. Foronda et al. [63] investigated the bending of the (0001) gallium nitride substrate, and found a gradient from compressive stress to tensile stress from bottom to top in the substrate. The crystal curvature is consistent with the stress gradient in the substrate, and both originate from the tilt of the edge dislocation, i.e., the dislocation line deviates from the [0001] direction. It is generally believed that different tilt directions of the dislocation half-atomic plane will lead to tensile and compressive stress, as shown in Fig. 1.18b.

The mechanism of dislocation bending is commonly believed to involve primarily tilting through climbing for threading dislocations. However, the growth temperature does not satisfy the climb movement involving bulk vacancy defects. Therefore, Follstaedt et al. [65] proposed a mechanism of surface-modulated climb, suggesting that dislocations climb by capturing vacancies on the surface, as illustrated in Fig. 1.19. J. Weinrich et al. used STEM to study the position and tilt direction of extra atomic planes when edge dislocations and mixed dislocations bend. The structure shows that Si doping causes a shortening of the half-atomic plane associated with dislocation, and tensile stress is observed in Si-doped GaN. In addition, for non-doped HVPE-GaN, the direction of dislocation tilt is $\langle 11\bar{2}0\rangle$, while in Si-GaN, there exists another tilt direction in along the $\langle 01\bar{1}0\rangle$ direction. Rohith Soman et al. [67] found that in gallium nitride thin films, Mg doping induces similar effects as Si doping, including dislocation bending and the generation of tensile stress. Although Si atoms have a smaller radius compared to Ga atoms and Mg atoms have a larger radius, both types of doping result in tensile stress and dislocation tilt, indicating that the size effect of atoms can be ruled out. Therefore, it can be speculated that the climb of dislocations is the main cause of tensile stress in thin films. Si and Mg doping promote the climb of dislocations, thereby introducing tensile stress. However, dislocation climbing can also be achieved by the diffusion of core atoms outward, and further research is needed to understand how doping atoms promote dislocation climbing.

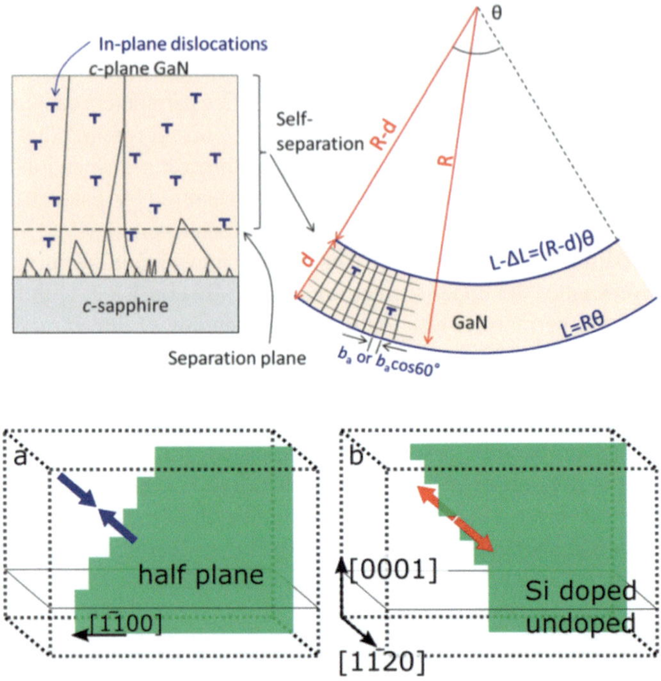

Fig. 1.18 Dislocation and curvature. **a** Crystal curvature caused by dislocation bending [63]. **b** Schematic diagram of stress introduced in the bending direction of dislocation: compressive and tensile stress cause the dislocation half-atomic plane to shorten or elongate [64]

1.2.3 Stacking Faults

Dislocations are the main type of defects in the (0001) plane growth of GaN, while stacking faults are the very important type of defects in semi-polar and non-polar planes of GaN. According to the crystal planes where the stacking faults are located, they can be divided into basal stacking faults (BSFs) and prismatic stacking faults (PSFs). In experiments, PSFs often appear along with BSFs, as shown in Fig. 1.20.

For basal stacking faults (BSFs), there are three common types of stacking faults according to the different stacking order of Ga-N atoms along the [0001] direction: I_1 type, I_2 type, and E type stacking faults, with the stacking order as follows [68]:

$$I_1 \text{ - type: } \cdots ABABA : CACACA \cdots$$
$$I_2 \text{ - type: } \cdots ABABAB : CACACA \cdots$$
$$E \text{ - type: } \cdots ABABA : C : BABABA$$

For PSFs, Hu et al. [69] used atomic-resolution STEM to verify that its atomic structure is similar to the relaxed Drum structure model, as shown in Fig. 1.21.

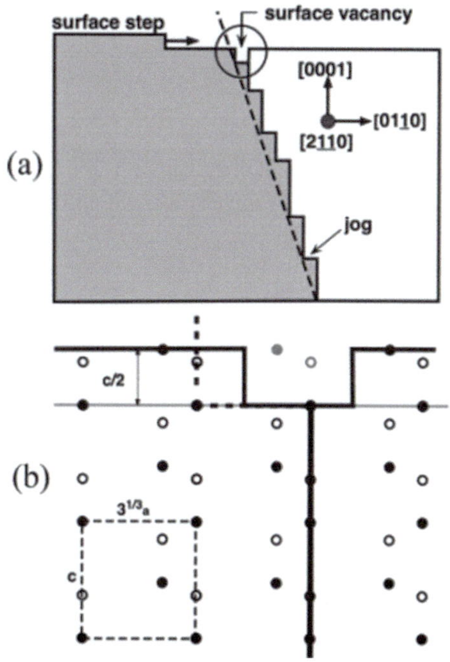

Fig. 1.19 Dislocation climb model (**a**), surface-modulated climb mechanism (**b**) [65]

Fig. 1.20 High-resolution transmission electron microscopy lattice image of stacking faults in GaN [66]

Numerous studies have shown that the generation of stacking faults in semi-polar and non-polar GaN is mainly related to the interaction between the N-polar face (0001) facet and the substrate interface. Song et al. [70] eliminated this N-polar facet by controlling the growth dynamics, thereby achieving the preparation of stacking fault-free semi-polar (20–21) substrates.

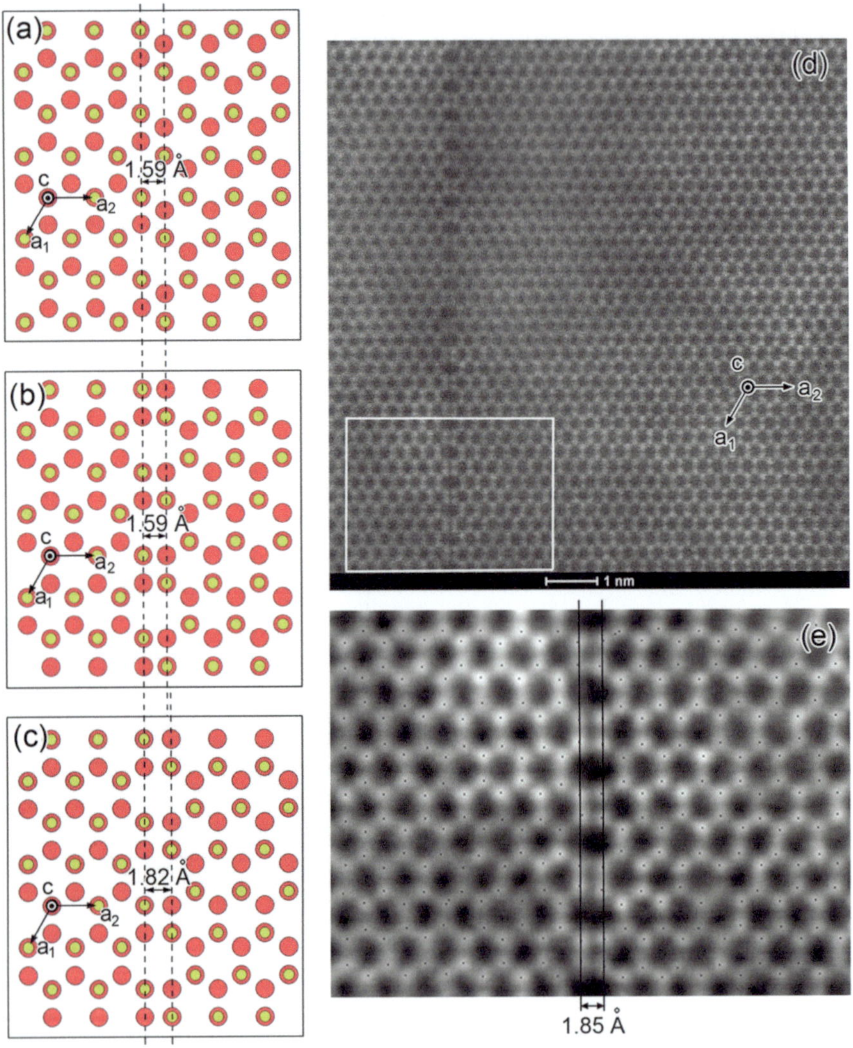

Fig. 1.21 Atomic resolution HAADF image of PSFs in GaN [69]

1.2.4 Inversion Domain Boundaries

The growth polarity of gallium nitride is a kinetically controlled process. The formation of reverse domains is related to substrate polarity, substrate surface treatment, growth conditions, and doping, etc. [71, 72]. The common models of inversion domain boundaries are shown in Fig. 1.22.

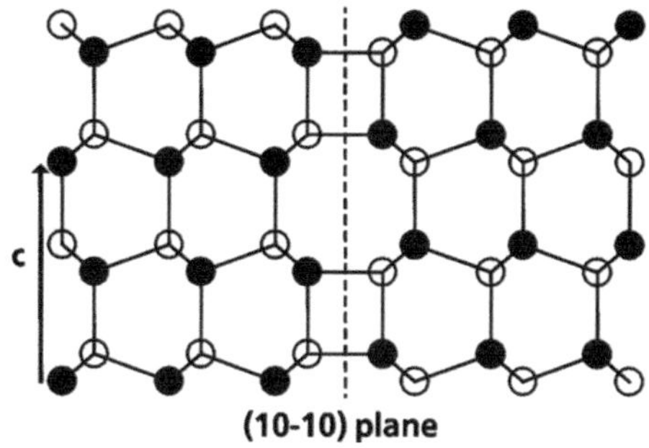

Fig. 1.22 Schematic diagram of the inversion domain boundary structure [73]

1.3 Material Characteristics of GaN

Compared with semiconductor materials like Si, GaAs, SiC, the excellent physical properties of GaN materials have led to their widespread use in areas such as blue light-emitting diodes (LEDs) and laser diodes (LDs), and they also have broad application prospects in RF microwaves and power electronic devices.

1.3.1 Optical Characteristics

1. Photoluminescence (PL) and Cathodoluminescence (CL)

Photoluminescence spectra (PL) are very useful means to study the optical properties of crystals. The non-radiative recombination in the sample at low temperatures will be greatly reduced, the luminescence spectral line is relatively narrow, which is conducive to the analysis of the luminescence peak fine structure. By changing the temperature and excitation power, etc., an analysis of the luminescence mechanism can be made. Figure 1.21 shows a luminescence spectrum of a GaN single crystal at low temperature. According to the distribution of luminescence peak energy, it can be roughly divided into near-band-edge luminescence peak, blue light, green light, yellow light, red light, and a series of defect-related luminescence peaks. According to the luminescence process, it can be divided into direct band-to-band recombination, radiative recombination between the band and impurity energy levels, donor–acceptor pair radiative recombination, free-bound exciton recombination, bound exciton recombination, and deep energy level impurity defect luminescence, etc.

Time-resolved photoluminescence (TRPL) is used to study the change of the spectrum with time, and the fluorescence intensity is a function of time and photon wavelength:

$$I = I(\lambda, t)$$

In the experiment, integration is performed near a certain central wavelength or delay time. This allows us to obtain time-dependent fluorescence decay spectra and fluorescence spectra at different delay times. By analyzing the fluorescence decay spectra, we can obtain information about the fluorescence recombination dynamics, where the carrier recombination lifetime τ is an important parameter. By analyzing the fluorescence spectra at different delay times, we can obtain information about the dynamics of photoexcited carrier energy relaxation and analyze the luminescence mechanism.

Cathodoluminescence (CL) is the luminescence produced by materials under high-energy electron beam excitation. Like photoluminescence, the cathodoluminescence of semiconductor materials is also the luminescence recombination of non-equilibrium carriers. The greatest advantage of cathodoluminescence is that it allows for the imaging of secondary electron microscopic morphology and the measurement of CL spectra in the same sample surface micro-area, enabling the analysis of micro-area structure and optical properties. During cathodoluminescence measurements, not only the cathodoluminescence spectra of the sample micro-area can be obtained, but also monochromatic microscopic images and full-color microscopic images of the micro-area can be obtained.

2. HVPE-GaN

Point defects in materials can introduce deep energy levels in the bandgap, which not only cause blue, green, yellow, and red light bands, but also lead to properties such as material breakdown and leakage. These deep energy level defect states are closely related to growth methods, growth conditions, and doping. Researches on point defects have a long history, but there is still much controversy about point defects related to deep energy levels. Combining first-principles calculations with optical measurement methods is an effective research method. Reshchikov et al. [74] used low-temperature PL and time-resolved PL to study the PL luminescence characteristics of undoped HVPE-GaN under the temperature of 18 K, as shown in Fig. 1.23a. The near-band-edge structure consists of free excitons (FE) ~ 3.480 eV and neutral donor-bound excitons (DBE) ~ 3.475 eV. The ultraviolet luminescence band (UVL) emission peak includes zero-order phonon sidebands (3.26 eV), two LO phonon sidebands. A broad red light band (RL), with a peak at 1.81 eV. A weak luminescence band located at 2.4–2.9 eV is the green light band (GL), with a peak at 2.5 eV. By measuring the changes in the peak position and intensity of the luminescence at different temperatures and different excitation powers, it is speculated that the RL luminescence band is related to a deep acceptor energy level located at (E_V + 1.13).

Reshchikov et al. [21] studied the effect of Mg doping concentration on the luminescent properties of GaN, as shown in Fig. 1.23b–d. For lightly doped HVPE-GaN with Mg, it is still n-type conductive, which is due to the presence of high concentrations of unintentional doping O_N and Si_{Ga} shallow donor impurities in the crystal. The low-temperature PL spectrum consists of near-band-edge (NBE ~ 3.471 eV), ultraviolet luminescence peak (UVL ~ 3.26 eV), yellow light band (YL ~ 2.20 eV), and red light band (RL ~ 1.67 eV), as shown in Fig. 1.23b–d. The NBE luminescence

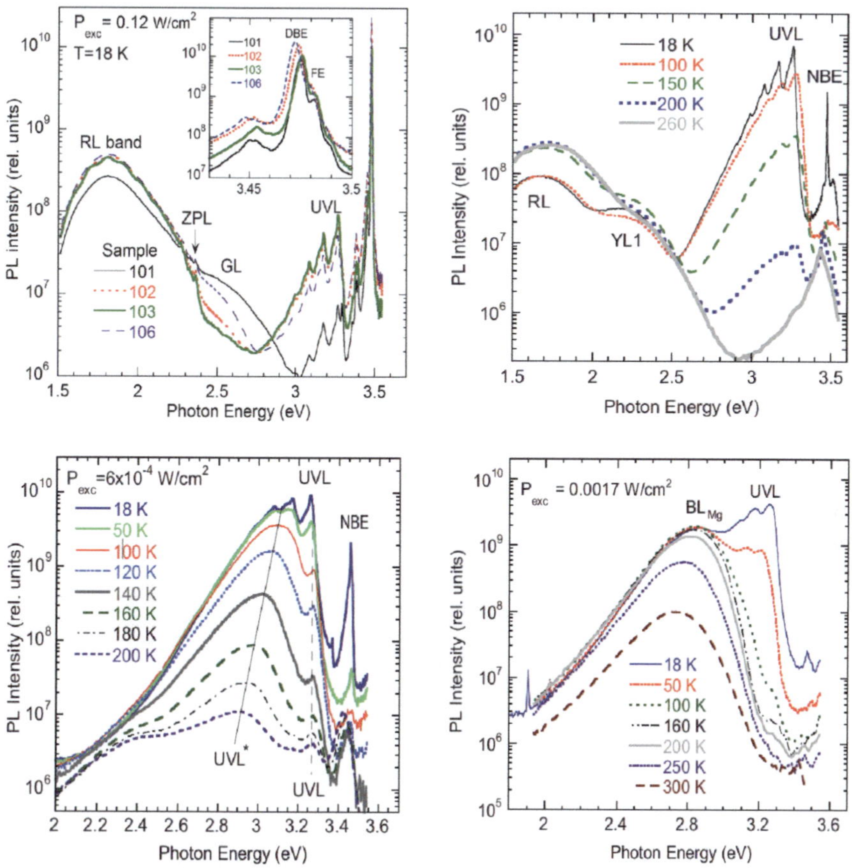

Fig. 1.23 **a** Luminescent peaks of Mg-doped HVPE-GaN N-type HVPE-GaN, with unintentionly-doped [Si] ~ 1E17. **b** N-type HVPE-GaN: lightly doped [Mg] ~ 1.3E17. **c** SI-type GaN: [Mg] ~ 1.5E18. **d** P-type GaN: [Mg] ~ 2E19

peak corresponds to the neutral shallow donor-bound exciton in unstressed GaN, a weaker peak of ~ 3.466 eV is considered to be the Mg_{Ga} acceptor-bound exciton luminescence peak. The ultraviolet luminescence peak (UVL ~ 3.26 eV) is identified as the luminescence peak related to the shallow Mg_{Ga} acceptor. Lyons et al. [18] found that Mg_{Ga} is related to the blue light peak, and the blue light peak BL can only be observed when the doping concentration is > 1E19. The Mg-H complex is the cause of UVL.

Bockowski et al. [75] studied the low-temperature optical properties of HVPE_GaN with different Si doping concentrations. The results of low-temperature PL measurements show that as the Si doping concentration increases, the intensity of the yellow light peak (YL) and the ultraviolet luminescent band (UVL) increase. The YL luminescent peak is related to the $V\text{-}_{Ga}$ defect, and Si doping could promote the generation of vacancy defects. Iwinska et al. [76] studied the effect of different Ge

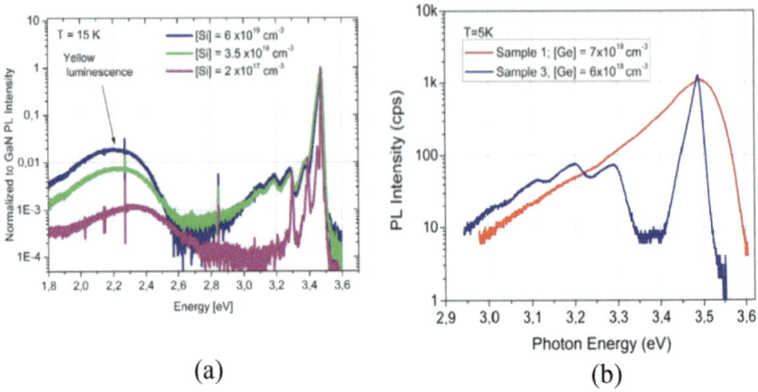

Fig. 1.24 N-type HVPE-GaN. **a** Si-doped GaN. **b** Ge-doped GaN

doping concentrations on the PL spectra. The results show that at low doping concentrations, the PL spectrum is mainly composed of near-band-edge luminescent peaks (NBE), donor–acceptor pairs (DAP), and YL peaks. As the Ge doping concentration further increases to 1E19, the DAP peak disappears (as shown in Fig. 1.24).

1.3.2 Electrical Characteristics

Power semiconductor devices play a key role in power conversion systems. The device efficiency, breakdown voltage and device size are closely related to the intrinsic electrical characteristics of materials. Table 1.3 compares the basic characteristics of four semiconductor materials: Si, GaAs, GaN and SiC. According to the application frequency, the comprehensive performance advantages of the material can be characterized by the low-frequency Baliga figure of merits (μ_n, ε_r, $E_c{}^3$) and the high-frequency Baliga figure of merits (μ_n, $E_c{}^2$).

Table 1.3 Electrical characteristic parameters of semiconductor materials

Parameter	Si	GaAs	4H-SiC	GaN
E_g/(eV)	1.12	1.42	3.26	3.39
E_c/(MV/cm)	0.23	0.4	2.5	3.3
μ_n/(cm^2/Vs)	1400	8500	950	1300
v_{sat}/(10^7 cm/s)	1.0	2.0	2.0	2.5
ε_r	11.8	13.1	9.7	8.9
k/(W/cm K)	1.5	0.5	3.8	1.3

1. Bandgap E_g

Hexagonal GaN is a wide bandgap, direct bandgap material and its bandgap width can reach 3.39 eV at room temperature, which is related to the strong chemical bond strength of ~ 2.20 eV between lattice atoms. The intrinsic carrier concentration in the material is related to the bandgap and temperature:

$$n_i = \sqrt{N_C N_V e^{-\frac{E_g}{2kT}}} \approx N_i(300K) \times \left(\frac{T}{300}\right)^{3/2} \cdot \exp\left(\frac{E_g(300K)}{0.0518} - \frac{E_g}{2kT}\right)$$

where N_C is the equivalent density of states for the material conduction and valence band bottom, and N_V is the equivalent density of states for the material conduction and valence band top. The intrinsic carrier concentration increases exponentially with temperature. Thus the devices made from intrinsic materials have difficulty in achieving stable operating characteristics. While the number of carriers provided by artificial doping coud remain essentially unchanged after all impurities are ionized. From the above formula, it can be seen that the larger the bandgap of the material, the higher the temperature required for the intrinsic carrier concentration to reach a certain level, so wide bandgap materials are suitable for the preparation of high-temperature resistant devices.

2. Critical breakdown field E_c

Stronger chemical bonds lead to a higher critical breakdown field when GaN devices undergo avalanche breakdown. There is an approximate proportional relationship between the critical breakdown field and the breakdown voltage of the device,

$$V_{BD} \approx \frac{1}{2} w_{Drift} \cdot E_c$$

where w_{drift} is the width of the device drift region. For devices with the same drift region thickness, the critical breakdown voltage of GaN devices is 14 times larger than that of Si devices; conversely, for devices with the same breakdown voltage, the drift region thickness of GaN devices can be 1/10 of Si devices.

Under high electric fields, carriers accelerate and the corresponding kinetic energy increases, and free carriers with sufficiently high energy could result in the ionization of lattice atoms through collision, generating new electron–hole pairs. The ionization coefficient characterizes the number of new electron–hole pairs produced by a carrier in a space charge region with unit length. Ji Dong et al. used homogeneously epitaxial avalanche diodes to characterize the collision ionization coefficients of electrons and holes at different temperatures. The experimental results show that the relationship between the ionization coefficients of electrons and holes in GaN and the electric field strength E are $2.1 \times 10^9 \exp(-3.7 \times 10^7/E)$ cm/s and $4.4 \times 10^6 \exp(-1.8 \times 10^7/E)$ cm/s, as shown in Fig. 1.25.

3. On-resistance

Low on-resistance means low device loss. For most carrier (taking n-type devices as an example) devices, the theoretical on-resistance R_{on} can be expressed as,

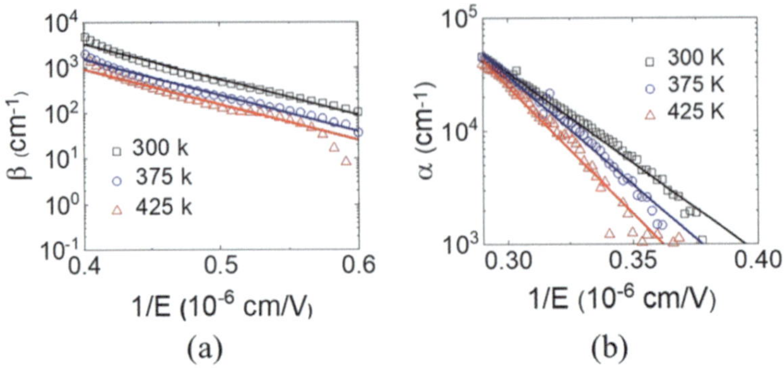

Fig. 1.25 Experimental measured values of GaN avalanche breakdown coefficient [77]

$$R_{on} = w_{\text{Drift}}/(q\mu_n N_D)$$

where μ_n is the electron mobility, N_D is the N-type doping concentration, the relationship between on-resistance and breakdown voltage, critical breakdown field is as follows,

$$R_{on} = 4V_{BD}^2/(\varepsilon_0 \varepsilon_r E_c^3)$$

where ε_0 and ε_r are the vacuum permittivity and relative permittivity, respectively. The on-resistance and breakdown voltage of the material restrict the actual use limit of power devices, as shown in Fig. 1.26.

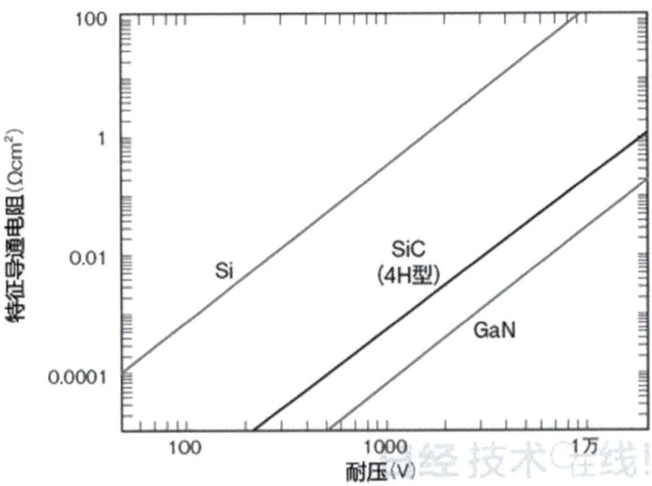

Fig. 1.26 The theoretical relationship of on-resistance and breakdown voltage for Si, GaAs, SiC, and GaN-based power devices

4. High field transport characteristics

The transport characteristics of carriers under high field are the main factors determining the frequency characteristics of devices. For high-frequency and high-power devices, drift velocity is a key parameter. Theoretically, the velocity-field relationship of electrons is usually obtained by Monte Carlo simulation. The characteristic of the electron velocity-field relationship under high field is that the drift velocity first increases with the increase of the external electric field strength, and then reaches a peak at a stronger electric field strength; as the electric field strength further increases, a wide negative differential resistance area appears where the speed gradually decreases, until the speed saturates under a sufficiently strong electric field. This phenomenon is related to electron inter-valley transfer, inter-valley deformation potential scattering, and polar optical phonon scattering. As shown in Fig. 1.27a, the drift velocity peak under steady-state in GaN crystals at room temperature is as high as 2.8×10^7 cm/s and the saturation drift velocity is as high as 1.3×10^7 cm/s, and the low-field mobility is in the range of 1600~1950 $cm^2/(Vs)$ [78]. Theoretically, the drift velocity peak under steady state in GaN crystals is 3.3×10^7 to 2.1×10^7 cm/s in the temperature range of 77~1000 K [79].

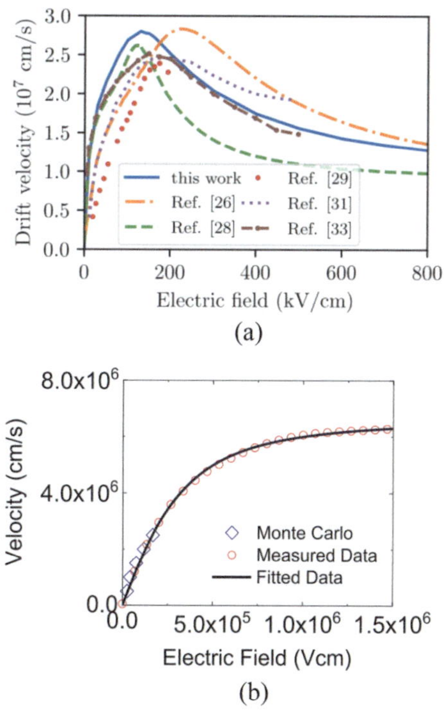

Fig. 1.27 Carrier drift velocity. **a** Electron drift velocity [78]. **b** Hole drift velocity [82]

The results of Hall effect test show that, at room temperature, the low-field mobility of the two-dimensional electron gas (2DEG density of ~ 3.3×10^{12} cm^{-2}) in GaN is 1300 cm^2/(Vs) under a background doping concentration of 10^{16} to 10^{17} cm^{-3} [80]. Through the measurement of I-V characteristics under high voltage pulse signal, the drift velocity peak of n-type GaN bulk single crystal at room temperature is obtained as 2.5×10^7 cm/s [81]. For the drift velocity of holes, it can be obtained experimentally by measuring the drift current. As shown in Fig. 1.27b, using light-assisted measurement to obtain drift current, and further using the Caughey-Thomas model to fit, the saturation drift velocity of holes in intrinsic GaN is about 6.63×10^7 cm/s, and the low-field mobility is 17 cm^2/Vs [81].

5. High temperature characteristics

The high temperature characteristics of GaN materials mainly refer to the variation rules of carrier concentration and electron mobility at high temperature, which have direct impacts on the device's on-state and off-state characteristics.

1. As known from semiconductor physics, the variation rule of carrier concentration n with temperature T is,

$$n \infty T^{3/2} \exp(-\Delta E_{\mathrm{D}}/kT)$$

where E_D is the ionization energy of the donor impurity. Assuming that all the donor impurities in the material are ionized at 300 K, the intrinsic carriers are the main factor causing the increase of carriers in the material. If the concentration of intrinsic carriers increases by an order of magnitude, it requires the temperature to increase to ~ 1400 K.

2. Electron mobility

Theoretically, the electron mobility is related to the scattering process, which can be represented as [83],

$$\mu(T) = \frac{1}{\mu_i} + \frac{1}{\mu_{dis}} + \frac{1}{\mu_{ap}} + \frac{1}{\mu_{op}}$$

The formula represents the electron mobility when only considering impurity ion scattering, dislocation scattering, acoustic phonon scattering, and optical phonon scattering. Temperature has different degrees of influence on the above four processes, and the change rule with temperature T can be simply summarized as,

$$\mu(T) \infty T^{-m}$$

In the formula, m is the exponent factor characterizing the change of mobility with temperature. As shown in Fig. 1.28, as to the high-quality GaN single crystal material

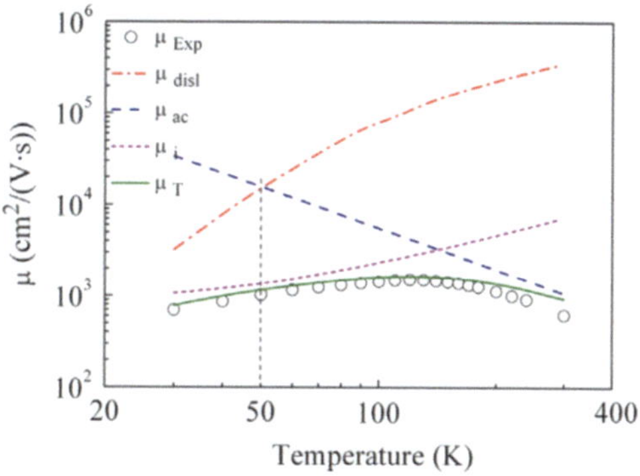

Fig. 1.28 Mobility change of the GaN single crystal with temperature [59]

with a dislocation density of 2.3×10^5 cm^{-2} and a background carrier concentration of 1.3×10^{16} cm^{-3}, its mobility reaches 1160 cm^2/(Vs) [59].

1.3.3 Thermal Characteristics

With the application of GaN-based high power density 5G microwave radio frequency, power electronic, and light-emitting devices, the heat dissipation problem and the thermal management of GaN have become research hotspots. The thermal conductivity is a key parameter, while the measured thermal conductivity of the material is far lower than the theoretical prediction. Studies have shown that the thermal conductivity is closely related to the dislocation density, point defects, doping, and temperature. Kihoon Park et al. [84] used the time-domain thermoreflectance method to study the influence of dislocation density σ_D on thermal conductivity κ_{GaN}, and the thermal conductivity decays according to the hyperbolic tangent function as the dislocation density increases. A new empirical formula was obtained by simulating experimental data,

$$k_{GaN} = 210 \tanh^{0.12}\left(\frac{1.5 \times 10^8}{\sigma_D}\right)$$

On this basis, the further revised Klemens's model is used to explain the decrease rate of thermal conductivity with the increase of dislocation density, highlighting the role of dislocation-induced scattering intensity on thermal conduction, as shown in Fig. 1.29a.

Rounds et al. [86] used SIMS and the 3ω method to compare the changes in thermal conductivity of GaN single crystals grown by ammono-thermal,, Na-flux, and HVPE in the temperature range of 30–295 K. The results show that phonons scattered by unintentional doping impurities is the main factor causing the decrease in thermal conductivity, as shown in Fig. 1.29b. Slomski et al. [85] further studied the influence

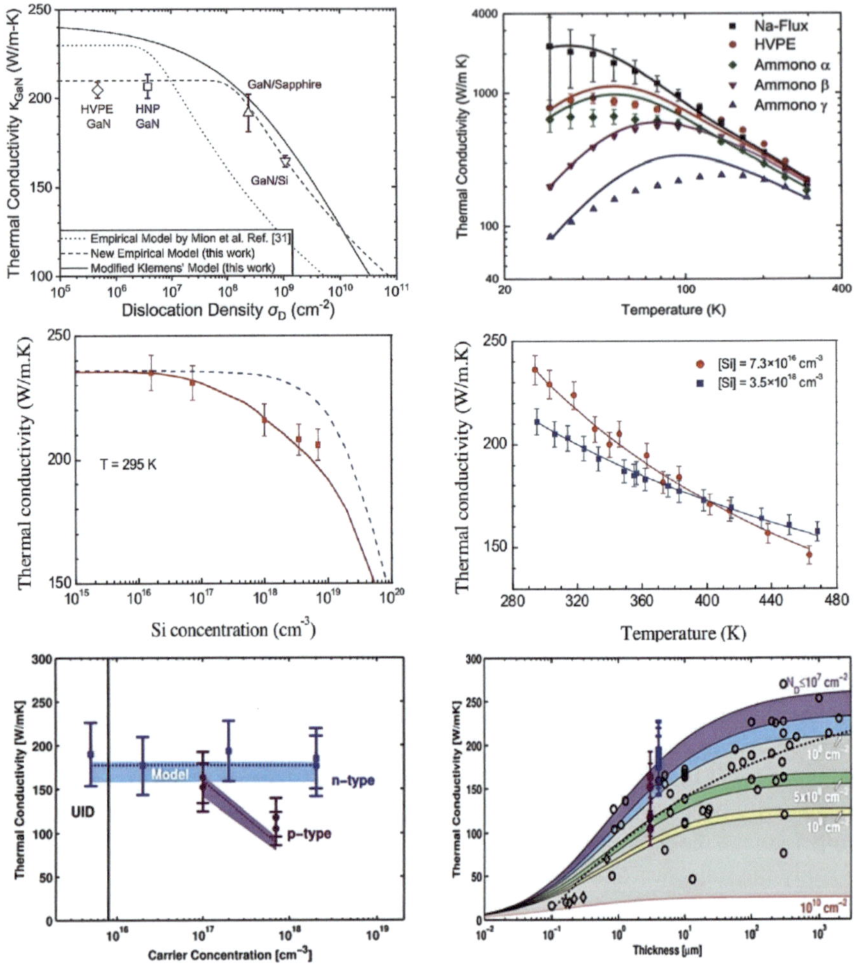

Fig. 1.29 Factors affecting GaN thermal conductivity. **a** Relationship with dislocation density, **b** unintentional doping impurities, **c** Si doping concentration, **d** Mg doping concentration, **e** Influence of Si impurity concentration and temperature on thermal conductivity, **f** Influence of film thickness *t* on thermal conductivity

of Si doping concentration on thermal conductivity at different temperatures. The results show that there are two mechanisms for the decrease in thermal conductivity with the increase in Si doping concentration: one is that phonon-impurity scattering and the other is phonon-free electron scattering. At high doping levels and high temperatures, phonon-free electron scattering dominates, as shown in Fig. 1.29c–d. Beechem et al. used the time-domain thermoreflectance method to study the effects of Mg doping concentration on thermal conductivity. The results show that as the Mg doping concentration increases, the thermal conductivity decreases, and the size effect is the main influencing factor, as shown in Fig. 1.29e. Further research indicates that the smaller the film thickness *t*, the more obvious effects on the GaN thermal conductivity. For a GaN film with a thickness of 1 μm, its thermal conductivity is half of that of the bulk single crystal material. Elbara Ziade et al. [86] used time-domain thermoreflectance and Monte Carlo simulation to analyze the dependence of thermal conductivity on film thickness. The results show that this dependence is related to phonon transport, as shown in Fig. 1.29f.

References

1. Ambacher O et al (1999) Two-dimensional electron gases induced by spontaneous and piezoelectric polarization charges in N- and Ga-face AlGaN/GaN heterostructures. J Appl Phys 85(6):3222–3233
2. Ren GB, Liu YM, Blood P (1999) Valence-band structure of wurtzite GaN including the spin-orbit interaction. Appl Phys Lett 74(8):1117–1119
3. Franz M et al (2019) Valence band structure and effective masses of GaN(10(1)over-bar0). Phys Rev B 99(19)
4. Yamaguchi AA et al (1998) Determination of valence band splitting parameters in GaN. J Appl Phys 83(8):4542–4544
5. Wei SH, Zunger A (1996) Valence band splittings and band offsets of AlN, GaN, and InN. Appl Phys Lett 69(18):2719–2721
6. Yeo YC, Chong TC, Li MF (1998) Electronic band structures and effective-mass parameters of wurtzite GaN and InN. J Appl Phys 83(3):1429–1436
7. Vurgaftman I, Meyer JR (2003) Band parameters for nitrogen-containing semiconductors. J Appl Phys 94(6):3675–3696
8. Reshchikov MA, Morkoc H (2005) Luminescence properties of defects in GaN. J Appl Phys 97(6)
9. Reshchikov MA (2015) Point defects in GaN, in defects in semiconductors. Elsevier, pp 315–367
10. Tuomisto F et al (2005) Effect of growth polarity on vacancy defect and impurity incorporation in dislocation-free GaN. Appl Phys Lett 86(3)
11. Chichibu SF et al (2018) Large electron capture-cross-section of the major nonradiative recombination centers in Mg-doped GaN epilayers grown on a GaN substrate. Appl Phys Lett 112(21)
12. Chichibu SF et al (2018) The origins and properties of intrinsic nonradiative recombination centers in wide bandgap GaN and AlGaN. J Appl Phys 123(16)
13. Polyakov AY, Lee I-H (2015) Deep traps in GaN-based structures as affecting the performance of GaN devices. Mater Sci Eng R-Rep 94:1–56
14. Reshchikov MA et al (2018) Thermal quenching of the yellow luminescence in GaN. J Appl Phys 123(16)
15. Nenstiel C et al (2015) Germanium—the superior dopant in n-type GaN. Physica Status Solidi-Rapid Res Lett 9(12):716–721
16. Ajay A et al (2016) Ge doping of GaN beyond the Mott transition. J Phys D-Appl Phys 49(44)
17. Lany S, Zunger A (2010) Dual nature of acceptors in GaN and ZnO: The curious case of the shallow Mg-Ga deep state. Appl Phys Lett 96(14)
18. Lyons JL, Janotti A, Van de Walle CG (2012) Shallow versus deep nature of Mg acceptors in nitride semiconductors. Phys Rev Lett 108(15)
19. Sun YY et al (2014) Anisotropic polaron localization and spontaneous symmetry breaking: comparison of cation-site acceptors in GaN and ZnO. Phys Rev B 90(16)

20. Miceli G, Pasquarello A (2016) Self-compensation due to point defects in Mg-doped GaN. Phys Rev B 93(16)
21. Reshchikov MA, Ghimire P, Demchenko DO (2018) Magnesium acceptor in gallium nitride. I. Photoluminescence from Mg-doped GaN. Phys Rev B 97(20)
22. Demchenko DO, Diallo IC, Reshchikov MA (2018) Magnesium acceptor in gallium nitride. II. Koopmans-tuned Heyd-Scuseria-Ernzerhof hybrid functional calculations of its dual nature and optical properties. Phys Rev B 97(20)
23. Kumar A et al (2019) Atomic-scale quantitative analysis of implanted Mg in annealed GaN layers on free-standing GaN substrates. J Appl Phys 126(23)
24. Wahl U et al (2017) Lattice location of Mg in GaN: a fresh look at doping limitations. Phys Rev Lett 118(9)
25. Zhang M et al (2012) The bound states of Fe impurity in wurtzite GaN. Appl Phys Lett 100(4)
26. Puzyrev YS et al (2015) Role of Fe impurity complexes in the degradation of GaN/AlGaN high-electron-mobility transistors. Appl Phys Lett 106(5)
27. Heitz R et al (1997) Excited states of Fe3+ in GaN. Phys Rev B 55(7):4382–4387
28. Wu S et al (2018) Unambiguous identification of carbon location on the N site in semi-insulating GaN. Phys Rev Lett 121(14)
29. Xu Y et al (2019) Influence of intrinsic or extrinsic doping on lattice locations of carbon in semi-insulating GaN. Appl Phys Exp 12(6)
30. Irmscher K et al (2018) Tri-carbon defects in carbon doped GaN. Appl Phys Lett 113(26)
31. Rhode SK et al (2013) Mg doping affects dislocation core structures in GaN. Phys Rev Lett 111(2)
32. Rhode SL et al (2015) Dislocation core structures in Si-doped GaN. Appl Phys Lett 107(24)
33. Elsner J et al (1997) Theory of threading edge and screw dislocations in GaN. Phys Rev Lett 79(19):3672–3675
34. Xin Y et al (2000) Direct experimental observation of the local electronic structure at threading dislocations in metalorganic vapor phase epitaxy grown wurtzite GaN thin films. Appl Phys Lett 76(4):466–468
35. Northrup JE (2001) Screw dislocations in GaN: the Ga-filled core model. Appl Phys Lett 78(16):2288–2290
36. Xin Y et al (1998) Direct observation of the core structures of threading dislocations in GaN. Appl Phys Lett 72(21):2680–2682
37. Usami S et al (2018) Correlation between dislocations and leakage current of p-n diodes on a free-standing GaN substrate. Appl Phys Lett 112(18)
38. Matsubara M et al (2013) Properties of threading screw dislocation core in wurtzite GaN studied by Heyd-Scuseria-Ernzerhof hybrid functional. Appl Phys Lett 103(26)
39. Northrup JE (2002) Theory of intrinsic and H-passivated screw dislocations in GaN. Phys Rev B 66(4)
40. Horton MK, Rhode SL, Moram MA (2014) Structure and electronic properties of mixed (a plus c) dislocation cores in GaN. J Appl Phys 116(6)
41. Arslan I, Stach EA, Browning ND (2005) Atomic and electronic structure of mixed and partial dislocations in GaN. Phys Rev Lett 94:025504
42. Xiong H, Wu J, Fang Z (2018) Atomic-scale studies of (a plus c)-type dislocation dissociation in wurtzite GaN. Appl Phys Exp 11(2)
43. Usami S et al (2019) Direct evidence of Mg diffusion through threading mixed dislocations in GaN p-n diodes and its effect on reverse leakage current. Appl Phys Lett 114(23)
44. Hawkridge ME, Cherns D (2005) Oxygen segregation to dislocations in GaN. Appl Phys Lett 87(22)
45. Kim B et al (2014) Investigation of leakage current paths in n-GaN by conductive atomic force microscopy. Appl Phys Lett 104(10)
46. Galiano K et al (2018) Spatial correlation of the E-C-0.57 eV trap state with edge dislocations in epitaxial n-type gallium nitride. J Appl Phys 123(22)
47. Mueller E et al (2006) Probing the electrostatic potential of charged dislocations in n-GaN and n-ZnO epilayers by transmission electron holography. Phys Rev B 73(24)

48. Cherns D, Jiao CG (2001) Electron holography studies of the charge on dislocations in GaN. Phys Rev Lett 87(20)
49. Cherns D et al (2002) Electron holography studies of the charge on dislocations in GaN. Physica Status Solidi B-Basic Res 234(3):924–930
50. Cai J, Ponce FA (2002) Determination by electron holography of the electronic charge distribution at threading dislocations in epitaxial GaN. Physica Status Solidi A—Appl Res 192(2):407–411
51. Robertson CA et al (2019) Modeling dislocation-related leakage currents in GaN p-n diodes. J Appl Phys 126(24)
52. Kyle ECH et al (2014) High-electron-mobility GaN grown on free-standing GaN templates by ammonia-based molecular beam epitaxy. J Appl Phys 115(19)
53. Fang ZQ et al (2005) Traps in AlGaN/GaN/SiC heterostructures studied by deep level transient spectroscopy. Appl Phys Lett 87(18)
54. Rackauskas B et al (2018) Leakage mechanisms in GaN-on-GaN vertical pn diodes. Appl Phys Lett 112(23)
55. Bian Z et al (2019) Leakage mechanism of quasi-vertical GaN Schottky barrier diodes with ultra-low turn-on voltage. Appl Phys Exp 12(8)
56. Shan Q et al (2011) Transport-mechanism analysis of the reverse leakage current in GaInN light-emitting diodes. Appl Phys Lett 99(25)
57. Besendoerfer S et al (2020) Vertical breakdown of GaN on Si due to V-pits. J Appl Phys 127(1)
58. Yakimov EB et al (2018) Recombination properties of dislocations in GaN. J Appl Phys 123(16)
59. Gu H et al (2016) The electrical properties of bulk GaN crystals grown by HVPE. J Cryst Growth 436:76–81
60. Pauc N et al (2006) Carrier recombination near threading dislocations in GaN epilayers by low voltage cathodoluminescence. Appl Phys Lett 89(16)
61. Kaganer VM et al (2019) Determination of the carrier diffusion length in GaN from cathodoluminescence maps around threading dislocations: fallacies and opportunities. Phys Rev Appl 12(5)
62. Yamane K et al (2016) Origin of lattice bowing of freestanding GaN substrates grown by hydride vapor phase epitaxy. J Appl Phys 119(4)
63. Foronda HM et al (2016) Curvature and bow of bulk GaN substrates. J Appl Phys 120(3)
64. Weinrich J et al (2019) Extra half-plane shortening of dislocations as an origin of tensile strain in Si-doped (Al)GaN. J Appl Phys 126(8)
65. Follstaedt DM et al (2009) Strain relaxation in AlGaN multilayer structures by inclined dislocations. J Appl Phys 105(8)
66. Soman R et al (2018) Dislocation bending and stress evolution in Mg-doped GaN films on Si substrates. J Appl Phys 124(24)
67. Kong BH et al (2012) Classification of stacking faults and dislocations observed in nonpolar a-plane GaN epilayers using transmission electron microscopy. Appl Surf Sci 258(7):2522–2528
68. Zakharov DN et al (2005) Structural TEM study of nonpolar a-plane gallium nitride grown on (1120) 4H-SiC by organometallic vapor phase epitaxy. Phys Rev B 71(23)
69. Hu Y-L et al (2012) Atomic structure of prismatic stacking faults in nonpolar a-plane GaN epitaxial layers. Appl Phys Lett 101(11)
70. Song J et al (2019) Elimination of stacking faults in semipolar GaN and light-emitting diodes grown on sapphire. ACS Appl Mater Interfaces 11(36):33140–33146
71. Stutzmann M et al (2001) Playing with polarity. Physica Status Solidi B-Basic Res 228(2):505–512
72. Zuniga-Perez J et al (2016) Polarity in GaN and ZnO: theory, measurement, growth, and devices. Appl Phys Rev 3(4)
73. Northrup JE, Neugebauer J, Romano LT (1996) Inversion domain and stacking mismatch boundaries in GaN. Phys Rev Lett 77(1):103–106
74. Reshchikov MA et al (2014) Fine structure of the red luminescence band in undoped GaN. Appl Phys Lett 104(3)

75. Bockowski M et al (2018) Doping in bulk HVPE-GaN grown on native seeds - highly conductive and semi-insulating crystals. J Cryst Growth 499:1–7
76. Iwinska M et al (2017) Crystal growth of HVPE-GaN doped with germanium. J Cryst Growth 480:102–107
77. Ji D, Ercan B, Chowdhury S (2019) Experimental determination of impact ionization coefficients of electrons and holes in gallium nitride using homojunction structures. Appl Phys Lett 115(7)
78. Fang J et al (2019) Electron transport properties of AlxGa1-xN/GaN transistors based on first-principles calculations and Boltzmann-Equation Monte Carlo simulations. Phys Rev Appl 11(4)
79. Bhapkar UV, Shur MS (1997) Monte Carlo calculation of velocity-field characteristics of wurtzite GaN. J Appl Phys 82(4):1649–1655
80. Webb JB et al (2001) Growth of high mobility GaN and AlGaN/GaN high electron mobility translator structures on 4H-SiC by ammonia molecular-beam epitaxy. Appl Phys Lett 78(24):3845–3847
81. Barker JM et al (2005) Bulk GaN and AlGaN/GaN heterostructure drift velocity measurements and comparison to theoretical models. J Appl Phys 97(6)
82. Ji D, Ercan B, Chowdhury S (2020) Experimental determination of velocity-field characteristic of holes in GaN. IEEE Electron Device Lett 41(1):23–25
83. Abdel-Motaleb IM, Korotkov RY (2005) Modeling of electron mobility in GaN materials. J Appl Phys 97(9)
84. Park K, Bayram C (2019) Impact of dislocations on the thermal conductivity of gallium nitride studied by time-domain thermoreflectance. J Appl Phys 126(18)
85. Slomski M et al (2017) Thermal conductivity of bulk GaN grown by HVPE: Effect of Si doping. Physica Status Solidi B-Basic Solid State Phys 254(8)
86. Ziade E et al (2017) Thickness dependent thermal conductivity of gallium nitride. Appl Phys Lett 110(3)

Chapter 2
Basic Characteristics of Gallium Nitride Single Crystal Materials

2.1 Growth Dynamics

The main research content of crystal growth dynamics is to clarify the crystal growth mechanism under different growth conditions, and the relationship between crystal growth rate and growth driving force. The crystal growth rate is governed by the growth driving force. When the heat or mass transport of the growth medium is changed, the crystal growth rate also changes. The structure of the crystal growth interface determines the growth mechanism, and different growth mechanisms show different growth dynamics. The morphology of crystal growth depends on the relative growth rates of the various crystal faces of the crystal, so the transport and other dynamic processes that affect the growth rate will also affect the morphology of crystal growth.

2.1.1 Thermodynamic Driving Force

The driving force for crystal growth refers to the thermodynamic driving force to restore balance when the system is in a non-equilibrium state.

1. Hydride Vapor Phase Epitaxy

In the growth area, the following chemical reactions occur between the reaction gases, forming epitaxial growth of GaN film on the substrate surface,

$$\mathrm{GaCl}(g) + \mathrm{NH_3}(g) \Leftrightarrow \mathrm{GaN}(s) + \mathrm{HCl}(g) + \mathrm{H_2}(g) \tag{2.1}$$

$$\mathrm{GaCL}(g) + 2\mathrm{HCL}(g) \leftrightarrow \mathrm{GaCL_3}(g) + \mathrm{H_2}(g) \tag{2.2}$$

The equilibrium constant of the above chemical reactions,

K. Xu et al., *GaN Single Crystal Growth and Application*, Wide Bandgap Semiconductors, https://doi.org/10.1007/978-981-96-7572-2_2

$$K_1 = \frac{P_{\mathrm{HCl}} P_{\mathrm{H_2}}}{P_{\mathrm{GaCl}} P_{\mathrm{NH_3}}} \tag{2.3}$$

$$K_2 = \frac{P_{GaCL_3} P_{H_2}}{P_{GaCL} P_{HCL}^2} \tag{2.4}$$

Including inert gas (IG) as a carrier gas, the total pressure of the reaction system is

$$\sum P_i = P_{\mathrm{GaCl}} + P_{\mathrm{GaCl_3}} + P_{\mathrm{NH_3}} + P_{\mathrm{HCl}} + P_{\mathrm{H_2}} + P_{\mathrm{IG}} \tag{2.5}$$

Further considering thermal decomposition of ammonia,

$$\mathrm{NH_3}(g) \rightarrow (1-\alpha)\mathrm{NH_3} + \frac{\alpha}{2}\mathrm{N_2} + \frac{3\alpha}{2}\mathrm{H_2}(g) \tag{2.6}$$

The decomposition ratio of ammonia in the formula is about $\alpha = 0.03$ at 1000 °C. According to the law of mass conservation, we know

$$P^0_{\mathrm{GaCL}} - (P_{\mathrm{GaCL}} + P_{\mathrm{GaCL_3}}) = P^0_{\mathrm{NH_3}} - P_{\mathrm{NH_3}} \tag{2.7}$$

where P^0_{GaCL} and $P^0_{\mathrm{NH_3}}$ are the input partial pressures of the source gas, considering that V/III > 1 in the HVPE crystal growth process, i.e., $P^0_{\mathrm{NH_3}} > P^0_{\mathrm{GaCL}}$, so the driving force Δf for HVPE growth is,

$$\Delta f = \Delta P = P^0_{\mathrm{GaCL}} - (P_{\mathrm{GaCL}} + P_{\mathrm{GaCL_3}}) \tag{2.8}$$

2. Ammonothermal Method

Unlike vapor phase epitaxy, the ammonothermal method is similar to the hydrothermal (temperature difference) method. The crystal growth reactor is divided into a dissolution area and a growth area. In the dissolution area, gallium nitride raw material is dissolved to form a saturated solution, which is transported to the growth area by natural convection. Due to the temperature difference ΔT between the growth area and the solubility, the saturated solution becomes supersaturated in the growth area, and crystallization grows on the seed crystal, as shown in Fig. 2.1. Therefore, the driving force Δf for ammonothermal growth is

$$\Delta f = \Delta\mu = RT\{\ln[c_1(T_1)] - \ln[c_2(T_2)]\} = RT\ln(\Delta c) \tag{2.9}$$

In the formula, μ is the chemical potential, and c_1 and c_2 are the solubilities of GaN material in the growth area (temperature T1) and dissolution area (temperature T2), respectively.

3. Growth of GaN by Na-flux Method

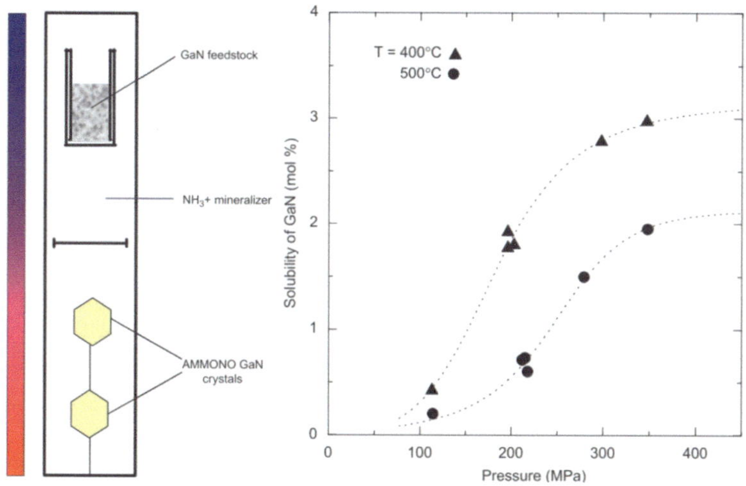

Fig. 2.1 Schematic diagram of ammonothermal growth principle and solubility curve [1]

The Na-flux method belongs to the high-temperature solution method. Since the supersaturation of the Ga source is large, the N source is the main factor limiting the crystal growth rate. The viscosity of the Ga-Na-X solution system is large, and convection can be basically ignored (except for forced convection, such as stirring, shaking, etc.), and mass transfer is mainly through diffusion. N_2 gas decomposes and dissolves at the gas–liquid interface, and under the drive of the concentration gradient, it diffuses to the surface of the GaN seed crystal. When the surface N source becomes supersaturated, the crystal grows (as shown in Fig. 2.2). Assuming that the system is at a constant temperature and there is no temperature difference, the driving force Δf for Na-flux growth is

$$\Delta f = \Delta\mu = RT\left\{\ln[c_N] - \ln[\overset{\circ}{c}_N]\right\} = RT\ln(\Delta c) \tag{2.10}$$

In the formula, μ is the chemical potential, and c_N and $c_N{}^0$ are the solubilities of the N source on the seed crystal surface and the equilibrium concentration, respectively.

2.1.2 *Mass Transport*

Diffusion and convection are the two main modes of mass transport in crystal growth, where diffusion involving the movement of molecules and convection transporting solutes through the macroscopic movement of fluids. The driving force of diffusion comes from the concentration gradient. Convection is divided into natural convection and forced convection, where natural convection being entirely caused by the gravitational field, divided into thermal convection and solute convection. The driving

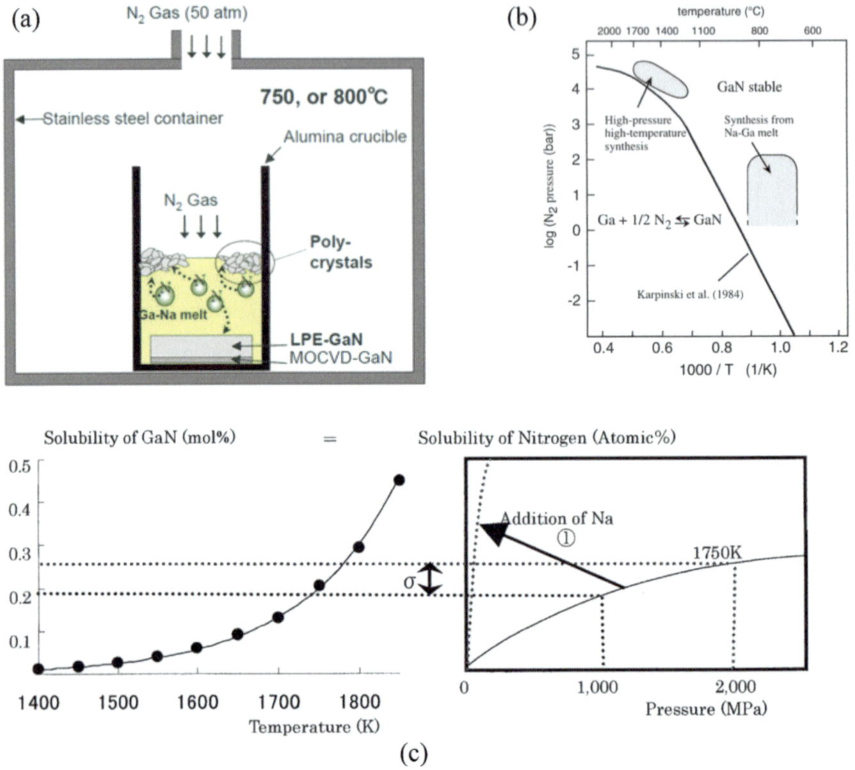

Fig. 2.2 Schematic of GaN growth by Na-flux. **a** Schematic of the growth process; **b** Ga-Na-N phase diagram[2]; **c** Solubility curve of N in the Ga-Na system [3]

force of thermal convection is the temperature gradient, while the driving force for solute convection is the solute concentration gradient. Forced convection is caused by the rotation of the crystal or the fluid surrounding the crystal.

For the HVPE vapor phase epitaxy method, mass transport is achieved by forced convection of the input gas. For the Na-flux method, the N source is mainly realized by volume diffusion. For the ammonothermal method, mass transport is mainly achieved by natural convection. Taking an alkaline mineralizer as an example, GaN has a negative solubility coefficient in the supercritical ammonia system, that is, the higher the temperature, the smaller the solubility, so the upper temperature zone T1 is the dissolution zone, and the lower temperature T2 is the growth zone (T1 < T2). In the dissolution zone, the solubility of the saturated solution is larger, and it is dominated by solute convection to transport downwards. When the saturated solution is transported to the growth zone, the temperature increases and the solution becomes supersaturated and crystallizes. At this time, it convects upwards under the drive of thermal convection. Ideally, under the combined action of solute convection

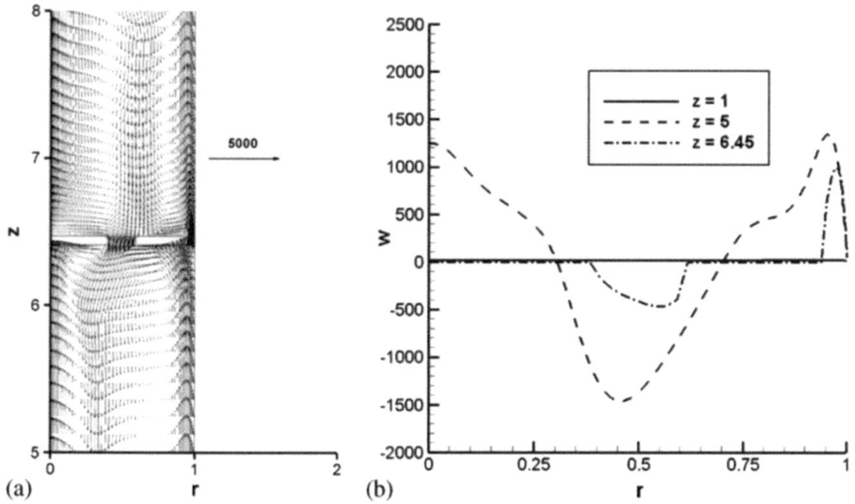

Fig. 2.3 Simulation diagram of the flow field pattern and flow rate distribution of the ammonothermal method [4]

and thermal convection, a stable circulating convection pattern is formed inside the reactor, as shown in Fig. 2.3.

2.1.3 Growth Mechanism

The crystal growth mechanism depends on the interface structure during the growth process. For HVPE vapor phase epitaxy and ammonothermal growth, the growth interface is a (complete and incomplete) smooth interface. There are three main growth modes during the epitaxial growth process: (1) Layered growth mode (Frank-van der Merwe mode, FVDM), when the interaction between the adsorbed atoms and the substrate is stronger than the interaction between the atoms, layered growth occurs; (2) Island growth mode (Volmer-Weber mode, V-W), when the interaction between the adsorbed atoms or molecules is stronger than the interaction between the adsorbed atoms and the substrate, the adsorbed atoms nucleate on the substrate surface in the form of atomic clusters, resulting in three-dimensional island growth; (3) Mixed growth mode (Stranski–Krastanov mode, S-K mode), a process between the two, first layered growth, and after exceeding a critical value, it transforms into island growth. Generally speaking, there are certain atomic steps and defect steps (screw dislocations, mixed dislocations, etc.) on the crystal surface. The ideal growth mode is the step flow growth mode, and the width of the steps is often controlled by controlling the oblique angle of the substrate in experiments.

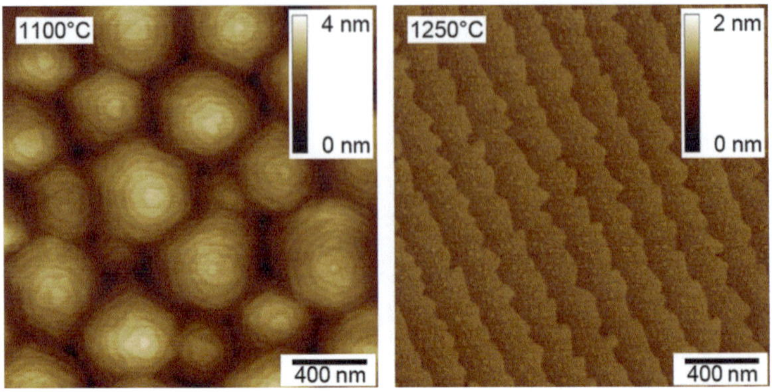

Fig. 2.4 AFM morphology of AlN epitaxial film [5]. **a** 3D island growth; **b** Double atomic layer step flow growth, accompanied by the formation of two-dimensional nucleation islands on the step surface

The transition between growth modes is not only related to the substrate used (heterogeneous or homogeneous substrate, oblique angle), but also closely related to the growth conditions (temperature T, V/III, growth rate), as shown in Fig. 2.4.

2.1.4 *Crystal Morphology*

The growth morphology of a crystal depends on the relative growth rates of its various crystal faces. The growth rate of each crystal face is not only related to the intrinsic properties of the crystal (dangling bonds, polarity, reconstruction), but also closely related to the physico-chemical conditions of growth. Despite different morphologies, the commonly exposed faces are $(000\underline{1})$, $\{10\underline{1}0\}$, $\{10\underline{1}1\}$ and (0001) crystal faces, as shown in Fig. 2.5. Whether it is Ga polarity or N polarity surface, through ammonia surface treatment of buffer layer surface, the conversion of GaN nanocolumns from pyramidal to prismatic has been achieved. Na-flux and hydrothermal methods, which are closer to thermodynamic equilibrium growth conditions, have intrinsic morphologies close to the ideal pyramidal shape.

2.2 Generation and Annihilation of Dislocations

Depending on the angle between the epitaxial film face and the c-axis, there are a variety of single-crystal substrate materials practically, such as polar, non-polar, and semi-polar faces, as shown in Fig. 2.6. Since there are no natural GaN single crystal materials in nature, heteroepitaxy is often used to artificially prepare large-size epitaxial films or single crystals. Sapphire, silicon carbide, silicon, lithium aluminate

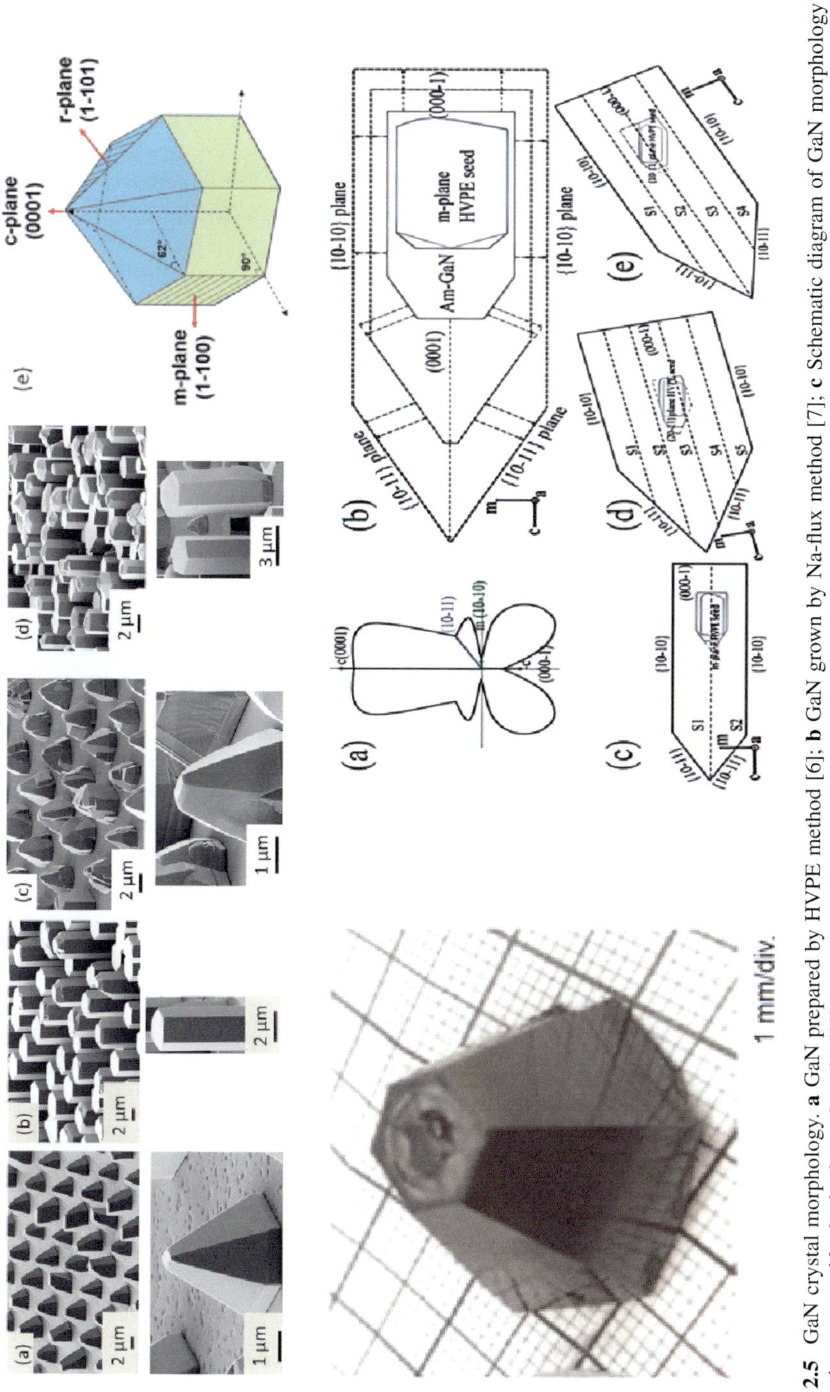

Fig. 2.5 GaN crystal morphology. **a** GaN prepared by HVPE method [6]; **b** GaN grown by Na-flux method [7]; **c** Schematic diagram of GaN morphology evolution prepared by hydrothermal method[8]

and other crystal materials are often used for the GaN heteroepitaxy. Heteroepitaxy causes a series of problems such as lattice mismatch, thermal mismatch, and chemical mismatch, introducing stress and dislocations and other defects in the epitaxial film. The generation and annihilation of dislocations are closely related to the relaxation and evolution process of stress in the epitaxial film.

Slip and climb are two modes of dislocation movement. The direction of dislocation slip movement (i.e., the Burgers vector b) and the slip plane form a slip system. Common slip planes include the basal plane {0001}, prismatic plane $\{10\bar{1}0\}$, and pyramidal plane $\{10\bar{1}1\}$, dislocation Burgers vectors include 1/3 $\langle 11\bar{2}0\rangle$, $\langle 0001\rangle$, 1/3 $\langle 11\bar{2}3\rangle$, 1/3 $\langle 10\bar{1}0\rangle$, 1/2 $\langle 0001\rangle$ and 1/6 $\langle 20\bar{2}3\rangle$, slip directions include $\langle 11\bar{2}0\rangle$, $\langle 0001\rangle$ and $\langle 11\bar{2}3\rangle$, as shown in Fig. 2.7. According to the relationship between the dislocation Burgers vector b and the lattice parameters, dislocations are divided into perfect dislocations and partial dislocations. Partial dislocations often accompany stacking faults.

Table 2.1 Burgers vector notation and miller-bravais indices in hexagonal close-packed crystals

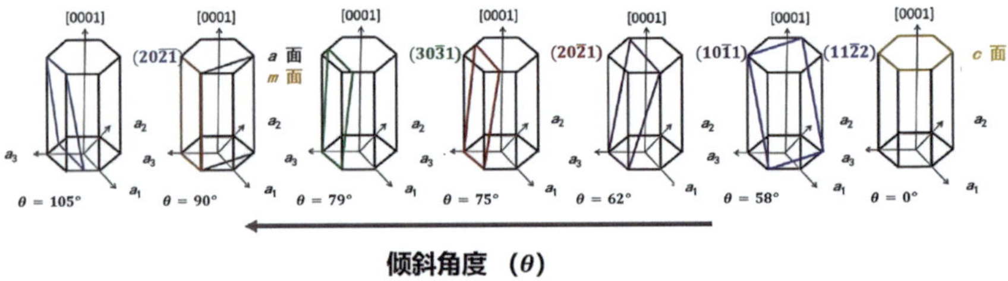

Fig. 2.6 Schematic diagram of different polar faces of GaN

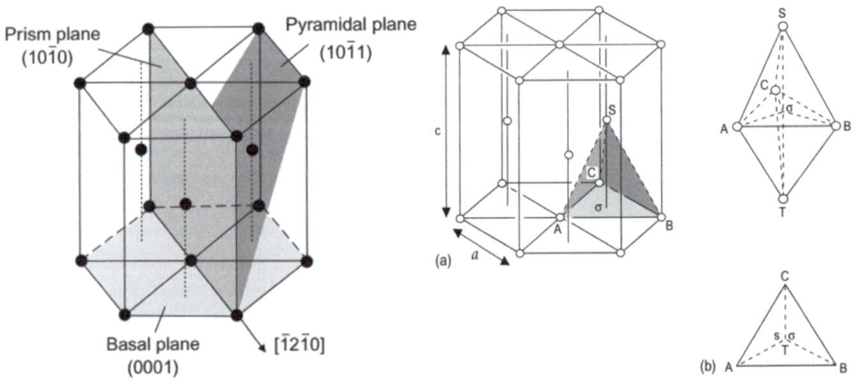

Fig. 2.7 Slip system of hexagonal lattice structure. **a** Slip plane. **b** Slip direction, also known as Burgers vector or slip vector

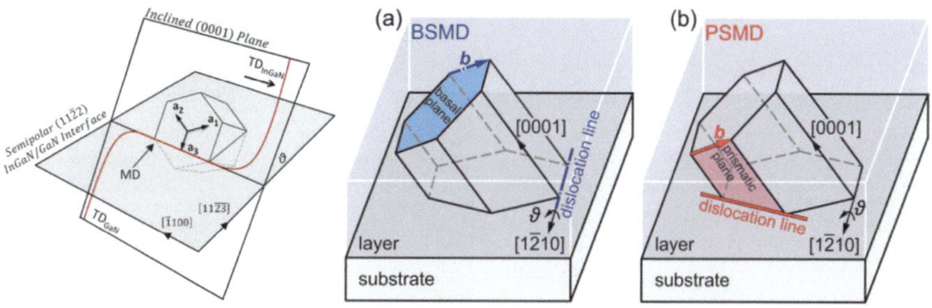

Fig. 2.8 Mismatch dislocations in semi-polar nitride semiconductor epitaxial materials. **a** Mechanism of basal plane slip generation [11]; **b** Mismatch dislocations generated by basal plane and prismatic slip [12]

2.2.1 Mismatch Dislocation Generation Mechanism

To relieve mismatch stress, mismatch dislocations with a line direction of <11$\bar{2}$0> and a Burgers vector of 1/3 <11$\bar{2}$0> are generated in the heteroepitaxial (0001) GaN film. The generation mechanism of mismatch dislocation is not very clear, it may originate from the fluctuations of local stress leading to the activation of the {0001} <11$\bar{2}$0> slip system, the slip of threading dislocation, and the dislocation multiplication mechanism, etc. Romanov et al. [9] studied the stress release mechanism in semi-polar GaN film, the results showed that the basal plane slip of dislocations, i.e., the <11$\bar{2}$0> (0001) slip system, generates 1/3 <11$\bar{2}$0> mismatch dislocations at the epitaxial interface, as shown in Fig. 2.8. Further research showed that when the semi-polar direction deviates from the c-axis direction to reach a critical angle of ~ 70°, the relaxation mechanism of mismatch stress changes from basal plane slip 1/3 <11$\bar{2}$0> (0001) to prismatic slip 1/3<11$\bar{2}$0> (01$\bar{1}$0) [10].

2.2.2 Threading Dislocation

1. Generation of dislocations in (0001) GaN

It is generally believed that lattice mismatch and thermal mismatch are the main causes of dislocation generation in heteroepitaxial GaN film. The typical two-step growth process is as follows: first, a GaN or AlN buffer layer is grown at low temperature on a heteroepitaxial substrate, then it is annealed at high temperature, followed by further high-temperature growth, and GaN nucleation islands merge into a film. It is generally believed that threading dislocation originates from the merging of nucleation islands, and thread dislocations are located at the Mosaic structure interfaces. Another view is that threading dislocations may originate from the GaN buffer layer, as shown in Fig. 2.9.

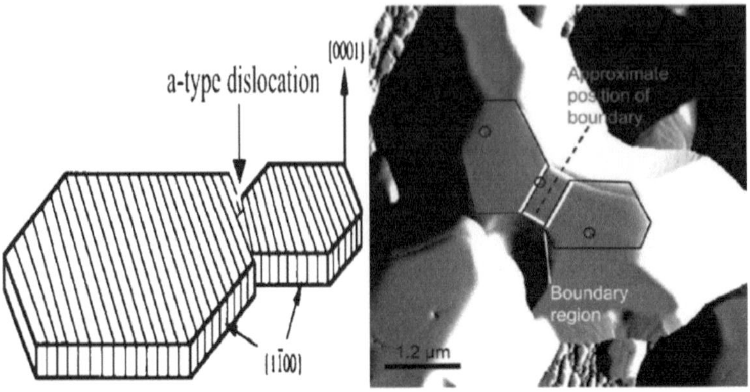

Fig. 2.9 Origin of threading dislocation. **a** Merging of nucleation islands; **b** GaN/sapphire interface

Fig. 2.10 Relationship between dislocation density and sample thickness [13]

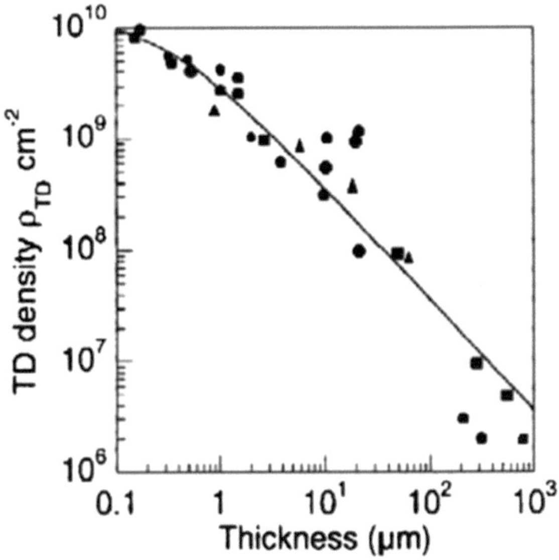

2. Evolution and Annihilation of Dislocation

Reducing the dislocation density has always been the key to obtaining high-quality materials by heteroepitaxy. The principles to reduce the dislocation density are mainly: (1) guiding the bending of dislocations through the mirror force, and then burying them in the material or reacting with each other to reduce the dislocation density. The main techniques include selective area growth (SAG), epitaxial lateral overgrowth (ELOG), in-situ SiN_x treatment, PSS substrate, etc.; (2) by increasing the crystal growth thickness, promoting the reaction or annihilation of dislocations to reduce the density of threading dislocations, as shown in Fig. 2.10.

References

1. Dwilinski R et al (2008) Excellent crystallinity of truly bulk ammonothermal GaN. J Cryst Growth 310(17):3911–3916
2. Yamane H et al (2000) GaN single crystal growth from a Na-Ga melt. J Mater Sci 35(4):801–808
3. Tandryo R et al (2020) Temperature dependence of nitrogen dissolution on Na flux growth. J Cryst Growth 535:125549
4. Chen QS, Prasad V, Hu WR (2003) Modeling of ammonothermal growth of nitrides. J Cryst Growth 258(1–2):181–187
5. Bryan I et al (2016) Surface kinetics in AlN growth: A universal model for the control of surface morphology in III-nitrides. J Cryst Growth 438:81–89
6. Avit G et al (2018) Crystal engineering by tuning the growth kinetics of GaN 3-D microstructures in SAG-HVPE. CrystEngComm 20(40):6207–6213
7. Mori Y et al (2019) Recent progress of Na-flux method for GaN crystal growth. Jpn J Appl Phys 58
8. Li T et al (2019) Growth behavior of ammonothermal GaN crystals grown on non-polar and semi-polar HVPE GaN seeds. CrystEngComm 21(33):4874–4879
9. Romanov AE et al (2011) Basal plane misfit dislocations and stress relaxation in III-nitride semipolar heteroepitaxy. J Appl Phys 109(10)
10. Yan X, Pennycook SJ (2003) High-resolution scanning transmission electron microscopy technology—atomic resolution atomic number contrast imaging. Science Press
11. Hsu PS et al (2011) Misfit dislocation formation via pre-existing threading dislocation glide in (11(2)over-bar2) semipolar heteroepitaxy. Appl Phys Lett 99(8)
12. Smirnov AM et al (2019) Stress relaxation in semipolar and nonpolar III-nitride heterostructures by formation of misfit dislocations of various origin. J Appl Phys 126(24)
13. Bennett SE (2010) Dislocations and their reduction in GaN. Mater Sci Technol 26(9):1017–1028

Chapter 3
Hydride Vapor Phase Epitaxy Method

3.1 Development History

In the 1960s, halide-based vapor phase epitaxy technology was developed for the growth of Ge [1], GaAs [2] and GaP [3] materials; based on this, the growth of GaAs, GaP materials was also achieved in a sealed chamber using HCl gas as the reaction precursor [4], which is the prototype of contemporary hydride vapor phase epitaxy (HVPE) growth technology. In 1966, J. J. Tietjen and J. A. Amick from the RCA laboratory further used HCl and metal Ga reaction as the group III source, and arsine and phosphine as the group V source, to achieve the growth of $GaAs_x P_{1-x}$ materials [5], marking the rapid development period of HVPE growth technology. In 1969, H. P. Maruska and J. J. Tietjen from the RCA laboratory used HCl gas and the metal Ga as the group III source, and ammonia as the group V source, to achieve the growth of GaN single crystal material [6]. This was the first time in human history that the method of vapor phase epitaxy was used to successfully prepare GaN single crystal material. Since then, GaN, a critical semiconductor material, has officially entered the historical stage.

However, due to multiple restrictions such as vapor phase epitaxy equipment, gas purity, and understanding of the growth mechanism, the development of GaN materials was relatively slow for more than 10 years. It was not until the late 1980s that three Japanese scientists, I. Akasaki, H. Amano, and S. Nakamura, achieved the growth of high-quality GaN materials using metal organic chemical vapor phase epitaxy growth technology on sapphire substrates, and further solved the difficult problem of P-type GaN material activation, greatly promoting the application of GaN light-emitting diodes and laser didoes (LD). Based on these key breakthroughs, the three Japanese scientists were awarded the 2014 Nobel Prize in Physics.

At the same time, the growth of GaN materials based on HVPE technology has been continuously developing. Thanks to the high growth rate of the HVPE method, which can reach more than 100 μm per hour, it is 1–2 orders of magnitude higher

K. Xu et al., *GaN Single Crystal Growth and Application*, Wide Bandgap Semiconductors, https://doi.org/10.1007/978-981-96-7572-2_3

than the MOCVD growth rate, making it one of the most ideal growth methods for GaN thick layer materials, especially for GaN single crystal substrates.

Many research institutes around the world have carried out a large amount of research on the growth of GaN single crystal substrates using HVPE technology. As the development of GaN single crystal substrates gradually matures, more and more companies are investing a lot of manpower and resources to gradually realize the commercial development of GaN single crystal substrates. Japanese companies such as Sumitomo Electric, Mitsubishi Chemical, SCIOS (formerly Hitachi Cable), and Furukawa are the main producers of gallium nitride single crystal substrates internationally, selling 2-inch GaN substrates prepared by HVPE with a thickness of about 450 μm and a dislocation density of about 10^6 cm^{-2}. At the same time, Sumitomo Electric of Japan has also carried out the development of large-size GaN free-standing substrates, achieving the research and development of 4-inch and 6-inch GaN single crystal substrates. The earliest company in the United States in this field was TDI, which was dedicated to the commercial development of HVPE equipment, achieving the growth of various nitride semiconductor materials such as GaN, AlN, AlGaN, InN and InGaN, and was acquired by Oxford in the UK in 2008; ATMI in the United States achieved the fabrication of 2-inch GaN single crystal substrates in 2003, and was acquired by CREE around 2004; Kyma in the United States is a company engaged in GaN substrates funded by venture capital, undertaking many projects supported by the US Department of Energy (DOE) and the Defense Advanced Research Projects Agency (DARPA), forming the commercial capabilities of GaN single crystal substrates and HVPE equipment. Lumilog, a French company established in 2001, was the first company in Europe to provide GaN substrates.

The main domestic research units that use HVPE technology to carry out the manufacture of GaN single crystal substrates include Suzhou Nanowin Technology Co., Ltd., Suzhou Institute of Nano-tech and Nano-bionics, CAS, Dongguan Sino Nitride Company, Peking University, Nanjing University, Shanghai Eta Research Ltd. and other units. Among them, Suzhou Nanowin has achieved mass production of 2-inch GaN single crystal substrates and completed the research and development of 4-inch GaN single crystal substrates. The product specifications have attain an internationally advanced level.

In order to achieve the growth of GaN single crystal substrates, the following key issues need to be resolved: (1) selection of starting materials; (2) stress control; (3) defect control; (4) substrate separation technology; (5) doping technology.

Looking at the growth technology of GaN single crystal substrates, Sumitomo Electric employs a distinctive growth technology that primarily utilizes GaAs substrates and SiO_2 as a nanometer-scale mask to fabricate 2-inch diameter free-standing GaN substrates with a defect density of 2×10^6 cm^{-2} . Using this method, the company has successfully achieved batch production of 2-inch free-standing GaN substrates. In the past 5 years, Sumitomo Electric has completed the development of 4-inch and 6-inch GaN single crystal substrates through further research and development.

Other organizations all use sapphire substrates to grow GaN single crystal substrates. In order to better control defects and achieve substrate separation, Hitachi

Cable uses the "Void Assisted Separation (VAS)" technology to prepare GaN single crystal substrates. This technology first inserts a nanometer-level mesh-shaped titanium nitride film between the sapphire substrate and the GaN formation layer, and generates a GaN single crystal film on this titanium nitride film, so that a large area of GaN crystal can be simply peeled off without causing any damage to the GaN wafer. Mitsubishi Chemical uses HVPE and self-separation technology to grow a pure and transparent GaN single crystal material with a diameter of 52 mm and a thickness of 5.8 mm, with no obvious V-pits and inverse pyramids defects on the surface. Suzhou Nanowin in China uses nanostructure to control stress and has achieved the manufacture of GaN single crystal substrates with lower defects density. In terms of substrate separation technology, in addition to self-separation technologies like VAS, Laser Lift Off technology can also be used to separate GaN bulk materials from sapphire substrates.

For doping in GaN materials, Si or Ge elements are generally used to achieve N-type doping, with a typical doping concentration of 10^{18} cm^{-3}. Fe or C elements are used to prepare semi-insulating GaN materials, with a resistivity generally greater than 10^6 Ω·cm. In recent years, Japanese SCIOS company has developed high-purity GaN materials, which is also a method to fabrication semi-insulating GaN single crystal substrates.

After more than 50 years of development, GaN single crystal substrates based on HVPE technology have achieved large-scale commercialization. Lasers around 405 nm have been developed, supporting the rapid development of the blue-ray DVD industry and forming a market in the data storage field; blue lasers around 450 nm and green lasers around 520 nm have been produced, creating a rapidly growing market space in the fields of laser display, laser lighting, and laser processing. In addition to lasers, radio frequency devices and power electronic devices based on GaN single crystal substrates are developing rapidly and are expected to play a more significant role in future 5G communications, electric vehicles, and smart grids, etc.

3.2 Growth Principles and Equipment

The methods for growing thin films include Hydride Vapor Phase Epitaxy (HVPE), Metal Organic Vapor Phase Epitaxy (MOVPE), and Molecular Beam Epitaxy (MBE), etc. These epitaxial methods all need to start growing from the high-temperature substrate surface. Depending on the similarities and differences between the substrate material and the epitaxial material, the epitaxial growth methods can be divided into heteroepitaxy and homoepitaxy [7]. For example, when growing GaN thin films by epitaxy, if the substrate used is not GaN material, it is called a heteroepitaxial substrate. If the substrate is also GaN material, it is called a homoepitaxial substrate, also known as a free-standing substrate.

Currently, the method of growing GaN thin films basically uses heteroepitaxy, and the commonly used substrate materials and their physical properties are shown in Table 3.1. Comparing the material properties in the table, it can be found that there is a large lattice mismatch and thermal mismatch between GaN and its substrate materials, which will lead to a decline in the quality of the epitaxial thin film and limit the improvement of the performance of GaN-based devices [8]. If GaN is used as a substrate for homoepitaxial growth, the lattice and thermal mismatches between the substrate and the epitaxial film are eliminated, which can significantly reduce the dislocation density of the material. As a result, this approach effectively reduces stress, thereby improving both the performance and lifetime of the devices. Consequently, the growth of GaN thin films on homoepitaxial substrates has garnered significant research interest [9].

The direct growth of high-quality crystals from the melt using techniques such as the Bridgman method and zone melting is a widely adopted approach for obtaining crystalline materials. The resulting crystal ingots are sliced, precision-polished, and ultimately fabricated into substrates. This method can effectively produce Si, Ge, GaAs and other materials, but it cannot be applied to the growth of GaN free-standing substrates. Due to the extremely high melting point of GaN (about 2300 °C), growth needs to be carried out under high pressure exceeding 30 KBar. The extreme growth conditions are not easy to obtain, and even if obtained, it will lead to excessively high crystal growth costs.

Among vapor phase epitaxy methods, the growth rate of the MOVPE method is typically 3–5 μm/h, while that of the MBE method is even slower. In contrast, the HVPE method enables the rapid growth of high-quality GaN thick films owing to the fast migration and coalescence of halides on the growth surface, with growth rates generally reaching around 100 μm/h and up to 800 μm/h in the best cases [11]. Also, because the structure of the HVPE reactor is relatively simple, it is considered to be the ideal method for growing GaN free-standing substrates [12]. The entire production process of HVPE growth of GaN thick films is shown in Fig. 3.1. First, a GaN buffer layer of about 2 μm is grown on the surface of the sapphire substrate

Table 3.1 Lattice parameters and thermal expansion coefficients of commonly used heteroepitaxial substrates in GaN growth [10]

Materials	Crystal structure	Lattice constant (nm)	Coefficient of thermal expansion (10^{-6} K^{-1})
GaN	Wurtzite	a = 0.3189 C = 0.5185	5.59 (a) 3.17 (c)
Al_2O_3	Hexagonal	a = 0.4758 C = 1.2991	7.5 (a) 8.5 (c)
6H-SiC	Wurtzite	a = 0.3081 C = 1.5112	4.2 (a) 4.68 (c)
Si	Diamond	a = 0.5430	3.59
GaAs	Zinc blende	a = 0.5653	6
γ-$LiAlO_2$	Tetragonal	a = 0.5159	7.1 (a)
$MgAl_2O_4$	Cubic	a = 0.808	7.45
ZnO	Hexagonal	a = 0.325	2.9

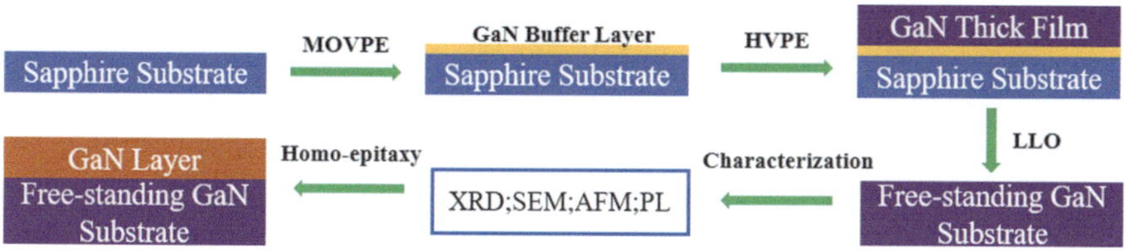

Fig. 3.1 Schematic of the GaN free-standing substrate manufaction

using the MOVPE method, and then the GaN thick film is grown on the surface of the buffer layer using the HVPE method; then the grown sample is put into the laser lift-off system to separate the GaN thick film from the sapphire substrate; usually after laser lift-off, the GaN surface will undergo an oxidation reaction to form an oxide layer, which needs to be removed by etching, and after polishing, a GaN free-standing substrate is obtained. The homo-epitaxy is used to continue growing high-quality GaN layer or GaN based device on the GaN free-standing substrate, which is well characterized by the XRD, SEM, AFM and PL.

3.2.1 *Basic Principle*

Hydride Vapor Phase Epitaxy (HVPE), was first proposed by Maruska et al. [13] in 1969 for the growth of GaN thin films. This method was used to grow the world's first GaN thin film. This technique involves transporting the halide gas containing Group III metal elements and the Group V hydride source gas to a high-temperature substrate, where a heterogeneous reaction occurs on the substrate surface to generate a GaN thin film. The HVPE growth system mainly consists of a furnace and reactor, gallium boat and gas delivery pipe, gas configuration system, and exhaust gas treatment system [14], as shown in Fig. 3.2.

The reaction chamber is usually made of quartz, which does not react with the source gas, and adopts a dual-temperature zone structure. Since metallic gallium (Ga) and ammonia (NH_3) cannot directly react to form GaN, a relatively low-temperature gallium boat area is set up in the reactor, generally maintained at around 800–900 °C, containing high-purity (better than 99.99999%) metallic gallium. HCl gas reacts with liquid metallic gallium to form GaCl gas, which serves as the Group III source gas. Carried by the carrier gas, it further enters the relatively high-temperature growth area with the Group V source gas NH_3, where the substrate surface temperature needs to be controlled at 1000–1100 °C. They mix and react on the substrate, thereby depositing GaN crystals. The reaction usually takes place at atmospheric pressure.

Since the reaction of HVPE growth GaN is a slightly exothermic reaction [15], to avoid parasitic reactions of GaN on the wall, resistance heating methods are often used, i.e., hot-wall heating. Also, since the temperature of the gallium boat and the growth area is different, the HVPE reactor must have at least two temperature zones to ensure that the reactions in the gallium boat area and the growth area can

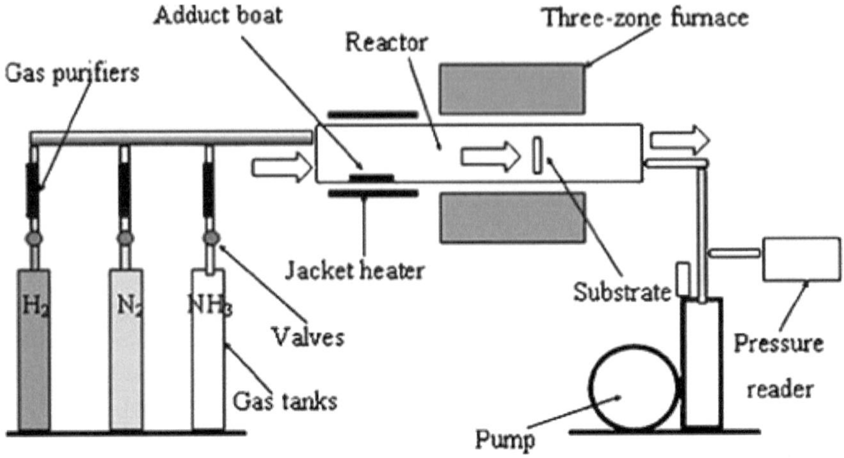

Fig. 3.2 Schematic of the HVPE growth system

proceed smoothly. To achieve precise temperature control, especially when the film has doping requirements, the reactor usually uses more heating zones, for example, the reactor in Fig. 3.2 uses three temperature zones.

To reduce the premature mixing and gas-phase parasitic reactions of the source gas after it is sprayed from the nozzle, an isolation gas, such as nitrogen, helium, argon, hydrogen, or their mixed gas, is often introduced between NH_3 and the halide gas to ensure that NH_3 and the halide gas do not react before reaching the substrate surface. The introduction of the isolation gas will affect the flow field distribution in the reaction chamber. Studies have shown that when using He or H_2 with a smaller molecular weight, it is easier to obtain laminar flow [16].

3.2.2 Research on the HVPE Growth Process

The HVPE growth process involves many fields such as thermodynamics, kinetics, fluid mechanics, heat and mass transfer, and crystal growth, each of which has a significant impact on film deposition. Among them, thermodynamics determines the driving force of the growth process, while kinetics determines the rate of different steps in the growth.

1. Thermodynamic theory of HVPE

In the HVPE growth process of GaN, thermodynamics determines the driving force of the entire growth process, such as the maximum growth rate and the components in the multi-component eutectic. In the HVPE growth system, the III group source is obtained by the reaction of HCl gas with liquid gallium, the reaction is as follows [17]:

$$\mathrm{Ga(l) + HCl(g) \rightarrow GaCl(g) + 1/2H_2(g)} \tag{3.1}$$

$$\mathrm{GaCl(g) + 2HCl(g) \rightarrow GaCl_3(g) + H_2(g)} \tag{3.2}$$

The subscripts "l" and "g" in the formula represent liquid and gaseous substances respectively. Ban [18] calculated that the conversion rate of formula (3.1) is as high as 99.5%. The V group source gas NH_3 has thermodynamic instability at high temperatures. However, only 3% of NH_3 decomposes into N_2 and H_2 gas at 950 °C [18]. The thermodynamic path of GaN deposition is as follows:

$$\mathrm{GaCl(g) + NH_3(g) \rightarrow GaN(s) + HCl(g) + H_2(g)} \tag{3.3}$$

$$\mathrm{3GaCl(g) + 2NH_3(g) \rightarrow 2GaN(s) + GaCl_3(g) + 3H_2(g)} \tag{3.4}$$

The thermodynamic constants of reactions (3.3)–(3.4) can be calculated by the following formula [19]:

$$K_p(T) = \exp(-\frac{\Delta G(T)}{RT}) \tag{3.5}$$

In the formula, ΔG(T) is the Gibbs free energy difference between the reactants and the products. The calculation formula is:

$$\Delta \mathrm{G(T)} = \Delta H_f^0(298) + S_{298}^0 \cdot C_p(T) \cdot T \tag{3.6}$$

where, $\Delta H_f^0(298)$ represents the reaction enthalpy at a temperature of 298 K, S_{298}^0 is the entropy difference at 298 K. The specific heat capacity of the reactants $C_p(T)$ varies with temperature as follows:

$$C_p(T) = a + bT + cT^2 + dT^3 + eT^4 \tag{3.7}$$

The parameters required to calculate the thermodynamic constant $\Delta H_f^0(298)$, S_{298}^0 a, b, c, d, and e are shown in Table 3.2 [19].

The relationship between the thermodynamic constants of reactions (3.1)–(3.4) and T calculated by formula (3.5) is shown in formulas (3.8)–(3.11):

$$\begin{aligned}\mathrm{K_1(T)} = \exp\big[-(&-17.92 + (2326.53/\mathrm{T}) + 1.08\ln\mathrm{T} - (2.73 \times 10^{-3})\mathrm{T} \\ &+ (1.46 \times 10^{-6})\mathrm{T}^2 - (4.70 \times 10^{-10})\mathrm{T}^3 + (5.84 \times 10^{-14})\mathrm{T}^4\big]\end{aligned} \tag{3.8}$$

$$\begin{aligned}\mathrm{K_2(T)} = \exp[-(&15.48 + (-21243.17/\mathrm{T}) + 0.9\ln\mathrm{T} - (3.61 \times 10^{-3})\mathrm{T} \\ &+(1.51 \times 10^{-6})\mathrm{T}^2 - (3.34 \times 10^{-10})\mathrm{T}^3 + (8.26 \times 10^{-14})\mathrm{T}^4\big]\end{aligned} \tag{3.9}$$

Table 3.2 Parameter values a, b, c, d, e, $\Delta H_f^0(298)$ and S_{298}^0

	$\Delta H_f^0(298)$ (kJ/mol)	S_{298}^0 J/(Kmol)	A J/(Kmol)	b J/(K^2mol)	c J/(K^3mol)	d J/(K^4mol)	e J/(K^5mol)
GaCl	− 70,542	240.25	29.585	0.0329	-5.01×10^{-5}	3.39×10^{-8}	-8.5×10^{-12}
H2	0	130.41	28.3	0.00301	-4.56×10^{-6}	4.72×10^{-9}	1.22×10^{-12}
HCl	− 92,312	186.6	30.9	− 0.011	2.04×10^{-5}	-1.06×10^{-8}	1.84×10^{-12}
GaC_{l3}	− 431,580	325.148	56.5	0.0679	-8.03×10^{-5}	$4.13 \times 10–^{8}$	7.71×10^{-12}
NH_3	− 45,940	192.451	26.1	0.0304	7.80×10^{-6}	-9.95×10^{-9}	2.15×10^{-12}
GaN	− 109,621	29.706	38.1	0.009	-7.35×10^{-9}	3.94×10^{-12}	-6.5×10^{-16}

$$K_3(T) = \exp[-(41.93 + (-11498.27/T) - 5.01\ln T + (3.74 \times 10^{-3})T - (1.17 \times 10^{-7})T^2 + (2.99 \times 10^{-10})T^3 - (5.65 \times 10^{-14})T^4] \quad (3.10)$$

$$K_4(T) = \exp[-(49.67 + (-22119.86/T) - 4.61\ln T + (1.94 \times 10^{-3})T - (4.09 \times 10^{-7})T^2 + (1.33 \times 10^{-10})T^3 - (9.78 \times 10^{-14})T^4] \quad (3.11)$$

By calculating formula (3.6), it is known that the Gibbs free energy difference ΔG of chemical reactions (3.1)–(3.4) is less than zero, indicating that these reactions can proceed spontaneously. However, the results of thermodynamic calculations only determine the driving force of the entire growth process, such as the maximum growth rate, and the rate of different steps is actually affected by kinetics.

2. Kinetic Analysis of HVPE

The chemical reaction process involved in the growth of GaN thin films by the HVPE method is significantly different from the growth of other III–V semiconductors. For example, when growing GaAs using the HVPE method, AsH_3 is thermally decomposed to generate As_4 and As_2 molecules, which have strong volatility and chemical activity and can participate in thin film growth. However, the N_2 molecules generated by the decomposition of NH_3 are very stable and cannot participate in the growth of GaN thin films at the growth temperature. The growth of GaN by HVPE mainly relies on the inert decomposition of NH_3 molecules on the substrate surface [20], therefore, the growth of large-size uniform thin films requires NH_3 effectively and evenly reaching the growth surface. In actual operation, to ensure that the growth of GaN is in the halide gas transport control area, an excess of NH_3 is often introduced.

Cadoret and others [21–23] proposed a detailed GaN surface reaction growth path by analogy with the GaAs growth path based on thermodynamic analysis.

1. First, NH_3 is adsorbed on the Ga surface, then the NH_3 molecule decomposes to produce active nitrogen atoms, and the nitrogen atoms adsorb GaCl molecules to form NGaCl. The corresponding reaction steps are as follows:

$$V + NH_3(g) \Leftrightarrow NH_3 \quad (3.12)$$

$$NH_3(g) \Leftrightarrow N + \frac{3}{2}H_2(g) \quad (3.13)$$

$$N + GaCl(g) \Leftrightarrow NGaCl \quad (3.14)$$

V represents the adsorption vacancy of NH_3.

2. The difference in the GaN generation path mainly lies in the process of NGaCl removing Cl atoms. Figure 3.3 is a schematic diagram of these two growth paths.

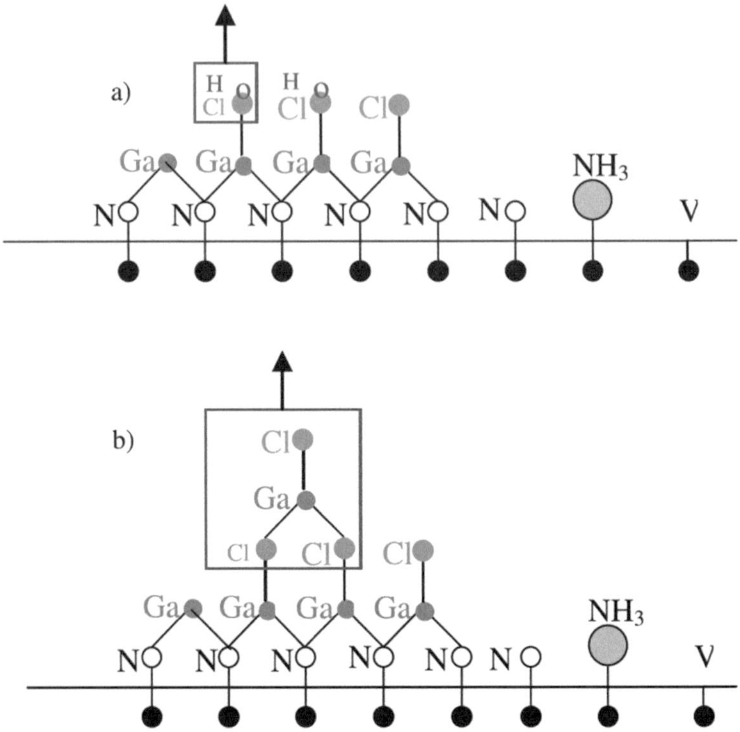

Fig. 3.3 Schematic diagram of GaN surface reaction. **a** H_2 mechanism; **b** $GaCl_3$ mechanism

3. H_2 path:

$$2\text{NGaCl} + \text{H}_2 \Leftrightarrow 2\text{NGa} - \text{ClH} \tag{3.15}$$

$$\text{NGa} - \text{ClH} \Leftrightarrow \text{NGa} + \text{HCl}(g) \tag{3.16}$$

Overall reaction:

$$\text{V} + \text{NH}_3(g) + \text{GaCl}(g) \Leftrightarrow \text{NGa}(s) + \text{HCl}(g) + \text{H}_2(g) \tag{3.17}$$

In this path, NGaCl mainly relies on adsorbed H_2 molecules to remove Cl atoms and generate HCl.

$GaCl_3$ path:

$$2\text{NGaCl} + \text{GaCl}(g) \Leftrightarrow 2\text{NGa} - \text{GaCl}_3 \tag{3.18}$$

$$\text{NGa} - \text{GaCl}_3 \Leftrightarrow 2\text{NGa} + \text{GaCl}_3(g) \tag{3.19}$$

Overall reaction:

$$2V + 2NH_3(g) + 3GaCl(g) \Leftrightarrow 2NGa(s) + GaCl_3(g) + 3H_2(g) \quad (3.20)$$

In this path, NGaCl removes Cl atoms by interacting with GaCl molecules to generate $GaCl_3$.

Trassoudaine and Aujol found through experiments that adding extra HCl to the main carrier gas N_2 can reduce parasitic reactions (but not completely eliminate them), and the substrate surface can still maintain a constant high growth rate [24]. Based on this, Cadoret proposed a $GaCl_2$–HCl mixed path to explain the above phenomenon.

$GaCl_2$–HCl mixed path:

$$NGaCl + GaCl(g) \Rightarrow NGa + GaCl_2(g) \quad (3.21)$$

$$NGa - HCl \Leftarrow NGa + HCl(g) \quad (3.22)$$

$$2NGaCl + H_2(g) \Leftarrow 2NGa - HCl \quad (3.23)$$

$$NGaCl + GaCl(g) \Leftarrow NGa + GaCl_2(g) \quad (3.24)$$

Gas phase equilibrium equation:

$$HCl(g) + GaCl(g) \Leftarrow GaCl_2(g) + \frac{1}{2}H_2 \quad (3.25)$$

In addition to the reaction mechanism proposed by Cadoret and others, Karpov and others [25] proposed another theory to describe surface reactions, that is, using the quasi-thermodynamic assumption [17, 20] on the substrate surface, various reactants reach thermodynamic equilibrium, and the substances after the reaction comply with the mass conservation equation. Adsorption and desorption use the Hertz-Knudsen equation to describe their reaction mechanism. Using this method, many difficult-to-verify reaction mechanisms can be avoided in the numerical simulation process.

3.2.3 *Classification and Research Status of HVPE Reactors*

At present, the growth of GaN free-standing substrates by HVPE has become a research hotspot at home and abroad. In the GaN HVPE growth system, the reactor is the most core part, and the gas transport, gas phase and surface reaction steps that play a decisive role in the growth of GaN thin films all occur in the reactor. HVPE reactors can be divided into two categories according to the flow direction of the main gas relative to the substrate, namely horizontal reactors and vertical reactors.

1. Horizontal reactor

The most commonly used reactor in the HVPE method is the horizontal reactor. As shown in Fig. 3.4a, the traditional horizontal reactor has a gas flow direction parallel to the substrate direction. There is a serious reactant depletion phenomenon along the path during the growth process, which will cause uneven thickness of the crystal thin film. Figure 3.4b is a variant improved from (a), that is, a tapered horizontal reactor obtained by tilting the substrate, and its gas flow direction is at a certain angle with the substrate, which can reduce the unevenness of thin film deposition to a certain extent.

In the horizontal HVPE reactor, regardless of whether the substrate is placed parallel or tilted, the uniformity of thin film deposition can be improved by rotating the substrate. In addition, increasing the gas flow rate is also a method to improve the uniformity of thin film deposition, but this will lead to a decrease in deposition efficiency (the deposition efficiency is defined as the molar ratio of Ga incorporated into the GaN lattice to the molar amount of Ga atoms in the intake).

2. Vertical reactor

Figure 3.5 is a schematic diagramof a vertical HVPE reactor. The direction of the gas flow in the reaction chamber can be the same as (Fig. 3.5a) or opposite to (Fig. 3.5b) the direction of gravity. When the two directions are the same, the substrate surface is placed upwards at the bottom of the reactor, and the gas flow is sprayed in from the top of the reaction chamber, that is, the reactor is downflow. When the two directions are opposite, the substrate surface is placed downwards at the top of the reaction chamber, and the gas flow goes from bottom to top to reach the substrate surface, that is, the reactor is upflow.

In the case where the direction of airflow is the same as the direction of gravity, there are several problems: (1) There are strong parasitic reactions in the HVPE growth process of GaN, which easily generate a large amount of parasitic particles on the walls of the reaction chamber and the air inlet structure. These large particles easily reach the substrate surface under the action of gravity, leading to a decrease in film quality; (2) Due to the opposite directions of buoyancy-driven convection and

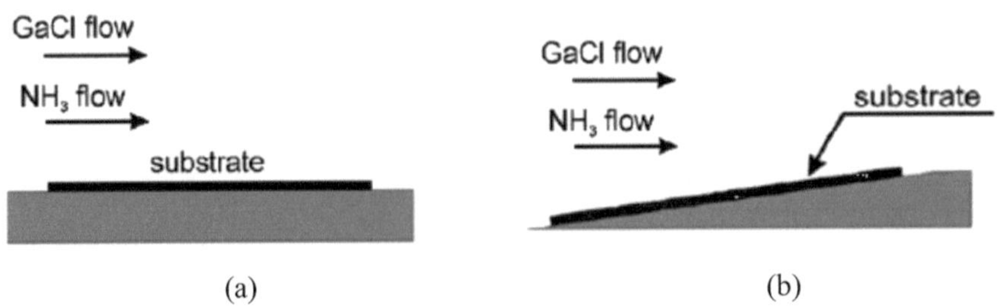

Fig. 3.4 Horizontal HVPE reactor: **a** traditional type; **b** flow reducer

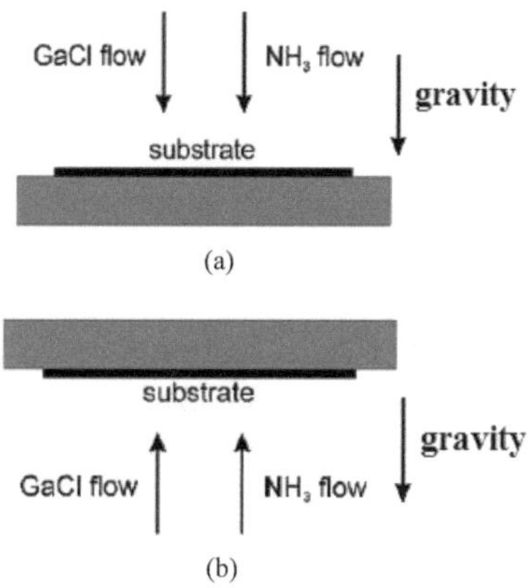

Fig. 3.5 Vertical HVPE reactor: **a** downflow; **b** upflow

thermal convection, the distribution of reactant concentration above the substrate is uneven, making it difficult to achieve uniform GaN growth. When the direction of airflow is opposite to the direction of gravity, the aforementioned problems in the upright vertical reactor will be overcome, and the quality of GaN thin films grown in the inverted vertical HVPE reactor will be improved to a certain extent, and a higher deposition uniformity can be achieved, but the film deposition rate is relatively low.

Because the nozzle of the vertical reactor is very close to the substrate, it can effectively reduce the pre-reaction of the gas, improve the utilization efficiency of the source gas, and achieve a higher growth rate than the horizontal reactor. At the same time, the uniformity of film growth can be further improved by rotating the substrate, therefore, the vertical reactor is more suitable for commercial production than the traditional horizontal reactor.

3. Current status of GaN-HVPE research

The key to achieving high-quality film growth is a carefully designed and reliably operated HVPE reactor. Due to the complex chemical reaction process in GaN-HVPE growth, there is no fully mature commercial GaN HVPE equipment so far. Most research groups use homemade or customized HVPE reactors for GaN growth research, and the optimization design of the reactor is still a hot topic in GaN HVPE research. The following introduces the research conducted by scholars at home and abroad on several common HVPE reactors.

Dam et al. [26] conducted a simulation study on the horizontal reactor shown in Fig. 3.6. This reactor reduces the unevenness of film deposition in the horizontal reactor caused by the reactant depletion phenomenon by changing the angle of the reaction gas inlet pipe. The simulation results found that changing the angle between the GaCl inlet nozzle and the horizontal direction has a great impact on the GaN growth rate and uniformity, and this result was experimentally verified and the same conclusion was obtained. They also studied the effect of different carrier gases on GaN growth and found that H_2 is more conducive to GaN growth than N_2. Based on this horizontal reactor that can grow 2-inch wafers, the group continued to use

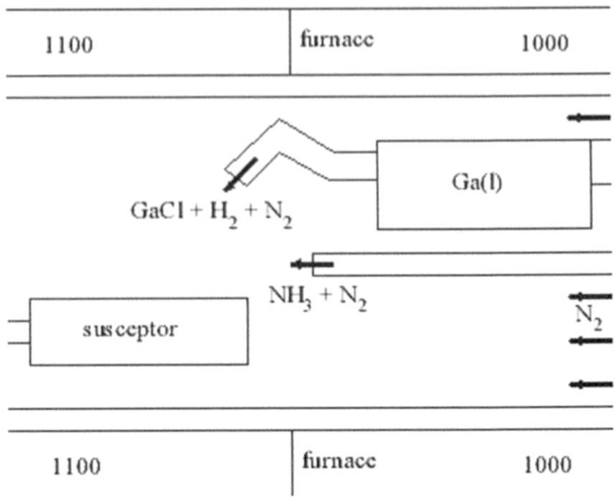

Fig. 3.6 Horizontal HVPE reactor

numerical simulation methods and similarity principles to expand the reactor to a horizontal reactor that can grow 4-inch wafers, and obtained similar flow and the same growth rate [27].

Meng [28] conducted relevant research on the impact of the relative positions of GaCl, NH_3 pipelines and the substrate in the horizontal HVPE reactor on the distribution of reactants. Numerical simulation research found that the distribution of GaCl above the substrate heavily depends on the vertical distance between the GaCl inlet and the substrate, while the horizontal distance has a smaller impact on the GaCl concentration distribution; the distribution of NH_3 above the substrate has a weaker dependency on the NH_3 inlet position.

Figure 3.7 is a commercial HVPE reactor developed by the German company Aixtron, which is characterized by using the tiny nozzles densely distributed on the showerhead to spray the GaCl gas is sprayed towards the substrate from a close distance, allowing the GaCl gas to be evenly distributed above the substrate, thereby obtaining a uniform GaCl distribution on the substrate, which also improves the utilization rate of GaCl gas. However, the parasitic reactions in the spray HVPE reactor are quite severe, with GaCl and NH_3 starting to mix on the left side of the substrate, and the parasitic particles produced by the reaction can easily block the spray head [29].

Rechter et al. [30] optimized the spray HVPE reactor shown in Fig. 3.7 by adding a gas regulation box to the NH_3 inlet pipe, moving the mixing position of GaCl and NH_3 to above the substrate, thereby optimizing the flow field in the reaction chamber and reducing parasitic deposition. By selecting appropriate operating parameters and epitaxial substrates, they grew crack-free and high-quality GaN crystals.

Lan et al. [31] used numerical simulation methods to study the effects of the direction of the spray head nozzle and the placement of the wafer in the horizontal spray HVPE reactor on material transport and GaN growth rate. They observed that changing the direction of the spray head nozzle is beneficial to the increase of GaN growth rate, but the uniformity of deposition has not been improved; they also found

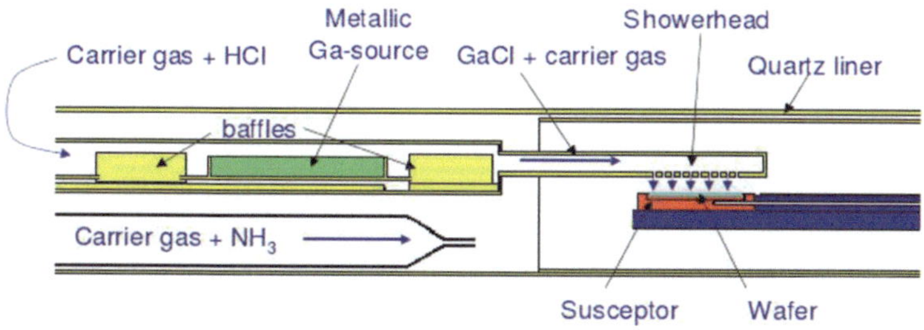

Fig. 3.7 HVPE reactor with showerhead

that the asymmetric placement of the wafer is an effective method to improve the uniformity of deposition.

Wu et al. [32] optimized the arrangement of the spray head in the horizontal reactor with GaCl on top and NH_3 on the bottom to NH_3 on top and GaCl spray head on the bottom, as shown in Fig. 3.8. After the improvement, they successfully grew smooth and uniform, crack-free GaN films without rotating the substrate.

The vertical HVPE reactor shown in Fig. 3.9 is also a commonly used reactor structure. This inverted reaction chamber can effectively prevent the uneven distribution of reactant concentration caused by thermal buoyancy, and can further improve the uniformity of film growth by rotating the substrate. Monemar et al. [34] simulated the flow field distribution in this reaction chamber and analyzed the influence of carrier gas on the flow.

Safvi et al. [35] introduced a HVPE reactor with a concentric ring inlet, as shown in Fig. 3.10. Due to the symmetrical structure of the air inlet, the radial growth rate above the substrate is the same, and the growth uniformity is good. They discussed the influence of the distance between the air inlet and the substrate on the growth rate through numerical simulation and experiments, and found that too large or too small distances may cause a decrease in crystal quality. In addition, they found that a low concentration of NH_3 corresponds to the production of polycrystalline films, and a low V/III ratio also corresponds to a decrease in the quality of GaN films.

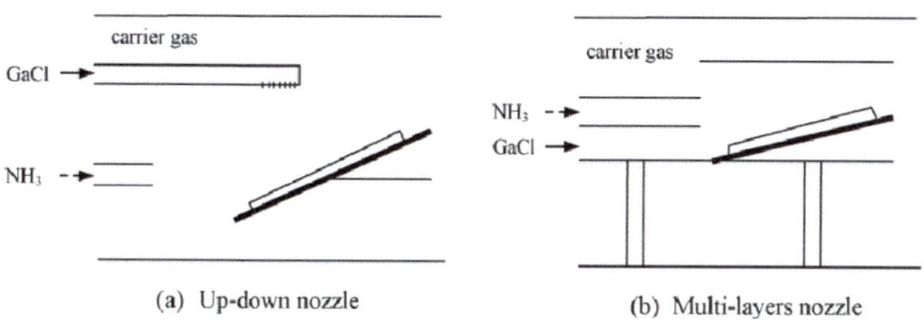

Fig. 3.8 Two kinds of nozzle structures used in HVPE reactor (a) the "old" up–down nozzle; (b) the "new" multi-layers nozzle [33]

Fig. 3.9 Vertical HVPE reactor with tube inlets on the bottom

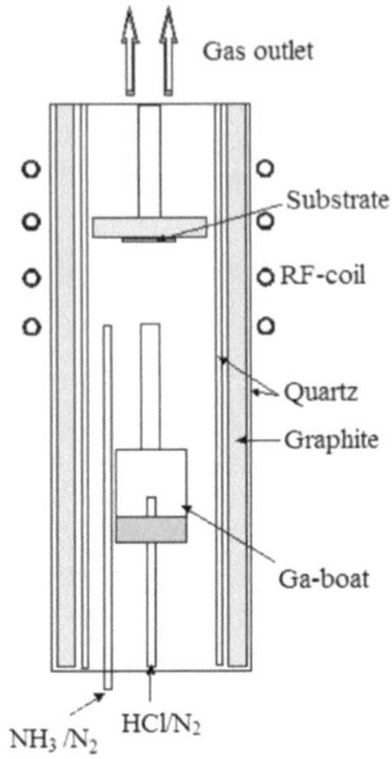

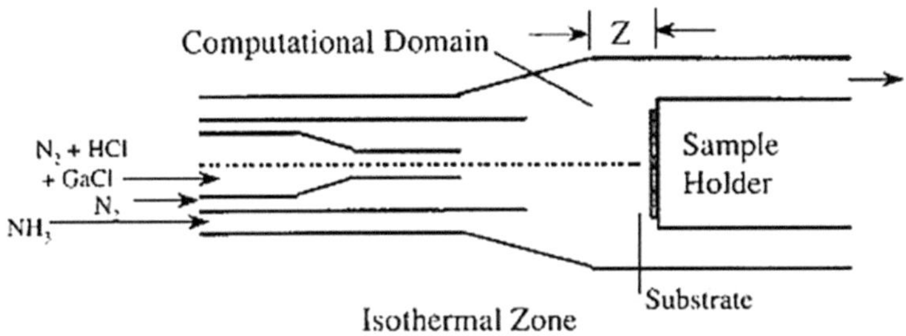

Fig. 3.10 Vertical HVPE reactor with annular inlets

4. Process Parameter Optimization

Han et al. [14] conducted numerical simulation studies on a multi-wafer large-size HVPE reactor that can grow 6-inch GaN films. They found that increasing the carrier gas flow rate within a certain range can improve the uniformity of deposition and change the location of the maximum deposition rate area.

Liu et al. [36] periodically controlled the inflow of source gas in a multi-plate reactor, which improved the uniformity of thin film growth and reduced the defect density.

Yu et al. [8] kept the flow of NH_3 constant, increased the flow of HCl (reducing the V/III ratio), successfully increased the growth rate, and also improved the crystal quality. This phenomenon may be due to the increased partial pressure of GaCl, which leads to an increase in its supersaturation, resulting in a smaller critical nucleation radius, manifesting as higher nucleation and growth rates.

Xiu et al. [9] found that introducing additional HCl gas downstream of the gallium boat in the reactor can improve the crystal quality and properties of GaN thin films, which may be due to changes in the supersaturation of the growth surface.

Momemar et al. [34] conducted a three-dimensional numerical simulation of a vertical HVPE reactor and found that increasing HCl would cause the formation reaction of GaN thin films to proceed in the reverse direction, thereby reducing the growth rate. In severe cases, this increase in HCl could even lead to the etching of GaN thin films. When the flow rate of the carrier gas N_2 is small (<11 slm), the flow is turbulent; when the flow rate is increased (3–10 slm), it is approximately laminar flow. Ideal HVPE growth is intended to produce GaN crystals solely on the substrate surface; however, in actual reactors, reactions can occur wherever NH_3 and GaCl gases coexist, particularly in the area between the source gas outlet and the substrate. Although parasitic reactions may also occur downstream of the tray, the impact on the growth of the substrate surface is small.

Karpov et al. [20] conducted a full three-dimensional simulation of a self-made horizontal reactor, using the surface adhesion coefficient to simulate the growth rate of the substrate surface, and the simulation results were in good agreement with the experimental values. The simulation results show that gas flow and material transport have a significant impact on the V/III distribution and growth rate on the substrate surface, and the growth rate is limited due to Cl atoms and H atoms occupying the adsorption sites of GaCl.

3.3 Stress and Defect Control

3.3.1 Stress Control

1. Source of Stress

The development of high-quality gallium nitride material growth technology is the key to promoting the application of gallium nitride-based materials. Due to the very high melting point and melting pressure of gallium nitride, the growth of gallium nitride single crystal material is very difficult. Since gallium nitride decomposes into metallic gallium and nitrogen gas before the melting point, it is difficult for gallium nitride to grow crystals like silicon by liquid phase pulling. The size of single crystals grown under high temperature and high pressure is small, which cannot meet the needs of device applications and is difficult to commercialize.

The absence of suitable bulk single-crystal substrates necessitates the heteroepitaxial growth of gallium nitride (GaN) materials. The crystalline quality of GaN is fundamentally determined by the choice of substrate and the optimization of the growth process. Mismatches in lattice constants and coefficients of thermal expansion between the heterosubstrate and GaN give rise to high densities of crystalline defects and significant internal stress within the gallium nitride material. A key challenge in GaN material fabrication is to improve crystal quality by selecting optimal foreign substrates and developing appropriate growth processes to mitigate the strain induced by lattice and thermal expansion mismatch.

2. Heteroepitaxial substrate materials and their mismatch with gallium nitride

In the heteroepitaxial growth of gallium nitride materials, the properties and structure of the substrate material have a significant impact on the quality of the epitaxy, and may even affect the quality of subsequent devices. Therefore, the choice of substrate material is crucial. Generally, the choice of substrate material involves consideration of the following factors:

3. Lattice matching between the substrate and gallium nitride

Lattice matching involves, first and foremost, the compatibility of crystal structures, for which substrates belonging to the same crystal system are preferentially selected. Secondly, it entails matching of lattice constants, both within the growth plane and along the direction normal to the surface, where a smaller lattice mismatch is highly desirable.

4. Matching of thermal expansion coefficients between the substrate and gallium nitride

A significant disparity in thermal expansion coefficients can induce excessive thermal stress during cooling post-growth, potentially leading to cracking or delamination. Additionally, during subsequent device operation, heat generated may adversely affect device performance or even result in failure.

5. Chemical compatibility between the substrate and gallium nitride

The substrate must exhibit chemical stability: it should not react chemically with GaN and should demonstrate good adhesion during epitaxy to prevent peeling. Moreover, it should maintain thermal stability at growth temperatures without undergoing decomposition throughout the process.

6. Complexity and cost of fabrication

To satisfy commercial development requirements, substrates should be relatively straightforward to manufacture, available in appropriate sizes, and offered at a reasonable cost.

Based on these criteria, commonly utilized substrates for the epitaxial growth of GaN include sapphire (Al_2O_3), silicon carbide (SiC), silicon (Si), gallium arsenide (GaAs), and lithium aluminate ($LiAlO_2$). Table 3.1 summarizes the crystallographic parameters of these substrates in comparison with those of GaN.

Measuring Stress in Gallium Nitride Using Raman Spectroscopy

In the heteroepitaxial thin film of semiconductors, due to the difference in lattice constants or thermal expansion coefficients between the epitaxial layer and the substrate, stress is generated in the epitaxial layer. Changes in stress can cause shifts in the position of the phonon Raman frequency. If the effect of biaxial stress (compressive stress and tensile stress) on the shift of the Raman spectrum peak frequency is known, the stress can be estimated based on the shift of the Raman peak frequency. Through this method, the value and distribution of stress in the epitaxial layer can be measured.

The main source of stress in gallium nitride heteroepitaxial structures at room temperature is the stress generated by thermal mismatch. Gallium nitride grown on a sapphire substrate is subjected to compressive stress, while gallium nitride grown on silicon carbide and silicon substrates is subjected to tensile stress. In the hexagonal gallium nitride epitaxial layer, apart from the E_2(low) branch phonon, compressive and tensile stresses can cause blue shifts or red shifts of each phonon peak. In the phonon Raman peaks of gallium nitride, the E_2(high) branch phonon mode Raman scattering has a narrower half-width and greater intensity when backscattered on the (0001) plane. Therefore, the frequency position of the Raman peak of the E_2(high) branch phonon mode is generally used to measure the stress in gallium nitride materials. The phonon transport is known to exhibit a linear correlation with the applied stress. To measure the stress magnitude through the E_2(high) branch phonon frequency, one must first know the position of the Raman peak at zero stress, and calculate the Raman stress factor according to the following formula:

$$\Delta\omega = k\sigma \tag{3.26}$$

where $\Delta\omega$ is the movement of the phonon energy of gallium nitride under the action of biaxial stress σ in the direction perpendicular to the growth axis (C axis). In the gallium nitride epitaxial layer, the calibration factor κ for the movement of the E_2(high) branch phonon mode Raman peak caused by biaxial stress has been measured to be 4.2 ± 0.3 cm^{-1}/GPa [37]. In heteroepitaxial gallium nitride on a sapphire substrate, due to the thermal expansion coefficient of sapphire (7.5×10^{-6} K^{-1}) being greater than the thermal expansion coefficient of gallium nitride (5.59×10^{-6} K^{-1}), at room temperature, the sapphire is subjected to the tensile stress, and the gallium nitride is subjected to the compressive stress. Through micro-Raman spectroscopy scanning, starting from the gallium nitride/sapphire interface and scanning to the surface of gallium nitride, the compressive stress magnitude at different positions of the gallium nitride film can be measured through the movement of the Raman E_2(high) branch phonon frequency (as shown in Fig. 3.11).

Double-layer film stress and curvature

The stress in the heterostructure was calculated using the Olsen and Ettenberg model [39], as shown in Fig. 3.12. We consider the structure with length L, width W, curvature κ, force F_i, and moment M_i, Young's modulus E_i, and film thickness t_i, where 1 and 2 represent sapphire and GaN respectively. When no external force is applied, the force and moment are in equilibrium:

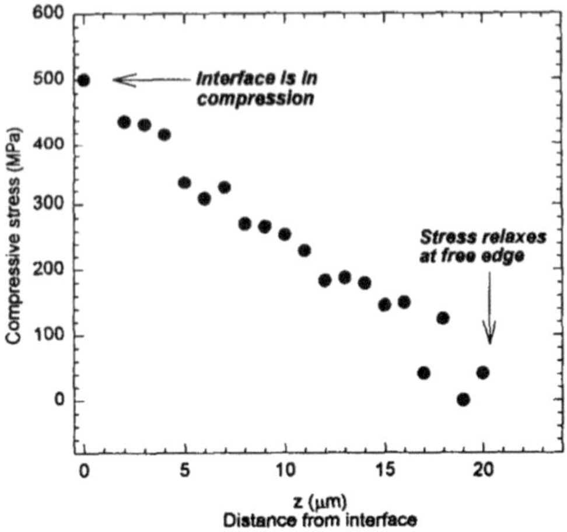

Fig. 3.11 Micro-Raman stress measurement of 20 μm thick HVPE GaN film [38]

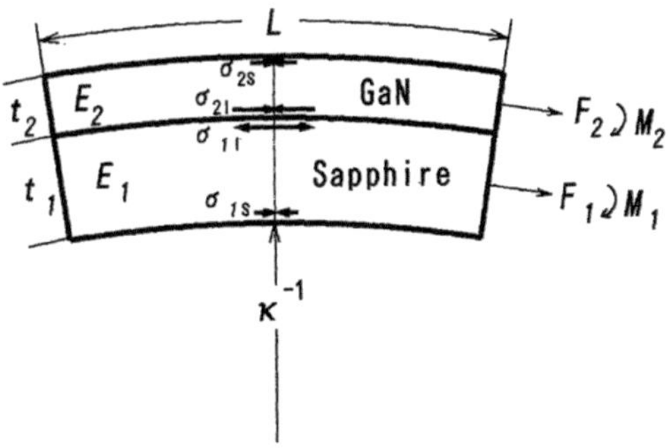

Fig. 3.12 Stress in the heterostructure (3.27)

$$\sum_{i=1}^{2} F_i = 0;$$
$$\sum_{i=1}^{2}\left[E_i\frac{t_i^3 L\kappa}{12} + F_i\left\{\sum_{j=1}^{i}(t_j) - \frac{t_i}{2}\right\}\right] = 0. \tag{3.28}$$

The positive direction of force and moment is the same as the positive direction of the vector shown in the figure, that is, tensile force is positive. Another equation is obtained by solving the strain $\varepsilon_{1,2}$ at the interface of sapphire and gallium nitride:

$$\varepsilon_{1,2} = \frac{F_2}{E_2 t_2 L} - \frac{F_1}{E_1 t_1 L} + \frac{(t_1 + t_2)\kappa}{2}. \tag{3.29}$$

Here, the strain $\varepsilon_{1,2} = \Delta T^*(\alpha_1 - \alpha_2)$, where α_1 and α_2 are the thermal expansion coefficients of sapphire and gallium nitride respectively, and ΔT is the temperature difference between the growth temperature and room temperature. The three equations are used to solve for the three unknowns F1, F_2, and κ.

To solve the equation, the stress $\sigma_i(z)$ of the i-th layer is given by F_i/t_iW and the bending stress $E_i\kappa(z - t_i/2)$, where z is the axial coordinate. Considering that the disc $L \approx W$ is spherically bent, the uniaxial stress is transformed into biaxial stress, $\sigma_i(z)|_{2D} = \sigma_i(z)|_{1D}/(1 - \nu)$, where ν is the Poisson's ratio.

The influence of sapphire substrate thickness on the curvature/bending of GaN/ Sapphire structure at room temperature (as shown in Fig. 3.13) (Table 3.3).

The thickness of the gallium nitride film and the thickness of the sapphire substrate have a significant impact on the warping of the wafer at room temperature. By fitting the results of the above formulas, the warping and strain of the wafer at different thicknesses of gallium nitride and sapphire are obtained. For a gallium nitride thickness of 100–300 μm and a sapphire thickness of 100–1000 μm, the actual results and simulation calculations match well [42] (as shown in Fig. 3.14).

Through the simulation calculation of the formula, the stress distribution from the interface to the surface in the gallium nitride film of different thicknesses at room temperature can be obtained [43] (as shown in Fig. 3.15):

In order to characterize the stress situation of the gallium nitride film during growth, the variable temperature XRD test method is used to test the relationship between the curvature radius of 50 μm HVPE GaN/Sapphire and temperature (as shown below). As the temperature increases, the compressive stress in the gallium

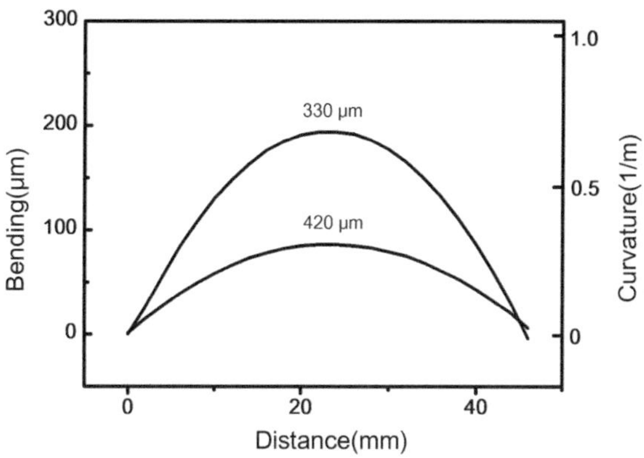

Fig. 3.13 Curvature of the wafer after growing 20 μm GaN on sapphire with different thicknesses by HVPE [40]

Table 3.3 Curvature radius at room temperature for GaN films of different thicknesses grown on sapphire substrates by HVPE [41]

Sample	A	B	C	D	E	F	G	H	I
Thickness (μm)	1.0	2.3	4.8	6.6	12	23	35	42	100
r (m)	16.4	7.2	5.0	3.5	3.3	1.7	1.4	1.1	0.7

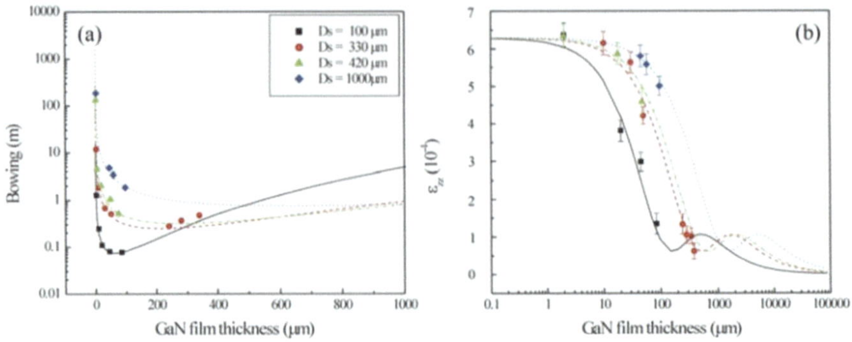

Fig. 3.14 Left—Warping at room temperature; Right—Strain at room temperature; Solid line represents simulation results, dots represent experimental results

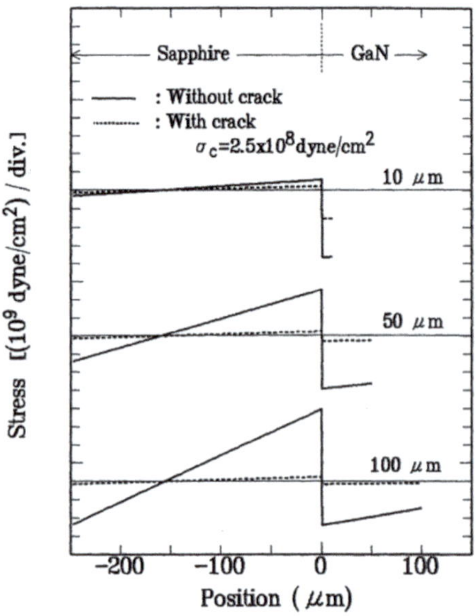

Fig. 3.15 Stress distribution from the interface to the surface in gallium nitride films of different thicknesses [43]

nitride film gradually decreases. By deriving the fitting curve, it can be judged that the stress received by the GaN film during growth is tensile stress [42] (as shown in Fig. 3.16).

GaN/Sapphire Stress Evolution

During the heteroepitaxial growth of gallium nitride (GaN) on sapphire substrates, the GaN layer experiences tensile stress induced by the sapphire, resulting in a concave warping of the sample. As biaxial tensile stress accumulates and surpasses the yield limit of the GaN film, cracks develop to alleviate the strain. These cracks manifest

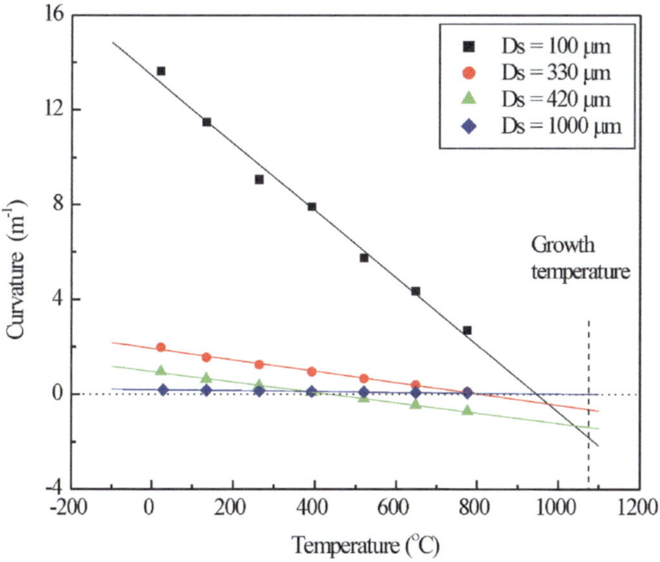

Fig. 3.16 Relationship between the curvature radius of 50 μm HVPE GaN/Sapphire and temperature

on the M-plane of the GaN crystal and are oriented at an angle of 60° relative to the gallium plane.

After the epitaxial growth is completed, the sample is cooled from the growth temperature to room temperature. Since the thermal expansion coefficient of sapphire is greater than that of gallium nitride, the sample gradually changes from concave warping to flat during the cooling process, and then to convex warping. At this time, the gallium nitride is subjected to compressive stress from the sapphire substrate, and the sapphire is subjected to tensile stress from the gallium nitride. Depending on the magnitude of the stress, the following situations may occur:

1. Compressive stress is below the yield limit of GaN, and tensile stress is also below the yield limit of sapphire. Under these conditions, when cooled to room temperature, the sample remains intact. Neither the sapphire nor the GaN requires crack formation to relieve strain. This scenario serves as a foundation for post-growth separation techniques, enabling methods such as laser lift-off, chemical etching for substrate removal, or mechanical grinding for substrate separation
2. Compressive stress exceeds the yield limit of GaN while tensile stress remains below that of sapphire. In this case, lateral cracks (parallel to the gallium-face) form within the GaN layer, leading to its separation from the sapphire substrate. This phenomenon represents a fundamental mechanism of cooling-induced self-separation. The interface at which separation occurs typically manifests in one of two ways: ① At a weakly bonded interface within the GaN structure, such as at the GaN/sapphire interface or through intentionally engineered weak connections (e.g. mask interfaces, interlayers, or buffer layers) (as shown in Fig. 3.17); ② At

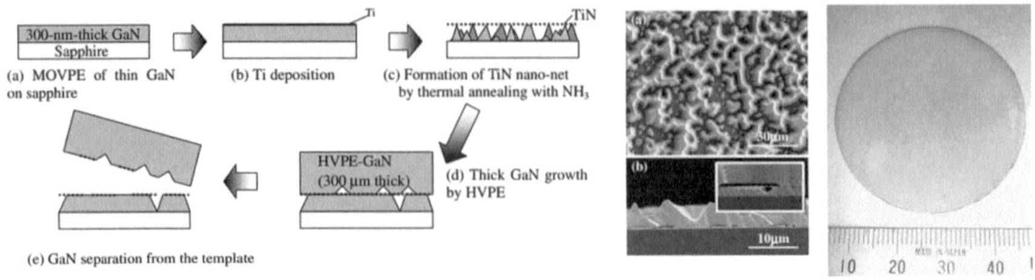

Fig. 3.17 Self-separation caused by compressive stress during cooling from the TiN weak connection [45]

a specific depth within the GaN film where separation takes place at locations that maximize strain relief and minimize total system energy [44] (as shown in Fig. 3.18). If new vertical cracks develop during this process, it may result in fragmentation of self-separated GaN (as shown in Fig. 3.19). However, if stress is well-controlled and only lateral cracks occur, it is possible to obtain a complete free-standing GaN wafer equivalent in size to that of its substrate. (as shown in Fig. 3.20).

3. Compressive stress remains below the yield limit of GaN while tensile stress exceeds that of sapphire. In this scenario, no lateral cracks are formed in the GaN film; instead, vertical cracks emerge within the sapphire substrate and may propagate into the underlying GaN layer resulting in fractures perpendicular to both growth surfaces. (as shown in Fig. 3.21).
4. Both compressive and tensile stresses exceed their respective yield limits for both materials involved: compressive stress surpasses that of GaN while tensile stress exceeds that of sapphire. Consequently, lateral cracks develop with in the GaN layer, and vertical cracks also appear in the sapphire substrate.

3. Common methods of stress control

The main source of stress in gallium nitride is lattice mismatch and thermal mismatch with the heterogeneous substrate. The main ways to relieve stress are:

1. Pretreat the heterogeneous substrate. Reduce interface stress to obtain a thicker gallium nitride film. For example, laser treatment of the sapphire substrate greatly increases the thickness threshold at which gallium nitride film cracks appear [50] (as shown in Fig. 3.22); wet etching of the sapphire substrate before growth has been shown to effectively relieve the stress during epitaxy.
2. Buffer layers and insertion layers. For example, high-temperature AlN buffer layers relieve stress [52, 53] (as shown in Fig. 3.23), AlGaN buffer layers reduce stress when growing GaN on SiC substrates [54], SiC is inserted to suppress cracks when growing GaN on silicon substrates [55], CrN buffer layers is employed to relieve stress [56];
3. Control at the beginning of growth. For example, initial gallium source pretreatment results in hundreds of micrometers of crack-free gallium nitride thick film

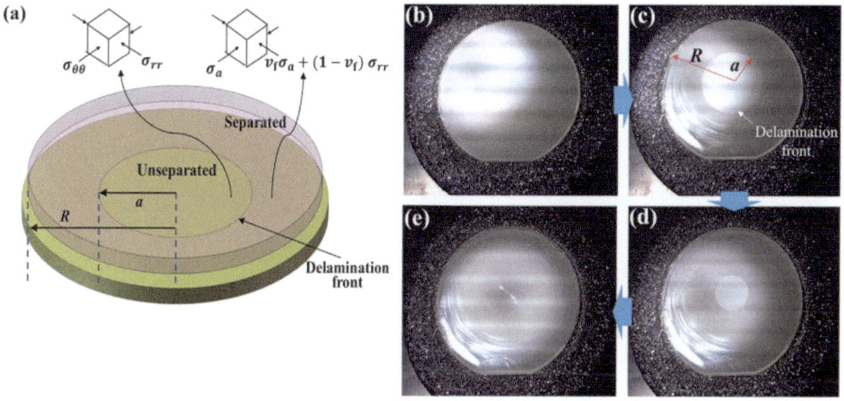

Fig. 3.18 (**a**) Model for stress-induced self-separation process (**b**)–(**e**) Stress-induced self-separation in thick GaN film, which completes following a circular delamination front beginning along the edge of wafer and shrinking into the center [44]

Fig. 3.19 Compressive stress generates transverse cracks, and longitudinal cracks result in incomplete gallium nitride [46]

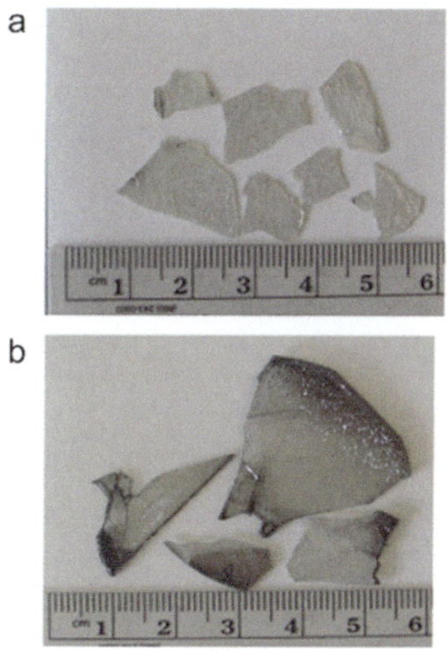

[57, 58] (as shown in Fig. 3.24), controlling the initial island nucleation density increases the crack-free thickness to 200 μm [59];

4. MOCVD template treatment. Many HVPE growths are not directly on heterogeneous substrates, but use a 0.5–4 μm gallium nitride template grown by MOCVD as the substrate. Pre-treating the template is also beneficial to reduce stress and obtain crack-free gallium nitride wafers. For example: using the template as a nanocolumn to obtain 400 μm crack-free gallium nitride [60–62] (as shown in Fig. 3.25), reducing stress by regrowing after phosphoric acid etching of the template [63] (as shown in Fig. 3.26).

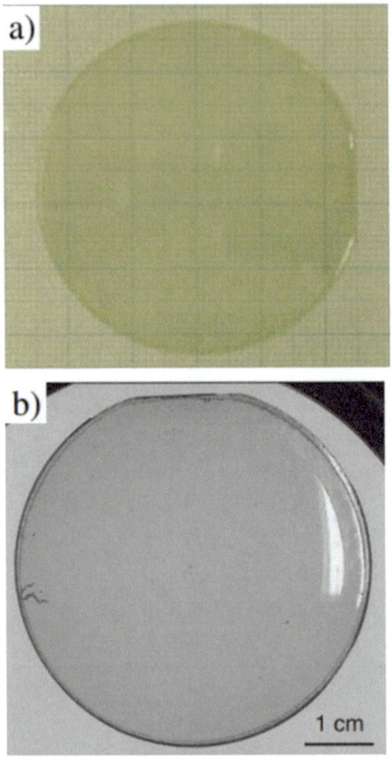

Fig. 3.20 After cooling, 140 μm. **a** [47] 250 μm. **b** [48] Crack-free GaN/Sapphire

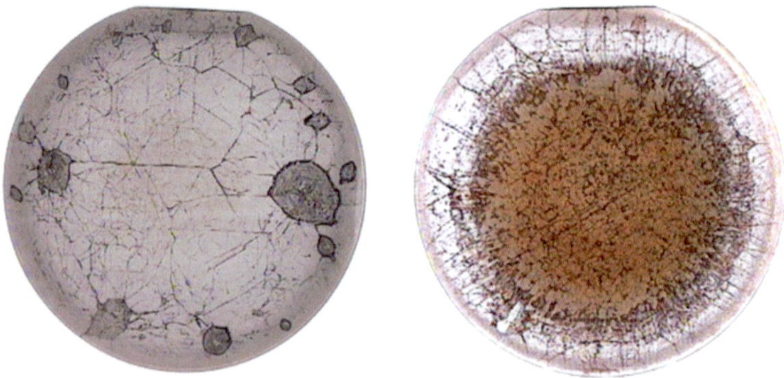

Fig. 3.21 Lateral cracks in gallium nitride and cracks in sapphire sheets after cooling [49]

5. Mask and ELOG. Through the mask and ELOG method, stress can be effectively released, and a complete wafer can be obtained, including SiO_2 mask [64, 65] (as shown in Fig. 3.27), SiN mask [66, 67] (as shown in Fig. 3.28), TiN mask [45, 68, 69], Pendeo GaN template [67] or air-bridged [70] (as shown in Fig. 3.29).

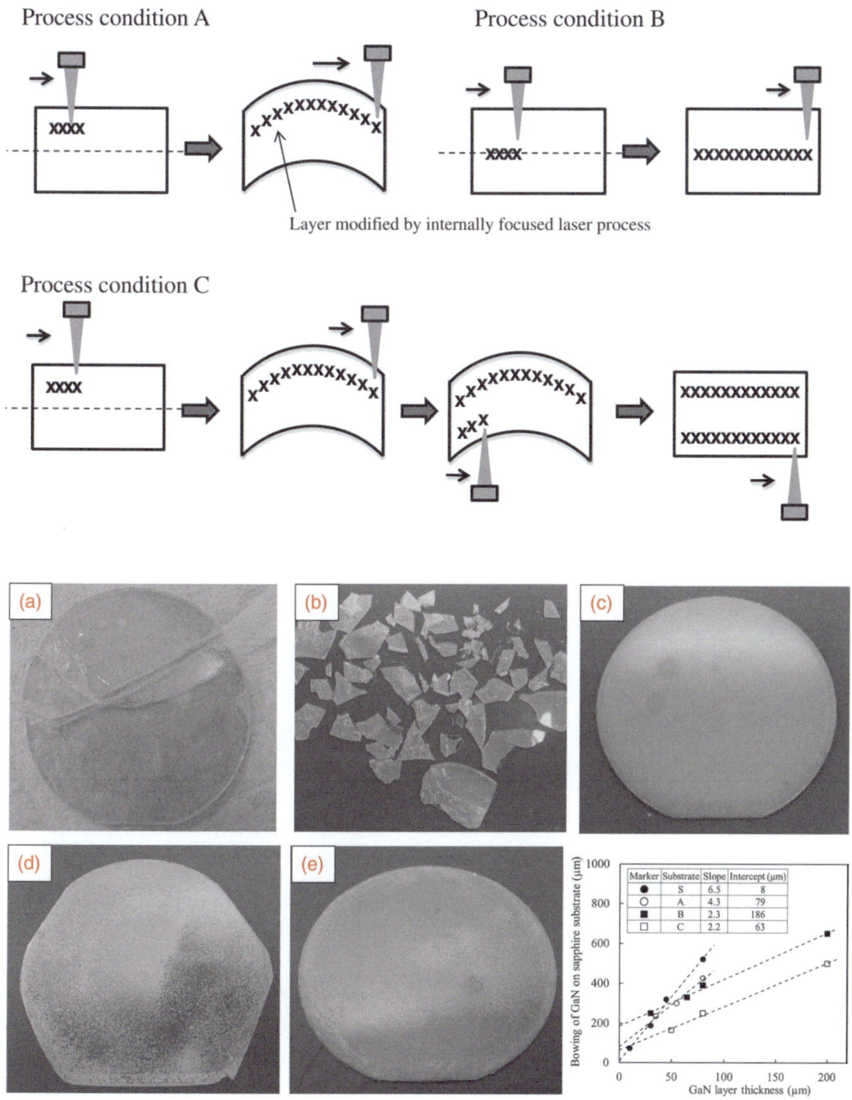

Fig. 3.22 1045 nm laser treatment of the sapphire substrate increases the crack-free gallium nitride thickness from 10–20 μm to 200–250 μm; For the two treatment methods B and C, the BOW at room temperature is smaller than the GaN thickness slope, indicating less stress [50]

Fig. 3.23 Smaller compressive stress with AlN buffer layer [52]

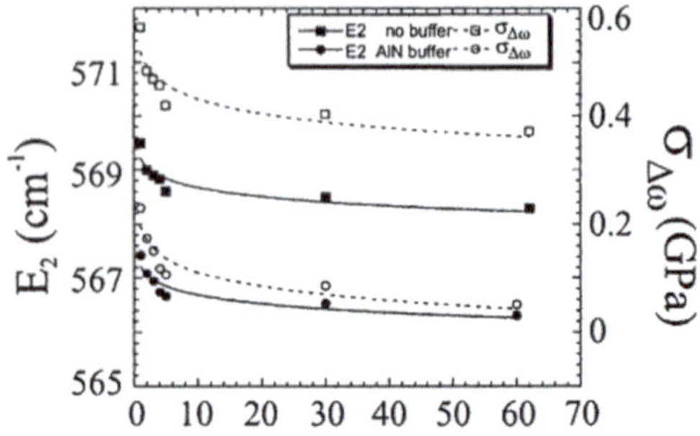

Fig. 3.24 Photograph of HVPE grown layer of 100 mm on a gallium treatment step (GTS) MOCVD template. No cracks are visible in either sapphire or GaN [58]

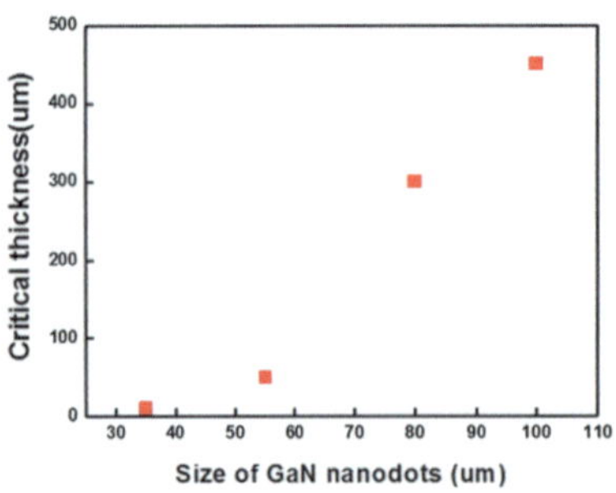

The dependence of the critical thickness of thick GaN layers on the average size of GaN nanodots.

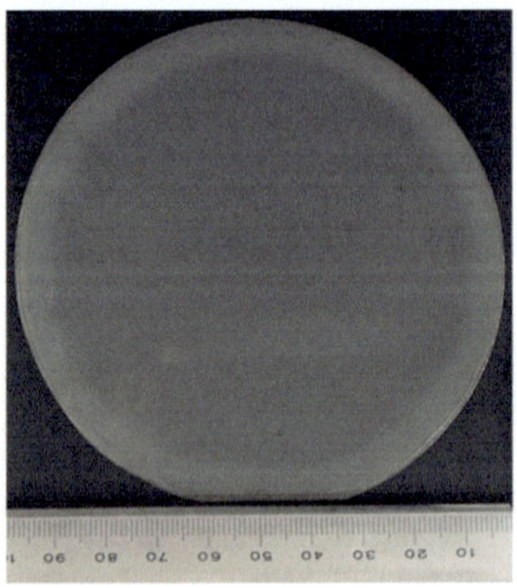

A thick GaN on an Al_2O_3 substrate with a diameter of 4 inches and a thickness of 400 μm.

Fig. 3.25 Left: The dependence of the critical thickness of thick GaN layers on the average size of GaN nanodots. Right: A thick GaN on an Al2O3 substrate with a diameter of 4 inches and a thickness of 400 μm [62]

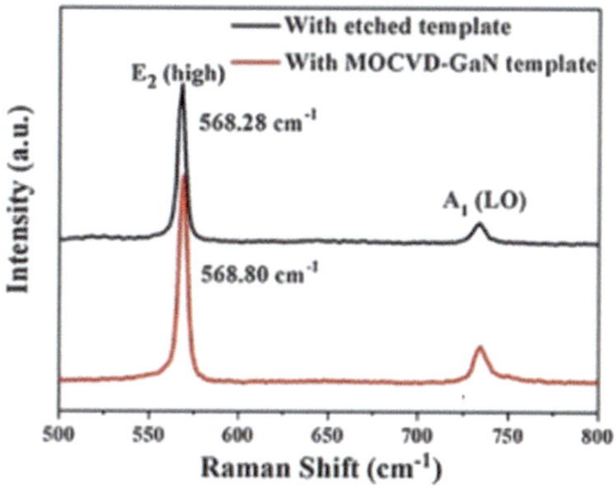

Fig. 3.26 Phosphoric acid corrosion of the template reduces compressive stress [63]

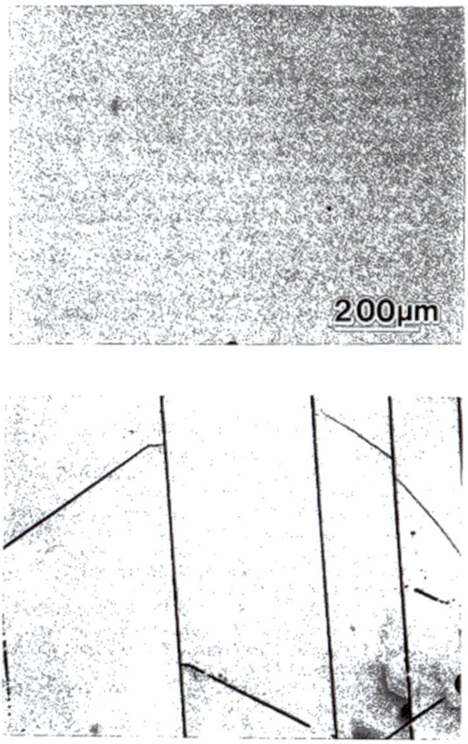

Fig. 3.27 SiO_2 mask relieves growth tensile stress [64]

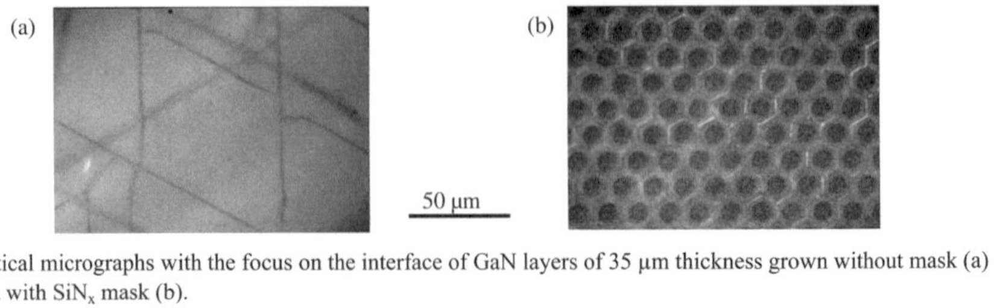

Fig. 3.28 Optical micrographs with the focus on the interface of GaN layers of 35 μm thickness grown without mask (a) and with SiNx mask (b) [66]

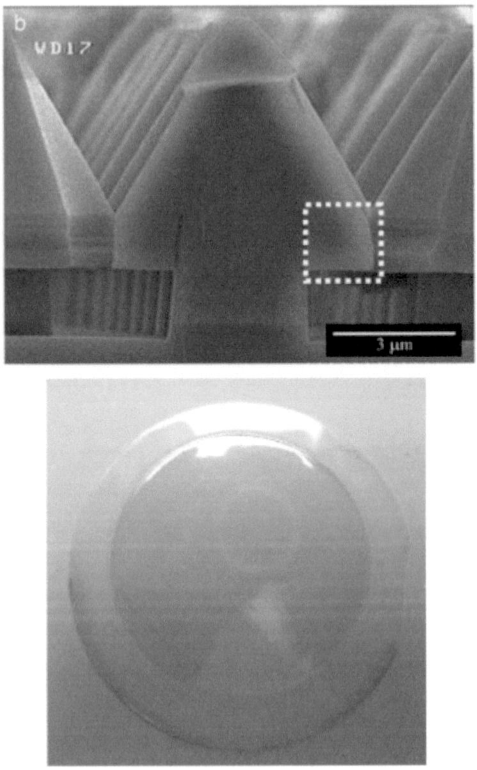

Fig. 3.29 Air-bridged structure grows 300 μm without cracks [70]

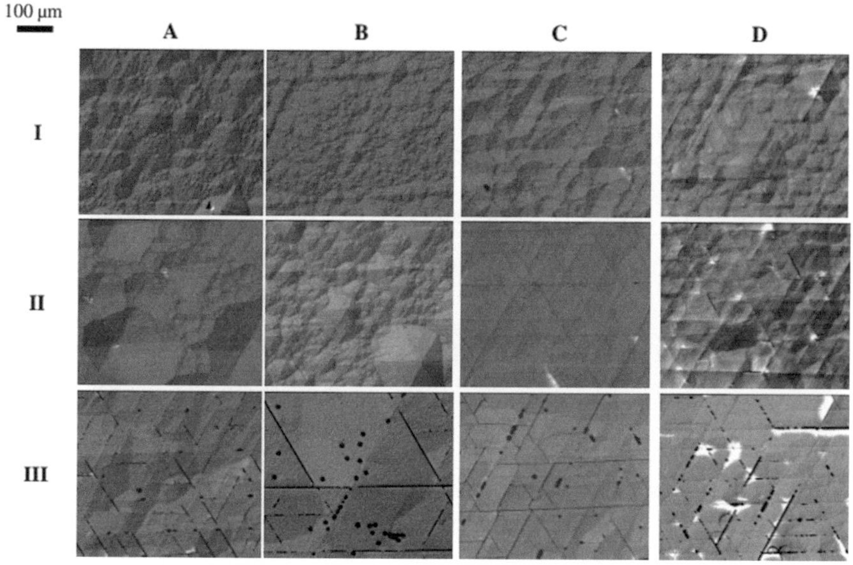

Fig. 3.30 The effect of V/III ratio and HCl flow rate on cracks [72]

Stress can also be adjusted by controlling growth parameters. For example, the choice of carrier gas, the appropriate ratio of nitrogen to hydrogen, and **V/III** [47, 70–72] (as shown in Fig. 3.30), growth pressure [73], pulsed ammonia flow rate [74];

6. Surface roughening or three-dimensional growth mode [75–81].

The coefficient of thermal expansion (CTE) of silicon is significantly lower ($3.59 \times 10^{-6}\ K^{-1}$) compared to that of gallium nitride ($5.59 \times 10^{-6}\ K^{-1}$). Consequently, during the cooling process, GaN contracts at a faster rate than silicon. This differential contraction induces tensile stress in the GaN film due to its attachment to the silicon substrate. Such tensile stress can result in the formation of cracks within the GaN layer, oriented perpendicular to its surface. As can be seen from Fig. 3.31, at the end of the growth process, the GaN/silicon wafer exhibits a convex bowing of 57 μm, with a maximum in-plane compressive stress of 0.07 MPa in the GaN layer. As the temperature decreases, the sample curvature gradually shifts from convex to concave, and the stress in the GaN film transitions from compressive to tensile. Ultimately, at room temperature, the GaN/silicon structure shows a concave bowing of 390 μm, with a maximum in-plane tensile stress of 173 MPa in the GaN layer.

The South Korean company Samsung Electronics [82] has developed a technology for in-situ removal of silicon substrates (as shown in Fig. 3.32). First, a 100 nm AlN buffer layer is deposited on the silicon substrate using MOCVD, followed by the growth of an AlGaN transition layer, and then the growth of a 500 nm thick gallium nitride film. Then, in the HVPE system, after growing 400 μm of gallium nitride, at a high temperature of 1000 °C, HCl is introduced to completely corrode the silicon substrate, leaving only gallium nitride. After cooling, free-standing gallium nitride

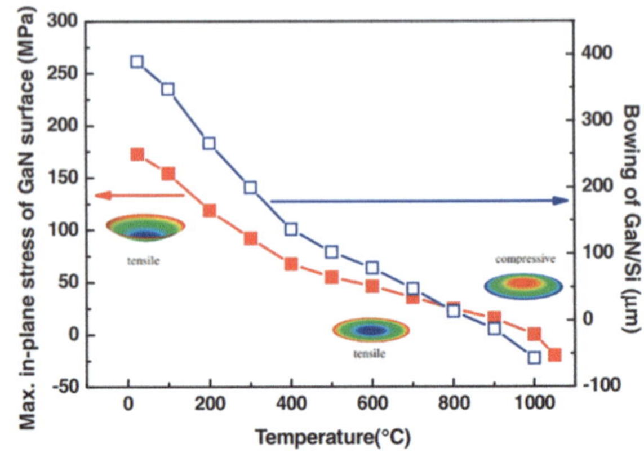

Fig. 3.31 400 μm GaN/400 μm Si stress and warpage from growth temperature to room temperature [82]

is obtained. This high-temperature method of removing the silicon substrate avoids the cracking caused by the thermal stress between silicon and gallium nitride.

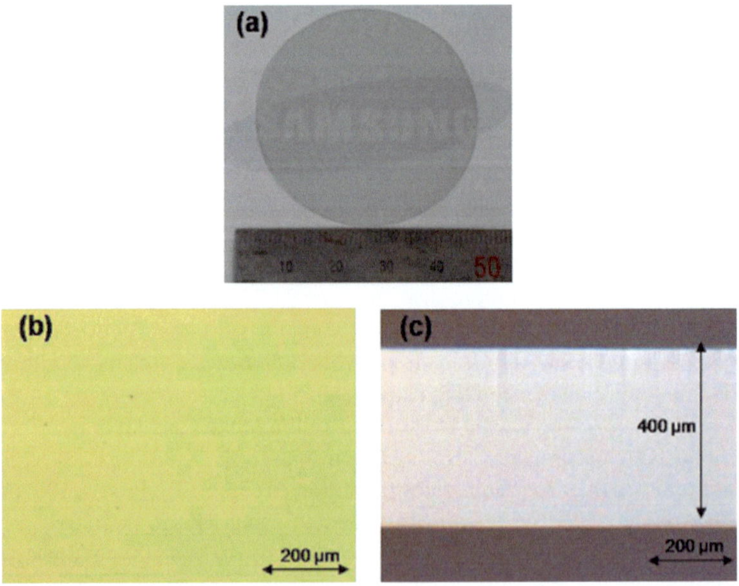

Fig. 3.32 In-situ etching of silicon substrate to obtain free-standing gallium nitride, **a** photo, **b** surface, **c** cross-section

3.3.2 Defect Control

1. Main defects of gallium nitride

Due to the lack of suitable homogeneous epitaxial substrates, the growth of gallium nitride materials usually uses heterogeneous substrates (sapphire, silicon carbide, gallium arsenide, silicon, etc.). The significant lattice mismatch and thermal mismatch between these hetero-substrates and GaN result in the formation of substantial crystalline defects within the GaN material.

0-Dimensional Defects

0-dimensional defects, also known as point defects, include intrinsic point defects (lattice vacancies, interstitial atoms), impurity atoms, etc.

There are mainly 6 types of intrinsic point defects, nitrogen vacancies V_N, gallium vacancies V_{Ga}, substitutional nitrogen N_{ant}, substitutional gallium Ga_{ant}, interstitial nitrogen N_{int}, interstitial gallium Ga_{int}.

1-Dimensional Defects

1-dimensional defects, also known as line defects, are mainly dislocations.

In contrast to conventional III-V compounds, gallium nitride (GaN) grown using standard methods typically with thicknesses on the order of several micrometers exhibits high dislocation densities ranging from 10^9 to 10^{10} cm^{-2}. This phenomenon is attributed to the significant lattice and thermal mismatch between GaN and its substrate.

2-Dimensional Defects

2-dimensional defects, also known as planar defects, such as grain boundaries, twins, stacking faults, etc. Gallium nitride has two structures, namely wurtzite structure and zinc blende structure. The atomic stacking method of the wurtzite structure is ABAB, and the atomic stacking method of the zinc blende structure is ABCABC. Stacking faults may arise when these two structures are intermixed or intergrown.

3-Dimensional Defects

3-dimensional defects, also known as bulk defects. Such as V-shaped pit [49], voids (internal), inclusions, inversion domains [83], etc.

2. HVPE bulk defects—V-shaped pits

Due to the generally substantial thickness of gallium nitride (GaN) grown via hydride vapor phase epitaxy (HVPE), typical layer thicknesses range from 250 to 500 μm for free-standing wafers, and from 2 to 7 mm for bulk crystals. Such significant thickness exacerbates defects on the crystal surface, including V-pits. In contrast, thin GaN and InGaN films produced by metal-organic chemical vapor deposition (MOCVD)

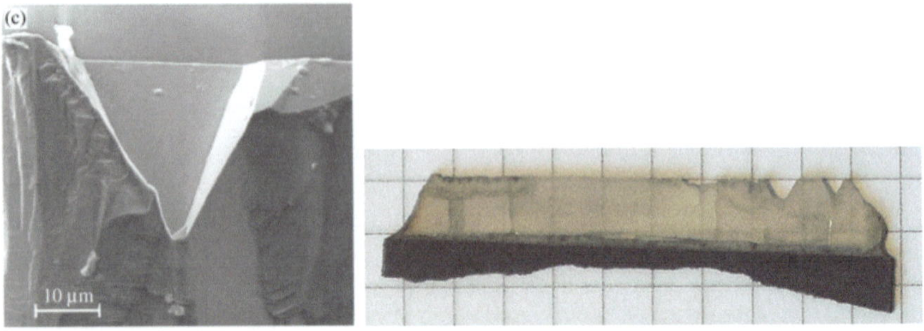

Fig. 3.33 V-shaped pits in HVPE samples. Left: V-shaped Pit on the surface of a 190 μm GaN film [49]; Right: V-shaped Pit with a diameter of ~ 2 mm on the M-plane of a 6 mm GaN crystal rod [84]

typically exhibit pit diameters of less than 1 μm. However, in millimeter-thick HVPE-grown crystals, V-pit diameters can vary from several tens of micrometers up to a millimeter scale. It is observed that as growth thickness increases, the diameter of the pits tends to enlarge correspondingly (as shown in Fig. 3.33).

V-pits can lead to several adverse effects:

1. They act as centers of stress concentration [85], which diminishes the yield during wafer processing;
2. They serve as sites for dislocation aggregation and impurity accumulation [86];
3. They reduce the effective usable area of the wafer;
4. During device epitaxy, stress concentrations and gradients may induce cracking; furthermore, the stress field can result in variations in ternary composition and doping concentration. Additionally, localized gas flow and surface atom diffusion may contribute to micro-scale surface roughening during epitaxial growth;
5. They can become initiation points for wafer cracking during processing;
6. In photolithography, spin-coated photoresist exhibits non-uniform thickness at the edges of V-pits, particularly along the direction of rotation;
7. During wafer dicing, V-pits may cause spontaneous cleavage or deviations in cleavage streaks;
8. Moreover, they can serve as potential origins of failure in fabricated devices [86].

Monemar, B. and others believe that there are two sources of Pits in HVPE. One is the pits formed by the adsorption of gallium nitride particles deposited on quartz to the sample surface, and the other is the pits produced by the continued growth of micro-cracks during the growth process [49].

Richter and others pointed out that the V-shaped Pits in HVPE growth may come from defects on the growth surface, such as nanotubes, or may be surrounded by local growth atmosphere. Improvements through reactor simulation may eliminate the production of V-shaped Pits [84]. Zhang et al. found that in germanium-doped gallium nitride, the Pit density is positively correlated with the germanium concentration, and metallic germanium droplets were found in the center of the pit in heavily doped germanium samples [87]. Wei et al. believe that the source of the pit is the

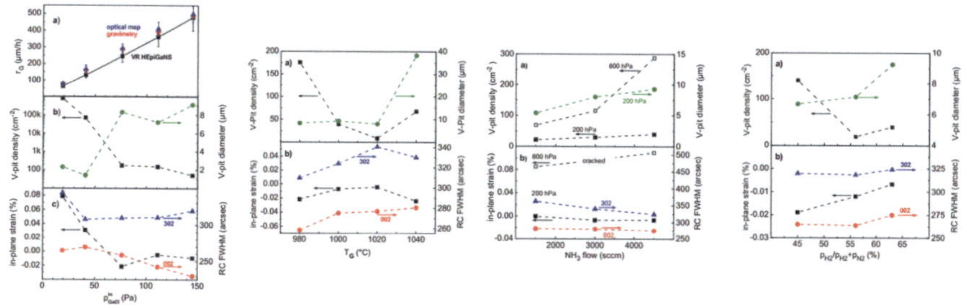

Fig. 3.34 The impact of HVPE growth conditions on the diameter and density of V-pits

inverse domain, and the V-shaped pit is produced because the growth rate of the nitrogen polarity is slower than the normal gallium polarity [88]. Wu et al. found that the structure design of the HVPE reactor will affect the formation of V-shaped pits on the surface of gallium nitride. By redesigning the reactor nozzle, a more uniform distribution of GaCl source can be obtained, which can eliminate the surface V-shaped pits to obtain a bright epitaxial surface [89].

Richter and others systematically studied the influence of HVPE growth conditions on the diameter and density of V-shaped pits. In order to balance cracks and crystal quality, it is necessary to choose the appropriate temperature, GaCl, NH_3 and carrier gas hydrogen ratio [90] (see Fig. 3.34).

Lee et al. proposed a method for the elimination of V-pits through cross-sectional cathodoluminescence (CL) characterization. Their findings indicate that when the crystallographic plane at the base of the pit transitions from (10-11) to (10-12), it promotes the coalescence of V-pits [91] (as shown in Fig. 3.35).

3. HVPE line defects—dislocations

There are mainly three types of dislocations in Group III nitride semiconductor materials, namely: edge dislocations, screw dislocations, and mixed dislocations, with their corresponding Burgers vectors as follows (Fig. 3.36):

$$b_e = \frac{1}{3}[11\bar{2}0], (b_e = a, b_e^2 = a^2) \tag{3.30}$$

$$b_m = \frac{1}{3}[11\bar{2}3], (b_{rmn} = \sqrt{c^2 + a^2}, b_m^2 = 3.66a^2) \tag{3.31}$$

$$b_s = [0001], (b_s = c, b_s^2 = 2.66a^2). \tag{3.32}$$

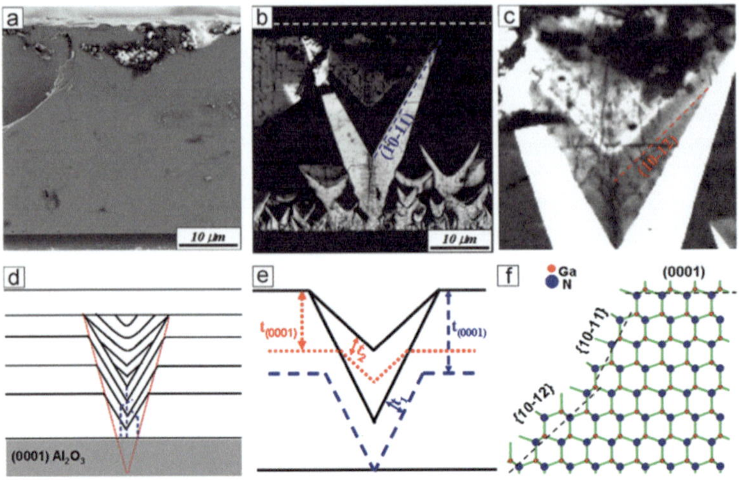

Fig. 3.35 V pit cross-section CL characterization

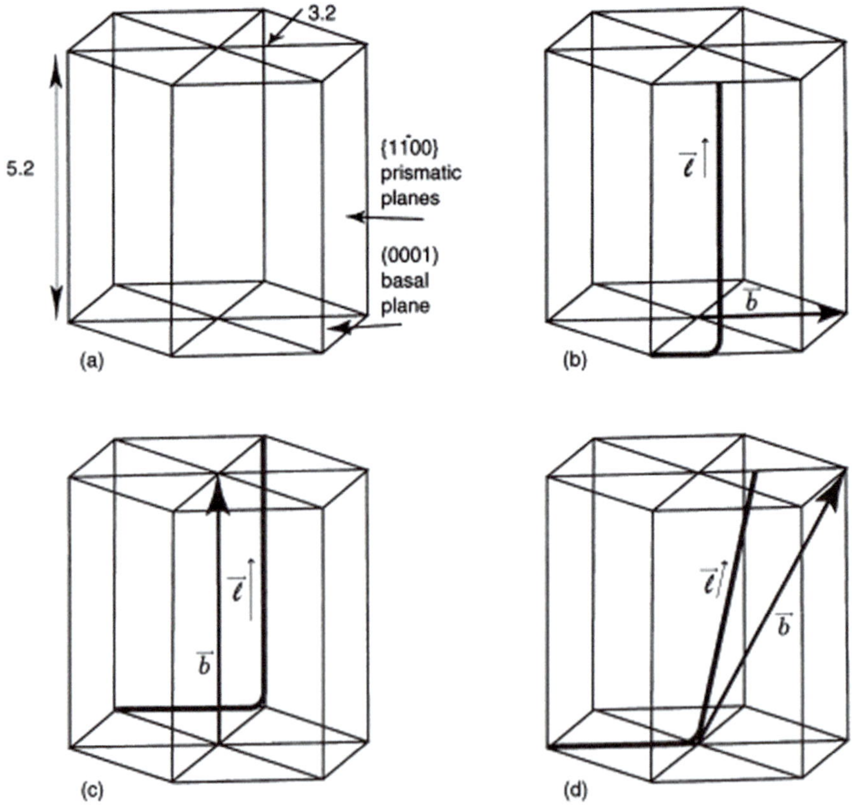

Fig. 3.36 **a** GaN unit cell. **b** Edge dislocation. **c** Screw dislocation. **d** Mixed dislocation

Characterization of dislocation density Using High-Resolution X-ray Diffraction Technology

Heteroepitaxially grown Group III nitrides have many crystal defects, which can be seen as many small grains embedded to form what is usually referred to as a mosaic structure crystal. Figure 3.37 is a schematic diagram of the structure of gallium nitride grown on a sapphire substrate. There are many small columns inside the gallium nitride, and there are differences in crystal orientation between each column in two dimensions, tilting along the C-axis and in-plane twisting. Each individual column can be considered as an almost perfect crystal. The dislocations in the material mainly originate from the overlap between the small columns, resulting in a significant difference in dislocation density in the directions perpendicular and parallel to the substrate. This can be described by the mosaic model [92, 93]. Combined with high-resolution X-ray diffraction technology, the characteristic parameters of the mosaic model can be obtained and then used to calculate the dislocation density in gallium nitride materials.

In the full width at half maximum (FWHM) of X-ray diffraction rocking curves obtained from practical crystalline materials, several factors contribute to broadening in addition to intrinsic broadening. These include instrumental broadening, dislocation-induced broadening, sub-grain size broadening due to mosaic structure, and sample curvature-induced broadening. It is important to note that intrinsic broadening is negligible, and the instrumental broadening associated with high-resolution X-ray diffractometers can also be disregarded. When the grain size exceeds 1 μm, grain size broadening becomes insignificant. Therefore, the primary considerations are dislocation-induced broadening and curvature-induced broadening. In practical characterization efforts, utilizing the (006) and (204) reflections instead of (002) and (102), along with employing narrower slits, can effectively mitigate broadenings caused by sample curvature [94].

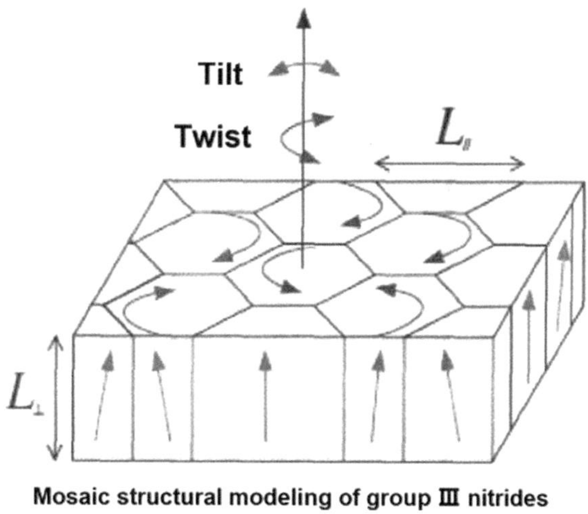

Fig. 3.37 Mosaic structure model

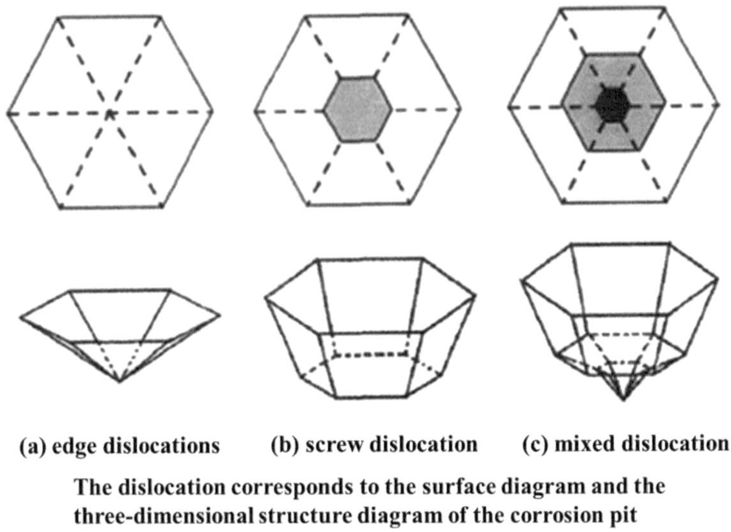

Fig. 3.38 Etch pits of different dislocations

Characterization of Dislocation Types by Etch Pits

Different types of dislocations exhibit distinct elastic strain energies. When gallium nitride (GaN) materials are subjected to etching, the characteristics—specifically the size and shape—of the resulting etch pits at these dislocations also demonstrate variability. These differences in etching behavior can be primarily attributed to variations in the Burgers vector [95] (as shown in Fig. 3.38). Common methods employed for etching include strong acid etching, molten alkali etching, and high-temperature annealing processes.

The investigation of various types of dislocations and their effects on device performance can be effectively conducted through selective etching techniques (as shown in Fig. 3.39). In a study focused on reverse-bias leakage in P-N junctions, Shigeyoshi Usami et al. observed that the leakage channels exhibited a certain degree of spatial correlation with dislocations; however, it was noted that not all dislocations were associated with leakage. Their research further indicated that only pure screw dislocations are responsible for contributing to reverse-bias leakage [96].

Dislocation Generation

Taking the growth of gallium nitride on sapphire as an example, when gallium nitride grows on the sapphire (0001) surface, the crystal orientation of gallium nitride rotates 30° relative to sapphire, thereby introducing shear strain and generating screw dislocations (as shown in Fig. 3.40); the lattice constants of sapphire and gallium nitride do not match, introducing tensile strain, leading to the generation of edge dislocations [92] (as shown in Fig. 3.41).

4. Common methods to reduce dislocation density

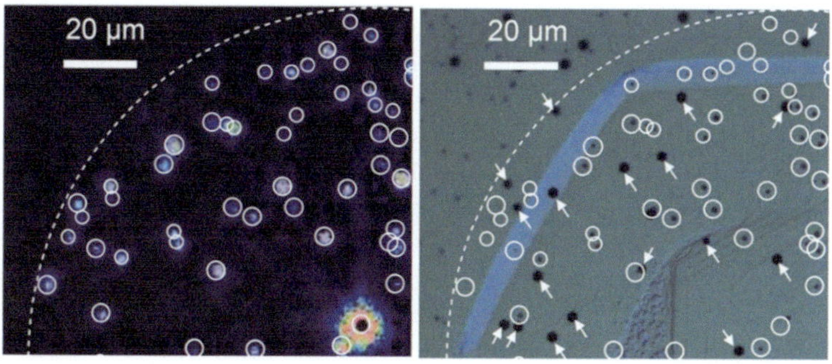

Fig. 3.39 Correspondence between leakage locations (circles) and corrosion pits [96]

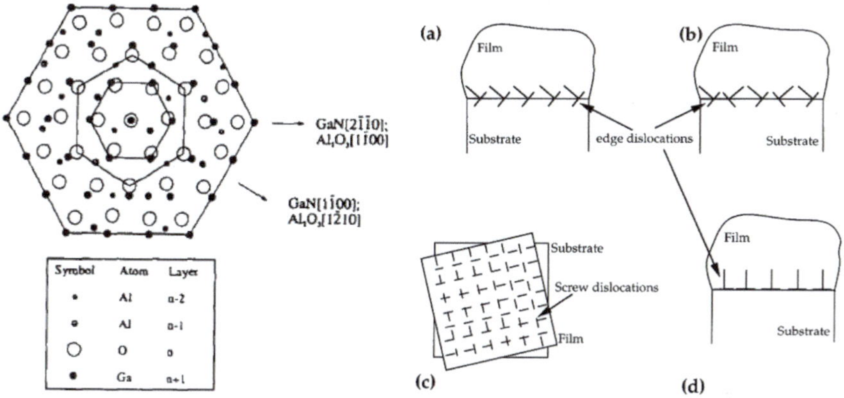

Fig. 3.40 Left: Schematic diagram of lattice matching between gallium nitride and sapphire; Right: 4 types of mismatch dislocations

Fig. 3.41 A large number of dislocations appear at the growth interface of gallium nitride grown on sapphire due to mismatch [97].

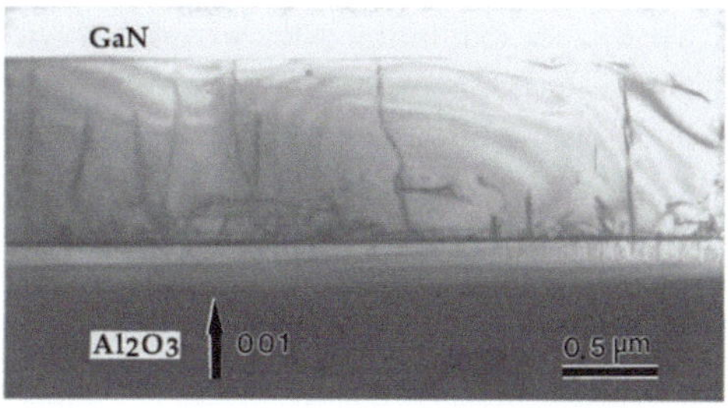

The predominant crystal defects in these materials are threading dislocations, which significantly degrade the crystalline quality and adversely affect the optoelectronic properties of the material. This degradation ultimately compromises the performance of associated devices.

Regarding material characteristics, dislocations lead to a reduction in carrier mobility and thermal conductivity. In device applications, dislocations function as non-radiative recombination centers and light scattering points, resulting in decreased luminescence efficiency for optoelectronic devices. Furthermore, they act as pathways for leakage currents, thereby diminishing the output power of the device.

The performance of nitride devices from prototype to commercial application depends on the improvement of the crystal quality of GaN base materials. The two-step growth method significantly reduced the dislocation density in gallium nitride films in 1986 [98], making the development of gallium nitride devices possible. The development of gallium nitride bulk materials and further reduction of dislocation density led to the fabrication of gallium nitride-based blue laser diodes [99].

MOCVD growth of gallium nitride has developed many methods to reduce the dislocation density of gallium nitride thin films, from nitridation of sapphire substrates, application of GaN and AlN buffer layers, AlN AlGaN insertion layers, buffer layer thickness and annealing conditions, control of initial nucleation density, to island coalescence control methods, reducing the dislocation density of 2 –5 μm gallium nitride thin films from 10^{10} to 10^8 cm^{-2}. Later, lateral epitaxy and PSS technologies were developed, further reducing the dislocation density to about 1×10^8 cm^{-2}.

The growth of HVPE has integrated techniques from Metal-Organic Chemical Vapor Deposition (MOCVD) to minimize dislocation density in thick GaN layers. Notable examples include: Molnar, R. J. et al. utilized an initial gallium source pretreatment on the substrate [100], achieving a dislocation density as low as 5×10^7 cm^{-2} in a 40 μm thick GaN layer; Tavernier et al. [101] implemented a two-step growth method that resulted in a dislocation density of 6×10^7 cm^{-2} in a 23 μm thick GaN layer; Kim et al. [102] introduced aluminum nitride (AlN) quantum dots to effectively reduce the dislocation density in HVPE-grown GaN.

Dislocation Reduction in GaN via Epitaxial Lateral Overgrowth (ELOG) Technology

A mask layer exhibiting poor wettability with respect to GaN is deposited on the substrate. Portions of this mask are subsequently removed through processes such as photolithography, thereby exposing the underlying substrate or the existing GaN layer. Following this step, HVPE growth commences within the exposed window regions and proceeds with lateral overgrowth, ultimately resulting in the formation of a smooth GaN film. This technique is referred to as epitaxial lateral overgrowth (ELOG). Figure 3.42 illustrates the schematic diagram of ELOG growth.

Why can ELOG growth reduce dislocation density? Firstly, because the surface of the substrate is partially covered and partially exposed, the dislocations under the covered part will be blocked and will not penetrate to the thin film above. Secondly, during the lateral growth process, the dislocations will turn parallel to the (0001) plane, and most dislocations will not extend to the surface. In this way, the dislocation density can be greatly reduced. Figure 3.43 is a TEM image of the HVPE film grown by ELOG [103], where the red box shows the dislocation lines turning and disappearing in large quantities.

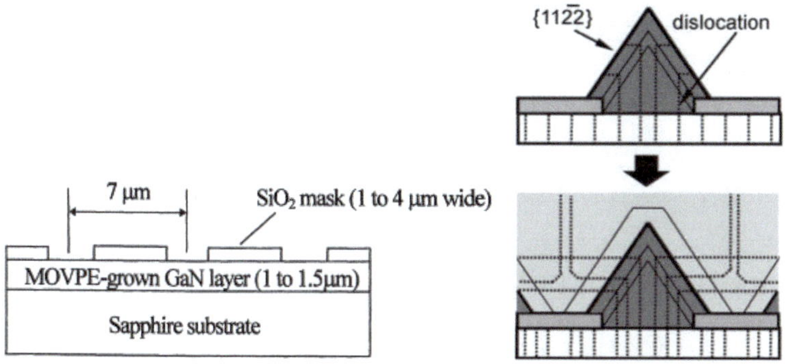

Fig. 3.42 ELOG technology

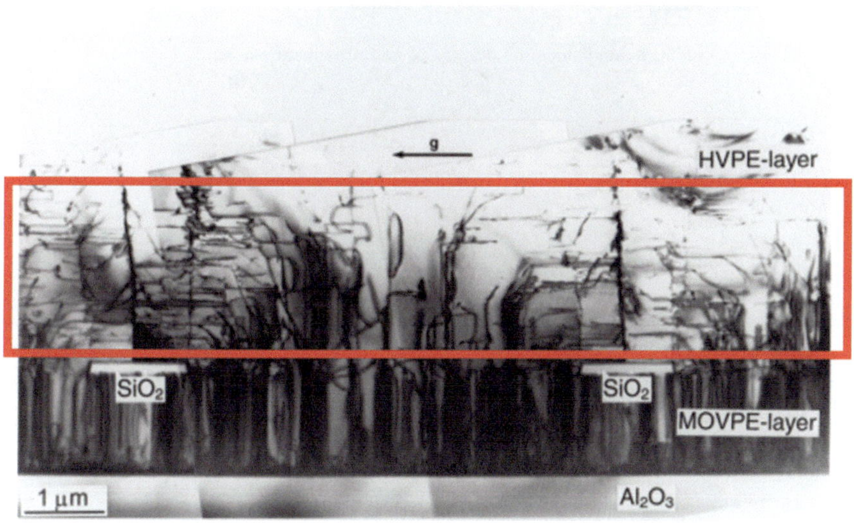

Fig. 3.43 Dislocation turning in ELOG growth

In 1997, Usui, A from NEC [64] first achieved ELOG growth in HVPE (as shown in Fig. 3.44), verifying that ELOG is not only effective in relieving stress and suppressing cracks, but can also reduce the dislocation density of gallium nitride materials. Finally, a 140 μm crack-free GaN sample was achieved, with a dislocation density as low as 6×10^7 cm^{-2}.

Linkoping University [48] grew 250 μm on an ELOG template using HVPE, reducing the dislocation density to 2×10^7 cm^{-2}. University of Ulm [104] deposited SiN on a MOVPE template to achieve ELOG growth, finally obtaining 1.5 mm self-separated GaN with an etch pit density of $< 1 \times 10^6$ cm^{-2} (as shown in Fig. 3.45).

Nearly all major manufacturers of free-standing GaN substrates currently utilize the Epitaxial Lateral Overgrowth (ELOG) technique to mitigate dislocation

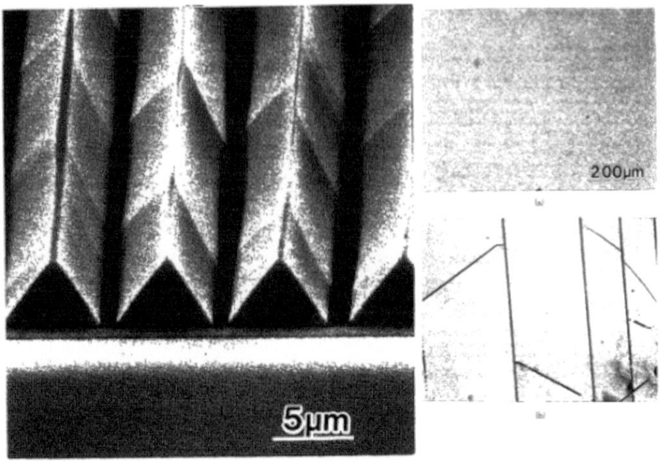

Fig. 3.44 ELOG reduces the dislocation density of gallium nitride materials

Fig. 3.45 ELOG grown GaN materials

density (as shown in Fig. 3.46). Sumitomo Chemical has developed a method known as Dislocation Elimination by Epitaxial-growth with Inverse-Pyramidal Pits (DEEP) [105, 106]. This approach involves growing on a GaAs substrate through designated window regions, leading to the formation of sloped facets that effectively eliminate dislocations. Variations in mask patterning techniques result in differences in the distribution of dislocation density (as shown in Fig. 3.47).

Hitachi Cable achieved ELOG growth by depositing a TiN mask on a gallium nitride film (as shown in Fig. 3.48). The weak connection formed by TiN allows gallium nitride to separate from the substrate after cooling, the gallium nitride wafer prepared by this method has a etch pit density of 5×10^6 cm^{-2} [107].

Furukawa deposited a TiC mask on sapphire, forming a tilted crystal face in the early stage of growth, and realized ELOG growth through the facet-initiated epitaxial lateral overgrowth method (FIELO). The dislocation density is less than

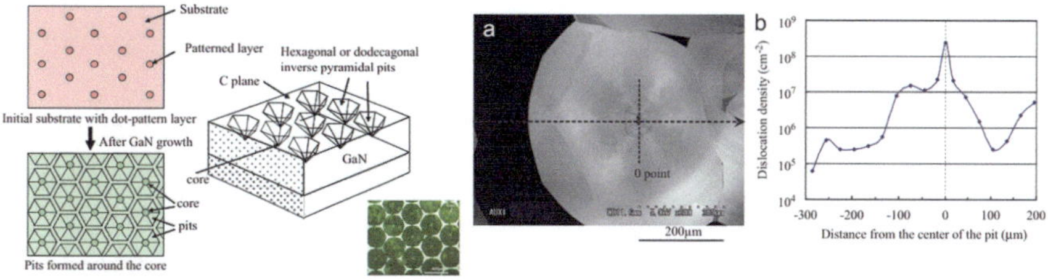

Fig. 3.46 Dot mask dislocation density from 3×10^4 to 1×10^7 cm^{-2}

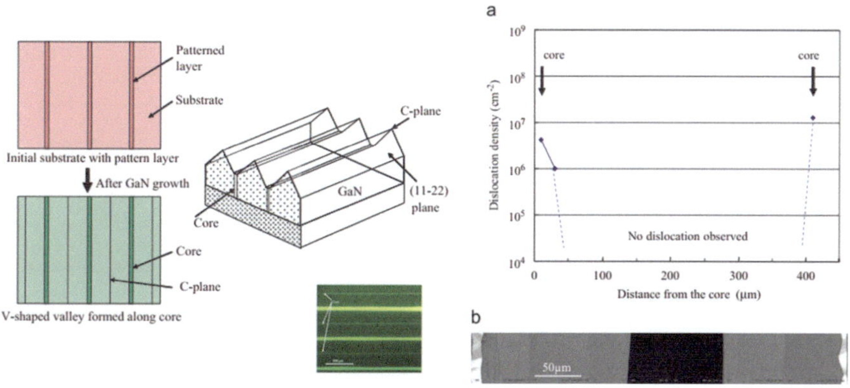

Fig. 3.47 Stripe mask, dislocation density from 1×10^7 cm^{-2} (core area) to $< 1 \times 10^4$ cm^{-2}

Fig. 3.48 Growth of GaN substrates using a TiN mask

3×10^6 [108]. As the thickness of the crystal increases, there is a corresponding decrease in dislocation density. With greater growth thickness, a higher number of dislocations experience phenomena such as bending, reactions, or annihilation. This leads to an inverse correlation between dislocation density and epitaxial thickness (as shown in Fig. 3.49) a relationship that has been extensively documented in various studies [83, 109]. For instance, Mitsubishi Chemical reported achieving a dislocation density as low as 1.2×10^6 cm^{-2} in a 5.8 mm-thick GaN ingrown crystal (as shown

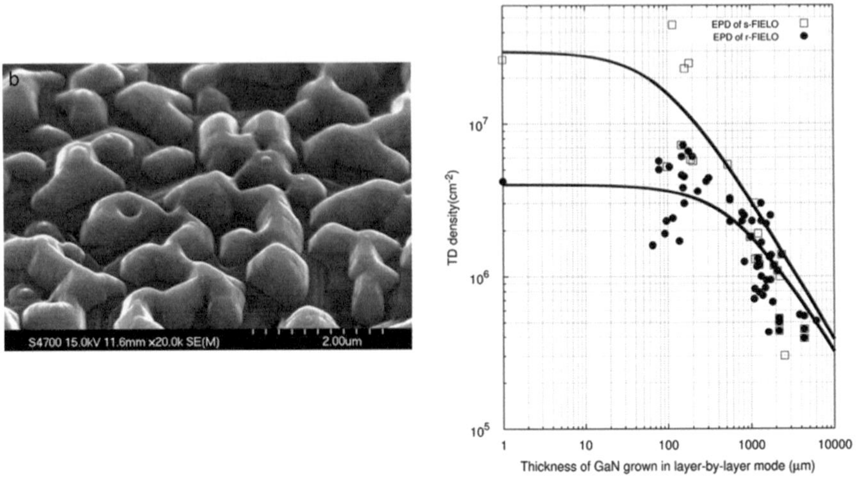

Fig. 3.49 Dislocation density changes with crystal thickness

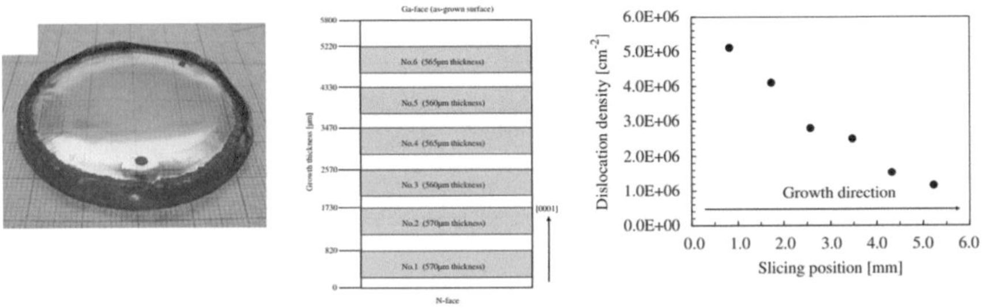

Fig. 3.50 Dislocation density under different thicknesses

in Fig. 3.50). They further examined the dislocation density at varying thicknesses subsequent to slicing the crystal into wafers [109] (as shown in Fig. 3.50).

3.4 Doping Technology

The electrical performance is a core parameter of GaN single crystal substrates. Effective doping and control of electrical characteristic are key for the widespread application of GaN single crystal substrates. In particular, high-conductivity n-type GaN materials are essential for high-power, high-current optoelectronic devices (such as LEDs and LDs). Additionally, high-resistance semi-insulating GaN materials are important for ensuring the high-reliability of power microwave devices (such as HEMTs). Furthermore, high-purity GaN materials are key to achieve high-voltage power electronic devices (such as SBDs and p-n Diodes). The research on doping and electrical characteristic control of GaN single crystal substrates not only provides significant guidance for scientific research, but also drives the development of related industries, possessing both theoretical and practical value.

3.4.1 Unintentional Doping

Unintentionally doped GaN materials generally exhibit N-type electrical characteristics, which are mainly caused by two factors. Firstly, some intrinsic defects in GaN (such as N vacancies) are donor-type charged centers, which affect the electrical characteristics of GaN [111]. Secondly, the reaction chamber for GaN material growth is generally made of quartz, which inevitably releases donor-type impurities such as Si, O during the growth process [112]. However, the carrier concentration of unintentionally doped GaN materials is generally in the range of 10^{16} –10^{17} cm^{-3}, and the resistivity is generally between 0.1 and 1 Ω cm. The fluctuation range is relatively large, making it difficult to use in semiconductor device.

In addition to point defects and unintentional impurities, the electrical properties of GaN materials are also affected by various scattering mechanisms. The main scattering mechanisms inside GaN materials include ionized impurity scattering, phonon scattering, and dislocation scattering. These scattering mechanisms greatly affect the mobility of the material, thereby affecting the electrical transport properties of GaN.

The mobility change caused by ionized impurity scattering can be expressed by the following formula:

$$\mu_i = \frac{64\sqrt{\pi\varepsilon^2(2kT)^{3/2}}}{N_i e^3\sqrt{m^*}}\left\{\ln\left[1+\left(\frac{12\pi\varepsilon kT}{e^2 N_i^{1/3}}\right)^2\right]\right\}^{-1} \tag{3.33}$$

where, ε is the dielectric constant of GaN, k is the Boltzmann constant, T is the Kelvin temperature, N_i is the ionized impurity density, e is the electron charge, and m* is the effective mass of electrons in GaN materials.

The mobility change caused by phonon scattering can be expressed by the following formula:

$$\mu_{ac} = \frac{e\sqrt{8\pi}\left(\frac{h}{2\pi}\right)^4 C_{11}}{3(E_{ds})^2(m^*)^{2.5}(kT)^{1.5}} \tag{3.34}$$

where, C_{11} is the average longitudinal elastic constant, and E_{ds} is the acoustic deformation potential.

The mobility change caused by dislocation scattering can be expressed by the following formula:

$$\mu_{disl} = \frac{30\sqrt{2\pi}\varepsilon^2 d^2 (kT)^{3/2}}{e^3 N_s f^2 \lambda_d \sqrt{m^*}} \tag{3.35}$$

where, f is the filling ratio of traps, N_s is the dislocation density, d is the C-axis lattice constant, and λ is the Debye screening length.

The total mobility can be expressed by the following formula:

$$\frac{1}{\mu_T} = \frac{1}{\mu_I} + \frac{1}{\mu_{disl}} + \frac{1}{\mu_{ac}} \tag{3.36}$$

GaN is generally grown on heterogeneous substrate materials such as sapphire and silicon, so its dislocation density is relatively high, and dislocation scattering has a large impact on the electrical properties of unintentionally doped GaN. For GaN materials, its dislocation density decreases with the increase of growth thickness [113], so increasing the growth thickness to reduce the dislocation density of GaN materials is an effective way to improve its electrical performance. For HVPE, its growth rate can reach several hundred micrometers per hour, which is much higher than other growth methods such as MOCVD and MBE, so it is easier to obtain GaN single crystal substrates with low dislocation density by HVPE.

Figure 3.51 shows the rules of different scattering mechanisms with temperature when the sample dislocation density is 2.4×10^6 cm^{-2} [114]. As can be seen from the figure, ionized impurity scattering dominates at low temperatures, and phonon scattering dominates at high temperatures. Therefore, when the dislocation density is at the level of 10^6 cm^{-2} or lower, dislocation scattering is not the main scattering mechanism of GaN.

Dislocations not only affect the mobility of GaN materials, but also have a significant impact on impurity adsorption and related optical properties. By using the HVPE method, thicker unintentionally doped GaN crystals can be prepared. As the dislocation density decreases with increasing thickness, the relationship between the concentration of unintentionally doped Si impurities and the dislocation density in GaN can be obtained by measuring the dislocation density and the concentration of unintentionally doped Si impurities at different positions, as shown in Fig. 3.52. Since

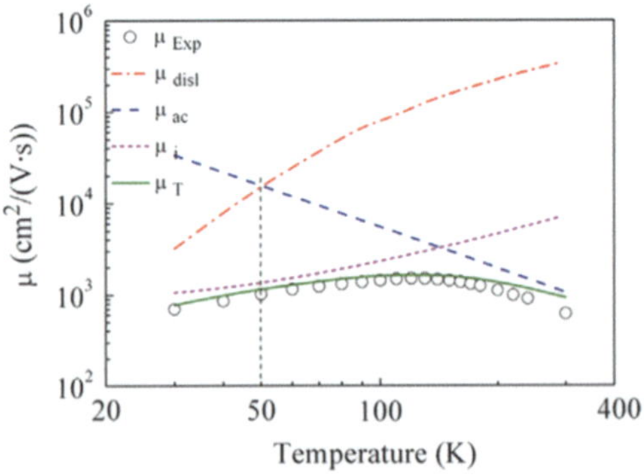

Fig. 3.51 The rules of different scattering mechanisms changing with temperature

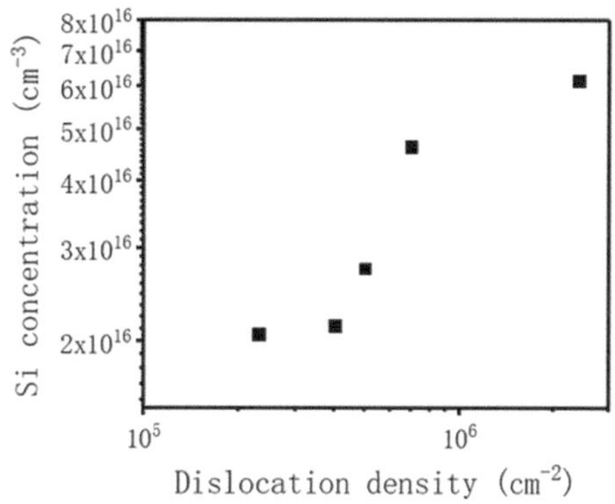

Fig. 3.52 Relationship between Si impurity concentration and dislocation density in intentionally doped GaN

the growth conditions did not change throughout the growth process, the changes in the concentration of unintentionally doped Si impurities are mainly affected by the dislocation density. As can be seen from the figure, the concentration of unintentionally doped Si impurities decreases with the decrease in dislocation density [114], mainly because there are a large number of dangling bonds along the dislocation line, which can bind charged impurity atoms (such as Si atoms) [115, 116]. At high temperatures, Si impurity atoms are easily desorbed from the quartz reactor and enter the lattice, where they are captured by the dangling bonds near the dislocation line. Therefore, the concentration of Si impurities is positively correlated with the dislocation density.

In GaN-based optoelectronic devices, the yellow luminescent band (YL) has a significant impact on the luminous efficiency of the device. It is generally believed that the yellow luminescent band is caused by the transition from the conduction band or shallow energy level to the deep energy level [117]. In GaN, as the dislocation density increases, the concentration of Si impurities adsorbed during the growth process also increases. These Si impurities provide the shallow energy levels required for YL luminescence, thereby increasing the intensity of YL luminescence [110].

3.4.2 Si Doping

High conductivity GaN single crystal substrates are mainly used for the fabrication of devices such as LEDs, LDs, etc., and have important applications in fields such as semiconductor lighting, laser display, etc. Among them, electrical parameters such as carrier concentration, resistivity, and mobility are the core parameters of GaN single crystal substrates, which have a significant impact on the ohmic contact, turn-on voltage, and other properties of GaN-based LEDs, LDs, etc.

The preparation of high conductivity GaN is generally achieved by doping shallow-level impurities such as Si, Ge, O, etc. [118]. These elements replace the Ga position in GaN and contribute electrons, making the entire semiconductor material exhibit low-resistance N-type electrical characteristics. Among them, Si doping is the main technical means to achieve N-type GaN materials and has been widely used in growth technologies such as MOCVD.

For the MOCVD process, the doping source of N-type GaN is generally SiH_4 gas. SiH_4 gas is transported to the substrate surface by the carrier gas, and in-situ doping is achieved by replacing the Ga position in the lattice with Si impurities during the high-temperature growth process. However, SiH_4 gas cannot be used for the HVPE process because the HVPE growth equipment is a hot wall system, and the growth temperature reaches above 1000 °C, while SiH_4 gas starts to decompose at around 500 °C and cannot reach the substrate surface.

The Si impurity sources currently used in HVPE equipment mainly include the following two types:

(1) HCl gas reacts with Si slices to generate $SiH_x\ Cl_y$. This method is simple and easy to implement, but as the growth progresses, the shape of the Si slice changes continuously, and the reaction surface area also changes accordingly. Therefore, as the reaction progresses, the concentration of $SiH_x\ Cl_y$ changes continuously, which is not conducive to stable doping control.
(2) $SiH_2\ Cl_2$ gas is used as the impurity source gas. The decomposition temperature of $SiH_2\ Cl_2$ reaches over 800°C, thus it can suppress high-temperature thermal decomposition, thereby achieving Si doping. At the same time, this method uses a flow meter to control the flow rate of $SiH_2\ Cl_2$ gas to regulate the doping concentration, thus it can achieve precise and stable control of the doping concentration.

A. V. Fomin demonstrated the use of Si slices and HCl gas to provide the impurity source gas, achieving the highest Si doping concentration of 2.5×10^{19} cm^{-3} [119]. In 1998, A. Usui proposed using $SiH_2\ Cl_2$ gas with better thermal stability to replace SiH_4 gas, and achieved the highest carrier concentration of 2×10^{18} cm^{-3} in the growth of GaN materials in HVPE equipment [120]. Currently, this method is adopted by most research institutions and companies, such as Hitachi Cable [121], Ferdinand-Braun-Institut [122], NaMLab [123], Institute of High Pressure Physics, Poland [124], Samsung Corning Precision Materials [125], etc.

The Si impurities in GaN will cause impurity scattering inside the GaN material. Figure 3.53a shows the relationship between the mobility of GaN material and temperature, comparing lightly doped samples with a Si impurity concentration of 1.1×10^{16} cm^{-3} and heavily doped samples with 6.5×10^{18} cm^{-3}. As can be seen from the figure, the mobility of the heavily doped samples is much lower than that of the lightly doped samples in the entire test temperature range (80–300 K), which also indicates that ionized impurity scattering has a significant impact on mobility. For both samples, their trends with temperature are opposite, because the dominant scattering mechanisms in the two samples are different. In lightly doped samples, the main scattering mechanism is phonon scattering. Phonon scattering strengthens with increasing temperature, so the mobility decreases with increasing temperature, which is basically consistent with the formula (3.34). In heavily doped

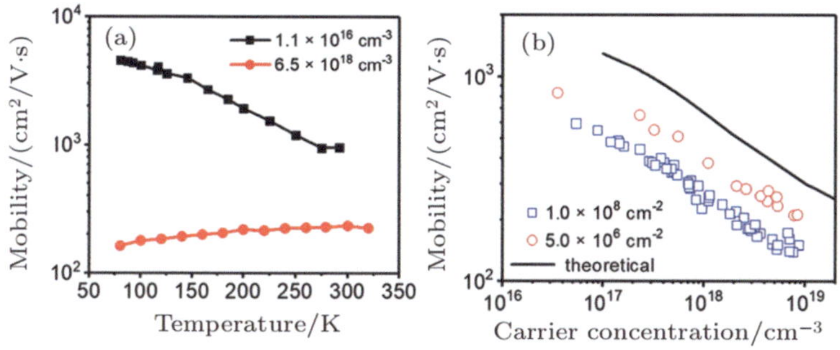

Fig. 3.53 Temperature-Dependent Mobility of GaN. **a** The mobility of lightly doped and heavily doped GaN materials changes with temperature. **b** The mobility of GaN with different dislocation densities changes with carrier concentration

samples, the main scattering mechanism is ionized impurity scattering. Ionized impurity scattering decreases with increasing temperature, so the mobility increases with increasing temperature, which is basically consistent with the formula (3.33).

Figure 3.53b hows the mobility of GaN as a function of carrier concentration for GaN samples with a dislocation density of 1.0×10^8 and 5.0×10^6 cm^{-3}. As can be seen from the figure, the mobility of both sets of samples decreases with the increase of carrier concentration, indicating that ionized impurity scattering has a significant impact on mobility. Moreover, throughout the entire carrier concentration range, the mobility of GaN thin film samples with high dislocation density is much lower than that of GaN single crystal samples with low dislocation density. This indicates that dislocation scattering has a significant impact on mobility, which is basically consistent with the formula (3.35).

The further increase of Si doping concentration in GaN is limited by the following two factors:

(1) Increase in tensile stress. Dadgar [126], Romanov [127], Cantu [128], and others found that Si doping can cause dislocations in GaN materials to tilt, and dislocation tilt can cause in-plane semi-atomic planes to be generated inside the material, thereby introducing tensile stress in GaN, leading to problems such as warping and cracking of GaN materials.
(2) Deterioration of surface morphology. In 2012, S. Fritze found that when the Si doping concentration reaches 1.9×10^{19} cm^{-3}, it will cause the growth mode to change to a 3D growth mode, leading to a deterioration in morphology [129]. In 2013, Markurt [130] found that the reason for the deterioration of morphology is that Si impurities are an antisurfactant during the growth process of GaN materials, causing the formation of a single atomic layer of $SiGaN_3$ on the GaN surface, and $SiGaN_3$ introduces a repulsive electric dipole moment on the GaN surface, hindering the growth of GaN on it, leading to a deterioration in surface morphology.

Limited by the above two factors, the carrier concentration of Si-doped GaN single crystal substrates on the market is generally less than 3×10^{18} cm^{-3}.

3.4.3 Ge Doping

In order to further improve device performance, it is necessary to further increase the carrier concentration of N-type GaN materials. Since the Si impurity concentration in GaN is limited by tensile stress and surface morphology deterioration, it is difficult to further increase. Hence, it is necessary to identify a more suitable doping source. In response to the problems of dislocation tilt and morphology deterioration introduced by Si doping with high impurity concentration, many institutes have tried to use Ge doping instead of Si doping to prepare GaN materials with high carrier concentration. C. Nenstiel's research found that Ge doping can solve the above two problems, and believes that Ge impurity is "the superior dopant in n-type GaN" [131]. Ge impurity is also a shallow level donor in GaN, which has the following characteristics:

(1) The electrical transport properties of Ge-doped GaN are basically the same as those of Si-doped GaN. The activation energy of Ge impurities in GaN is 19 meV [132], while it is 17 meV for Si impurities in GaN [133]. Both are lower than the thermal energy of electrons at room temperature, i.e. 26 meV, so they can be fully activated at room temperature. Moreover, the electrical transport properties of Si-doped GaN and Ge-doped GaN prepared by the MOCVD method are basically the same, and the relationship curve between their mobility and carrier concentration is also basically consistent [134].
(2) Literature reports that Ge doping does not introduce tensile stress in GaN. Even when the doping concentration is greater than 10^{19} cm^{-3} level, Ge doping does not introduce significant tensile stress [135].
(3) Ge doping does not cause changes in the growth mode. When Si doping is high, a thin layer of silicon nitride tends to grow on the surface of GaN, leading to a deterioration in the surface morphology of GaN at high doping concentrations. For Ge doping, Ge_3 N_4 is thermally unstable and cannot exist stably above 1000°C, so it will not cause deterioration of the surface morphology [136].
(4) The covalent radius of Ge atoms is close to that of Ga atoms. The covalent radii of Si, Ge, and Ga atoms are 116, 124, and 124 pm, respectively. Therefore, the covalent radius of Ge is basically the same as that of Ga, while the covalent radius of Si is significantly different from Ga, so the influence of Ge impurities on the lattice structure and stress is less than that of Si impurities.

Therefore, Ge doping is expected to solve the problems of stress and morphology deterioration caused by Si doping, and further increase the carrier concentration of GaN materials. At the same time, Ge doping can have similar electrical transport properties to Si doping, hence it is an important method to achieve high carrier concentration N-type GaN materials and has attracted widespread attention [131, 137].

There are mainly two types of impurity sources used for Ge doping:

(1) The reaction of HCl gas with Ge metal, controlling the doping concentration by controlling the flow rate of HCl gas reacting with Ge metal and the exposed area of Ge metal [137]. However, the area of Ge metal changes with the reaction, so this method is not stable enough.
(2) Bubbling N_2 in $GeCl_4$ liquid, carrying out the impurity source gas containing Ge elements [134]. Oshima prepared high-quality Ge-doped GaN by this method, with a Ge impurity concentration as high as 2.4×10^{19} cm^{-3} [134].

Figure 3.54 is a comparison of the relationship curves between the mobility and carrier concentration of Ge-doped and Si-doped GaN materials. The dislocation density of both sets of samples is at the 10^8 cm^{-2} level. As can be seen from the figure, under the same dislocation density, doping concentration, and test temperature, the mobility of Ge-doped samples is almost the same as that of Si-doped samples.

A common defect in Ge-doped GaN is the V-pit [134, 138]. Iwinska and others found that H_2 has a great influence on the formation of V-pits [138]. Thermodynamic calculations show that in H_2 atmosphere, the equilibrium vapor pressure of Ge is higher than its saturated vapor pressure, so Ge vapor tends to condense into droplets, leading to the formation of V-pits. However, under pure N_2 conditions, its equilibrium vapor pressure is lower than the saturated vapor pressure, so increasing the proportion of N_2 in the carrier gas is beneficial to reduce the pit density.

Although the doping limit of Ge in GaN is higher than that of Si, the doping concentration of Ge is generally less than 2.9×10^{20} cm^{-3} [135]. Further increasing the doping concentration will not lead to an increase in carrier concentration, which is mainly determined by the following two factors:

(1) With the increase of Ge doping concentration, the concentration of compensatory defects will also increase, which will reduce the carrier concentration in GaN.
(2) Ge impurities in GaN have a solubility limit. Exceeding the solubility limit will cause new phases to appear, leading to deterioration of morphology.

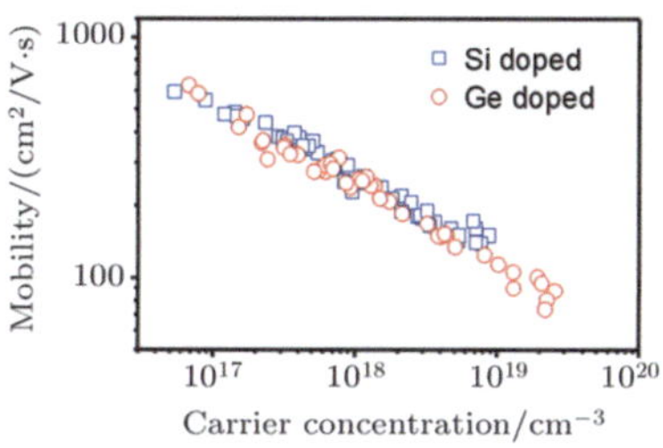

Fig. 3.54 Relationship curve between mobility and carrier concentration

3.4.4 Semi-insulating Doping

Semi-insulating GaN single crystal substrates can overcome the signal loss caused by parasitic capacitance related to substrate materials in devices such as HEMTs, which has important applications in fields such as 5G communication and phased array radar. Semi-insulating GaN materials are generally achieved by doping deep-level Fe impurities into GaN to compensate unintentionally incorporated shallow-level donor impurities in the substrate. The doping source is usually generated by the reaction of HCl gas with metallic Fe [139], or directly using ferrocene (Cp_2 Fe) as the doping source [140].

The relationship between the resistivity of Fe-doped GaN and the concentration of Fe impurities is shown in Fig. 3.55. From this, we can see that the resistivity of Fe-doped GaN is mainly affected by two factors:

(1) Concentration of Fe impurities.

Fe is a deep-level impurity in GaN, which achieves semi-insulating electrical characteristics by compensating for the electrons contributed by donor-type impurities inside the material. The resistivity of Fe-doped GaN almost linearly increases with the increase of Fe impurity concentration, and there is no saturation phenomenon of resistivity even when the Fe impurity concentration reaches 1×10^{19} cm^{-3}.

(2) Concentration of unintentionally doped Si impurities in the substrate.

The first group of samples with a Si impurity concentration of 1×10^{16} cm^{-3} has about one order of magnitude higher resistivity than the second group of samples with a Si impurity concentration of 1×10^{17} cm^{-3} under the same Fe impurity concentration. Si impurity in GaN is a shallow donor, it provides electrons, leading to an increase in carrier concentration and a decrease in resistivity.

The resistivity of Fe-doped GaN is influenced by both the concentration of Fe impurities and the background Si impurity concentration. Therefore, to increase the resistivity of Fe-doped GaN, one can either increase the Fe doping concentration or decrease the concentration of background donor-type Si impurities. However, when the Fe doping concentration is too high, the GaN single crystal substrate can

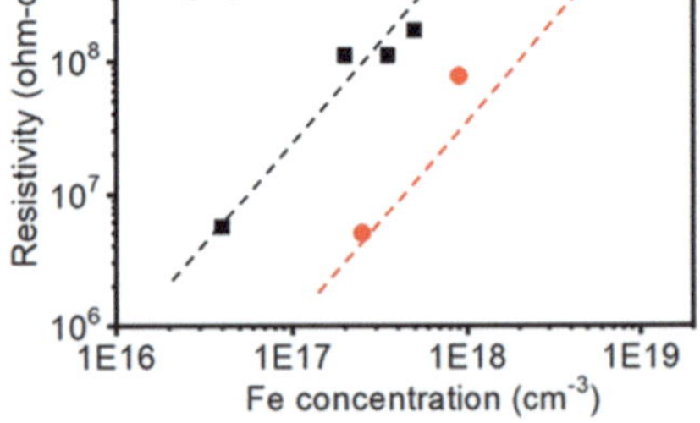

Fig. 3.55 Relationship between the resistivity of Fe-doped GaN and Fe impurity concentration

have a negative impact on device reliability. Therefore, to ensure the reliability of Fe-doped GaN substrates, we can prepare semi-insulating GaN [141] by reducing the background Si impurity concentration, which has provided a direction for the development of high purity semi-insulating GaN (HPSI-GaN).

In 2003, Vaudo prepared semi-insulating GaN [139] with a resistivity of $2 \times 10^9\ \Omega$ cm at room temperature using the HVPE growth method. In 2009, M. Kubota prepared Fe-doped GaN materials by reacting HCl gas with metallic Fe to generate impurity source gas. The Fe impurity concentration is proportional to the partial pressure of HCl gas, but the resistivity saturates when the Fe impurity concentration reaches 4×10^{18} cm^{-3} [142].

Fe-doped GaN has rich optical properties. Vaudo [139] observed the internal transition luminescence of Fe^{3+} ions at 1.3 eV through the PL method. Gladkov [143] systematically studied the spectrum of Fe-doped GaN and found that its luminescence spectrum, transmission spectrum, reflection spectrum, and ellipsometry spectrum are all affected by the Fe doping concentration. Based on this dependency, the impurity concentration of Fe in GaN can be approximately estimated. Fang [144] studied the deep energy levels inside GaN materials and found that there are at least 6 types of deep energy level traps inside Fe-doped GaN, with the main trap located at 0.56–0.60 eV. Jarasiunas [145], Fang [146] and others have systematically studied the optical nonlinear phenomena and the dynamics of photogenerated carriers in GaN crystals, providing a basis for the design of ultrafast optoelectronic devices based on Fe-doped GaN.

The magnetism of Fe-doped GaN has also been widely studied. Dalpian [147] and Theodoropoulou [148] observed from experiments that Fe-doped GaN exhibits ferromagnetism at room temperature and explained it through theoretical calculations. Akinaga [149] prepared GaN with a Fe impurity concentration of up to 3×10^{19} cm^{-3} at a growth temperature of 400°C using the MBE growth process, and found that it has ferromagnetism with a Curie temperature of 100 K. However, when the growth temperature is raised above 500°C, the material exhibits paramagnetic characteristics. At present, many optical, electrical, and magnetic phenomena of Fe-doped GaN have not reached a consensus and need further research.

One significant defect in Fe-doped GaN is the pit. Semi-insulating GaN has properties such as high resistivity and high breakdown voltage. However, pits have a severe impact on the devices grown on semi-insulating GaN substrates. The center of the pit contains a large number of defects, all of which are non-radiative recombination centers for carriers. The pit has a high carrier concentration [150, 151], making it an important leakage channel for semi-insulating GaN single crystal substrates [152]. Since semi-insulating GaN is generally used as a current blocking layer, the pit, as a leakage channel, has a significant impact on the reliability of the semi-insulating GaN substrate. The origins of pits in HVPE-grown GaN are generally considered to be dislocations [153], reverse domains [154], substrate defects [155], surface contamination [155], heterogeneous particles [155], cracks [155], etc. The pits on the substrate surface can be smoothed after polishing, but the internal defects still exist. By naked eye or optical microscope, a pit halo can be seen left at the original pit location. Detailed measurement characterization of the Fe-doped GaN pit halo is shown in Fig. 3.56. The optoelectronic properties of the pits inside Fe-doped GaN were studied by micro-Raman, AFM, and other methods. It was found that the carrier concentration inside the pit is much higher than in the pit-free area, and the carrier concentration gradually increases from the center of the pit to the edge of the pit. This is because the growth surface inside the pit is a semi-polar face such as (10–11), and the doping efficiency of shallow donor impurities (Si, O, etc.) on these semi-polar faces is much higher than those on the C-face, which leads to an increase in the donor impurity concentration inside the pit, and thus an increase in the carrier concentration. The high carrier concentration causes a significant decrease in the resistivity inside the pit. Scanning spreading resistance testing shows that the resistivity at the edge of the pit is 5 orders of magnitude lower than in the normal area. Therefore, when the device is fabricated on the pit, the current will mainly flow inside the pit, leading to the problem of device leakage [156]. The high carrier concentration further causes changes in the optical properties of GaN inside the pit, with band-edge luminescence broadening, NBE peak position first red-shifting then blue-shifting with the increase of carrier concentration. These phenomena are due to the influence of the Burstein-Moss effect and band renormalization effect on the band when the carrier concentration is higher than the Mott transition point.

With the development of devices, it is required to reduce the Fe impurity concentration in semi-insulating GaN single crystal substrates. Semi-insulating GaN single crystal substrates are mainly used in fields such as 5G communication, phased array radar, etc., and these fields have high requirements for device reliability. Currently, semi-insulating GaN is generally achieved by in-situ doping with deep acceptor Fe impurities during growth, but excessive Fe impurities may have adverse effects on the

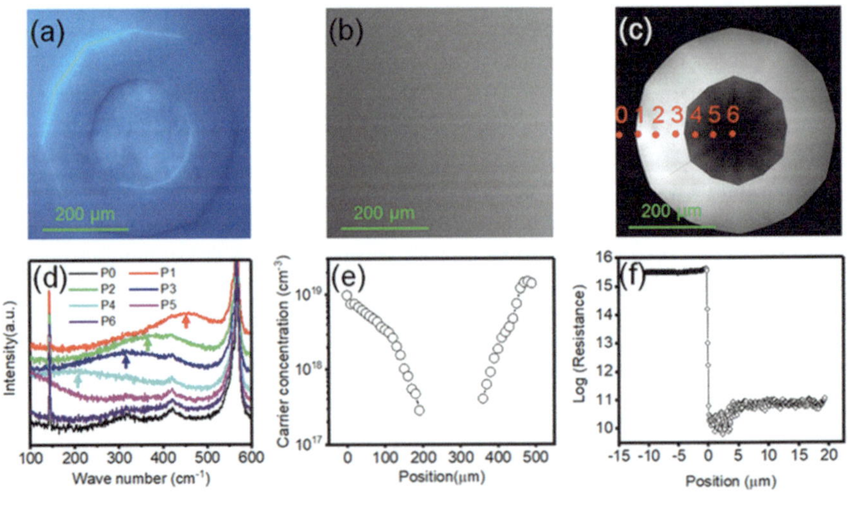

Fig. 3.56 Fe-doped GaN pit halo: **a** Optical microscope image. **b** SEM image. **c** CL image. **d** Raman spectra at different positions inside the pit halo. **e** Carrier concentration distribution inside the pit calculated from Raman spectra. **f** Scanning spreading resistance test results at the edge of the pit

performance and life of the device, mainly in the following aspects: ① Fe impurities are an important cause of current collapse in AlGaN/GaN HEMT devices. Studies have shown that an increase in the concentration of deep-level traps at 0.5–0.6 eV below the conduction band will lead to current collapse, and these trap defects are mainly Fe_{Ga} -V_N [157]. ② The trapping effect of Fe impurities on electrons leads to a longer recovery time after device response. AlGaN/GaN HEMTs made with Fe-doped substrates require a recovery time of 20 ms after being stimulated at 28 dBm [158], which cannot meet the response speed requirements of devices such as radars and electronic warfare receivers. ③ Fe impurities in GaN are prone to diffusion, which could affect device performance. In HEMT devices, Fe impurities in the substrate can easily diffuse into the channel, leading to an increase in channel resistance, thereby reducing the drain current [159], affecting device performance.

In order to reduce the impact of Fe impurities on device reliability, it is necessary to reduce the concentration of Fe impurities as much as possible while ensuring the electrical characteristics of semi-insulation. High-purity semi-insulating GaN material is an important technical approach to achieve this goal. It reduces the concentration of donor-type impurities in the substrate, so only a small amount of deep-level impurities need to be doped to prepare semi-insulating GaN. In 2017, SCIOCS company in Japan [68, 112] believed that Si and O impurities in GaN mainly come from quartz in the reaction chamber. They prepared high-purity GaN material with a Si impurity concentration as low as 2×10^{14} cm^{-3} using a quartz-free HVPE device, and its resistivity reached 1×10^9 Ω cm.

3.5 Gallium Nitride Single Crystal Substrate Separation Technology

Hydride Vapor Phase Epitaxy (HVPE) is the preferred method for preparing large gallium nitride crystals. In order to separate the thick gallium nitride grown by HVPE from the original substrate (sapphire), a series of separation technologies have been developed, such as stress-induced self-separation, interface weak connection, laser lift-off, and mechanical grinding.

3.5.1 Self-separation Process

1. Stress-induced self-separation

Stress-induced self-separation is essentially the process of crack propagation along the interface between gallium nitride and sapphire from the edge to the center. For brittle material crack propagation, according to the Griffith criterion, if the energy release rate is greater than the fracture energy, the crack will propagate. In this criterion, the fracture energy refers to the energy consumed to separate a unit area of fracture surface, which is a characteristic parameter related to fracture resistance. The energy release rate is the strain energy released per unit depth when a crack in the material extends by a unit length, which is closely related to the stress state before and after separation.

Under the approximation of the double-layer model and biaxial stress, the in-plane stress σ_{rr} (z) is,

$$\sigma_{rr}(z) = \begin{cases} \frac{E_s}{1-\nu_s}\left[\varepsilon_0 - \kappa\left(z + \frac{h_s}{2}\right)\right], \text{ when } -h_s < z < 0 \\ \hline \frac{E_f}{1-\nu_f}\left[\varepsilon_0 - \kappa\left(z + \frac{h_s}{2}\right) + \varepsilon_m\right], \text{ when } 0 \le z < h_f \end{cases} \tag{3.37}$$

where E_s and E_f are the elastic modulus of the substrate and film, v_s and v_f are the Poisson's ratio of the substrate and film, ε_0 and κ are the strain and curvature of the midplane (dotted line in the figure) of the substrate, h_s and h_f are the thickness of the substrate and film, ε_m is the thermal mismatch strain between GaN and the substrate, as shown in Fig. 3.57.

Generally speaking, due to the thermal mismatch between gallium nitride and sapphire, the cooled gallium nitride film is compressed, but the actual GaN/Sapp. thick film substrate has a certain degree of warping, and the stress in the gallium nitride film gradually decreases from the interface and even transforms into tensile stress. According to formula 3.1, the total stress $\boldsymbol{\sigma_{rr}}$ in the film is at its maximum compressive stress at the GaN/Sapp. interface. As the distance z from the interface increases, the tensile stress caused by the warping of the wafer linearly increases with the film thickness z, and there is a critical thickness $\mathbf{h_0}$ where the stress is zero, as shown in Fig. 3.58a.

During the separation process, the energy release rate **G** is the sum of the positive work done by the release of compressive stress and the negative work done by the release of tensile stress,

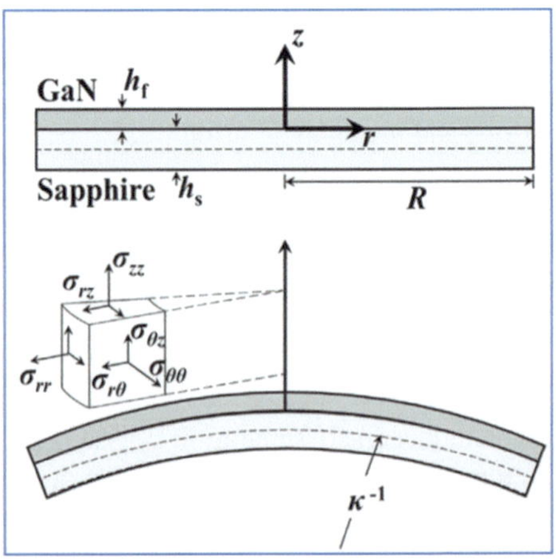

Fig. 3.57 Stress analysis of GaN/Sapp. Substrate [100]

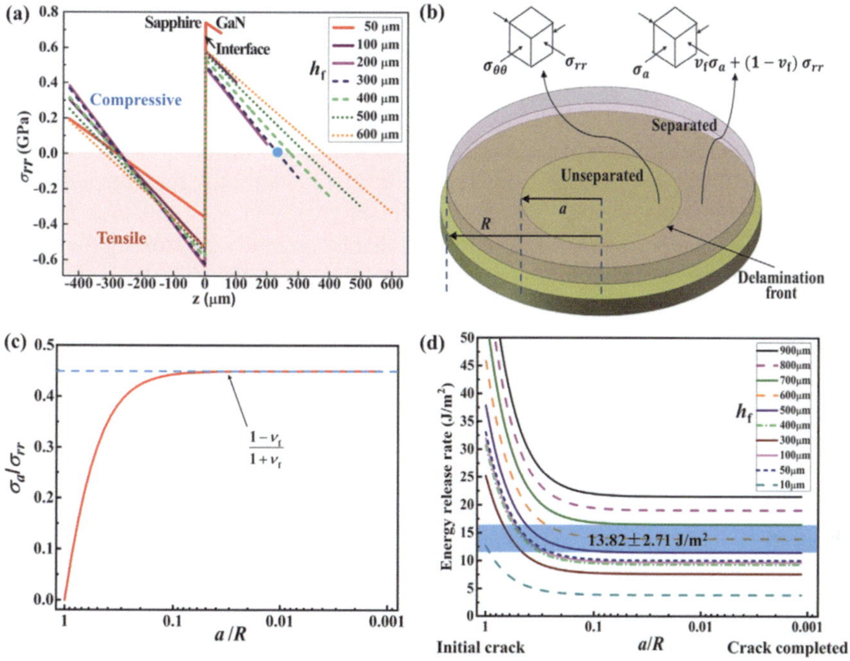

Fig. 3.58 Stress-induced self-separation technology. **a** Stress distribution in GaN/Sapp. film before separation. **b** Schematic of self-separation. **c** Change in the ratio of unrelaxed stress σ_a to pre-separation stress σ_{rr} during the separation process. **d** Energy release rate for different film thicknesses

$$G = \int_0^{h_0} \frac{1 - v_f^2}{2E_f}(\sigma_{rr} - \sigma_a)^2 dz - \int_{h_0}^{h_f} \frac{1 - v_f^2}{2E_f}(\sigma_{rr} - \sigma_a)^2 dz \tag{3.38}$$

where $\boldsymbol{\sigma_a}$ is the stress in the GaN film corresponding to the undetached radius a, as shown in Fig. 3.58b. From the above formula, it can be seen that as the separation progresses, $\boldsymbol{\sigma_a}$ tends towards a constant value, as shown in Fig. 3.58c. For GaN films of different thicknesses, the stress release rate gradually increases with the increase in film thickness. For brittle materials, crack propagation, according to the Griffith criterion, if the energy release rate is greater than the fracture energy, the crack will propagate. For GaN materials, the fracture toughness along the C-plane is 2.12 ± 0.21 MPa $m^{1/2}$, and the corresponding energy release rate is 13.82 ± 2.71 J/m^2, therefore there is a critical thickness $\boldsymbol{h}_c$, such that when the film thickness is greater than this value, complete separation will occur, as shown in Fig. 3.58.

Williams and Moustakas [160] used low-temperature gallium nitride as buffer layer and grew gallium nitride thick films ranging from 200 μm to 3.8 mm on sapphires. After the growth was completed, the GaN thick film naturally peeled off from the sapphire substrate during the cooling process. Further research found that when the thickness of the GaN film was reduced (< 200 μm), a large number of cracks appeared in the GaN self-separation substrate, and increasing the thickness of the epitaxial film not only led to the generation of heterogeneous cracks, but also as the thickness of the gallium nitride increased, the area of self-separation also increased, eventually leading to the complete separation of the free-standing gallium nitride substrate. The research team at Peking University further studied the mechanism of stress self-separation. The theoretical model study found that there is an optimal thickness of 500–700 μm, which can greatly increase the area ratio of self-separation. Experiments have confirmed that by precisely controlling the thickness of the gallium nitride film near the critical thickness, the complete separation ratio of 2-inch GaN can be greatly increased to 74%, as shown in Fig. 3.59 [161].

2. Void assisted separation

The key factors of void assisted separation include the selection of mask material, the optimization of the geometry and duty cycle of the mask. Due to the presence of a large number of voids at the growth interface, the voids expand and connect along the growth interface under the stress of thermal mismatch and wafer bending during the cooling process, ultimately achieving complete substrate separation. In 2008, Yoshida et al. [162] used a TiN mask to assist in forming voids and successfully prepared a 3-inch crack-free, free-standing gallium nitride substrate for the first time. The difficulty of separating a 3.2-inch gallium nitride thick film is not much greater than that of a 2-inch GaN, and it has good repeatability. Hennig et al. [163] used WSiN mask and adjusted the duty cycle of the mask to successfully separate a 2-inch crack-free free-standing GaN substrate from sapphire. Lee et al. [164] applied in-situ growth of NH_4 CL layer and low-temperature buffer layer technology on the surface of sapphire and obtained porous buffer layer technology through high-temperature annealing. Sato et al. [165] further carried out high-temperature growth of thick films on this porous buffer layer, achieving a 91 mm diameter free-standing GaN

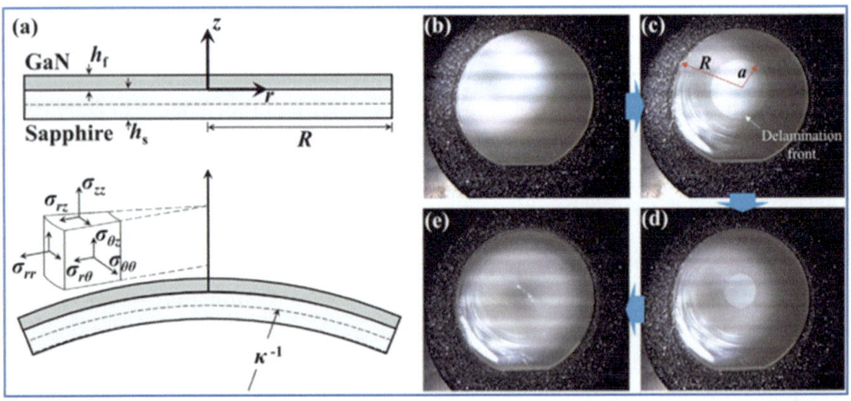

Fig. 3.59 Process of stress-induced self-separation preparation of GaN substrate. **a** Schematic diagram of GaN/Sapp. growth temperature and room temperature force. **b** Optical photo during the separation process

substrate. Zhang et al. [166] annealed MOCVD grown GaN thin films in situ in N_2 atmosphere, forming a porous structure, and used this substrate as the substrate for HVPE growth, achieving GaN self-separation (as shown in Fig. 3.60). Nikolaev et al. [167] used Ni as a mask to prepare a GaN nanocolumn buffer layer and achieved a 2-inch free-standing substrate.

3. Interface weak connection self-separation technology

Interface weak connection self-separation includes two key steps: first, forming a weak connection at the GaN and sapphire interface, and then realizing the separation through cooling or external stress. The main idea to achieve weak connection is based on the two-dimensional material interlayer, including quasi-van der Waals epitaxial growth on graphene, boron nitride, carbon nanotube arrays, etc.

Chung et al. [169] firstly reported the use of high-density, vertically aligned zinc oxide nanowires as an interlayer for heteroepitaxial growth of GaN thin films and LED structures on a graphene layer, as shown in Fig. 3.61a. Kim et al. [170] further used epitaxial graphene on the surface of a SiC substrate as the epitaxial starting interface, realizing direct van der Waals epitaxial growth of GaN thin films on graphene,

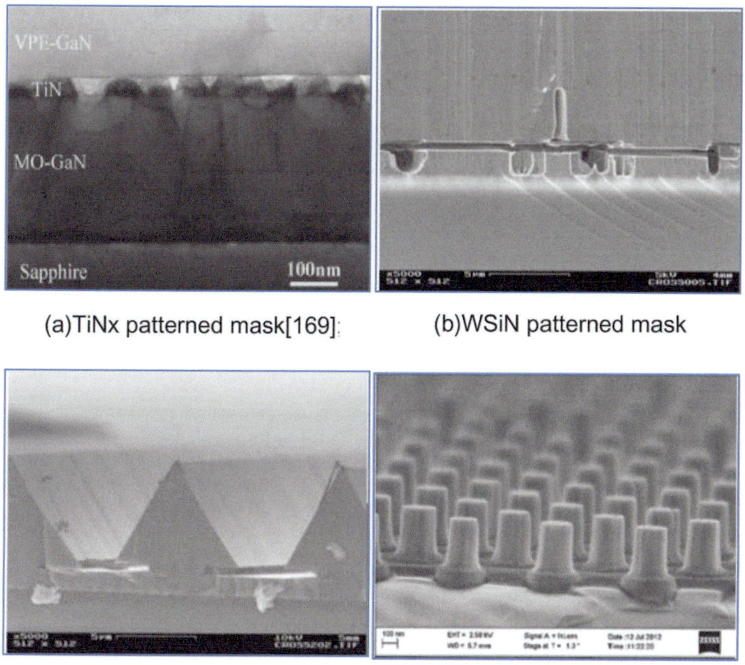

(a)TiNx patterned mask[169]; (b)WSiN patterned mask

(c)SiO 2 patterned mask (d)GaN nanocolumn buffer layer

Fig. 3.60 Void-assisted-separation method (VAS)

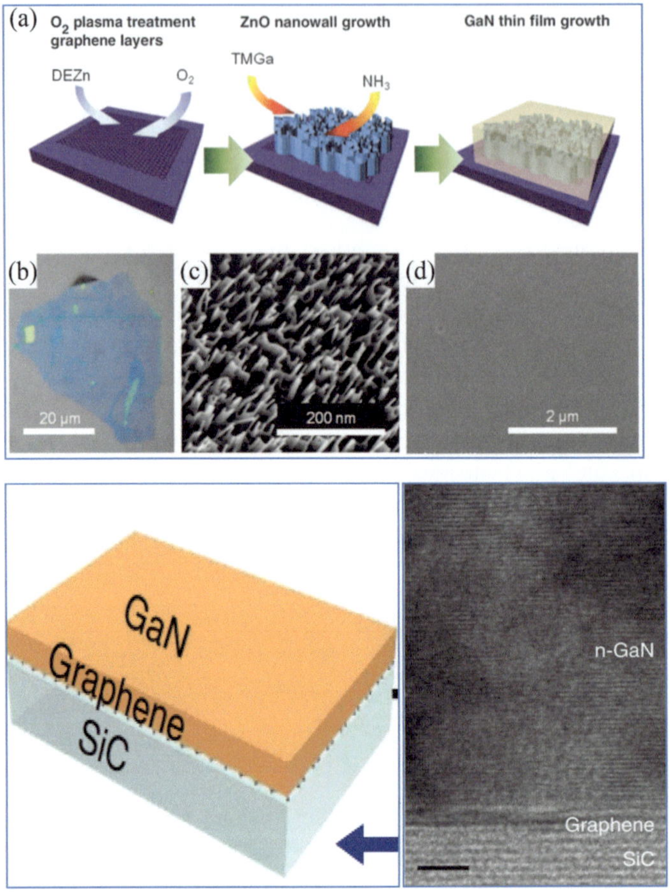

Fig. 3.61 Epitaxial GaN on graphene. **a** Schematic diagram and surface morphology of GaN/ZnO/graphene epitaxial structure. **b** Schematic diagram and cross-sectional TEM high-resolution image

and the gallium nitride thin film can be transferred to any substrate, as shown in Fig. 3.61b.

3.5.2 Laser Lift-Off

Laser lift-off (LLO) of GaN substrates refers to the technique of thermally decomposing GaN material at the interface using the thermal effect of high-energy lasers, thereby separating the GaN thick film from the sapphire substrate [171]. Figure 3.62 is a schematic diagram and energy band diagram of laser lift-off. A UV laser beam with energy higher than the bandgap width of GaN material (3.4 eV) but lower than the bandgap width of sapphire (9.9 eV) (248 nm or 5.0 eV), is absorbed by a thin

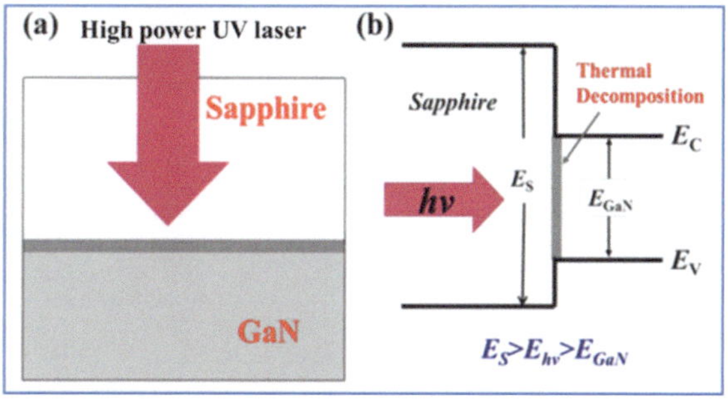

Fig. 3.62 Schematic diagram of LLO peeling principle. **a** Schematic diagram of laser lift-off. **b** Energy band principle diagram

layer of GaN material at the interface with an absorption depth of about 100 nm. In a very short time, this thin layer of GaN material is heated to a high temperature. When the temperature reaches or exceeds the decomposition threshold of GaN, the GaN at the interface decomposes to form liquid Ga and nitrogen gas, achieving the separation of GaN and sapphire.

The laser lift-off device mainly includes three parts: the light source, the mechanical control platform, and the heating platform, as shown in Fig. 3.63. The light source part mainly includes a laser, a homogenizer, a slit, a reflector, and a focusing mirror, aiming to obtain a laser beam with uniform energy distribution and adjustable spot size; the mechanical control platform mainly includes a translation stage (controlling X, Y direction stepping) and a rotation stage (controlling the stepping of rotation: constant linear speed or constant angular speed); the heating platform is mainly used to heat GaN/sapphire substrates to release thermal mismatch stress. First, the GaN/sapphire thick film is inverted on the BN heating platform and heated to 750–800 °C. Under the coordinated control of the rotation stage and the translation stage, the laser spot is scanned in a spiral manner from the edge to the center at a constant linear speed. To ensure that all places are irradiated by the laser during the scanning process, all experiments use a 30% repetition rate. The UV pulsed KrF laser (COHERENT LSX 200 K) used has a central wavelength of 248 nm, a pulse half-peak width of 25 ns, and the highest power density is 600 mJ/cm^2. The original laser beam forms a uniform energy distribution spot after the action of the beam limiter and homogenizer, and the spot size is 3 × 3 mm ^{2}. After the laser lift-off, it is kept warm to more than 30 °C above the melting point of Ga metal to remove the sapphire, and then the free-standing GaN substrate is put into a 50% hydrochloric acid solution ($HCl:H_2O$ = 1:1 volume ratio) to remove the Ga metal remaining on the N-polar face.

Controlling the generation of cracks is the key to laser lift-off technology. The generation of cracks has two sources, one is the release of thermal stress, and the other is laser-induced impact force. Su et al. [172] found that there are many cracks on the N-polar surface after laser lift-off, which can be divided into two major types of cracks according to the distribution characteristics and extension paths of the cracks: lateral cracks (LCs) located on the basal plane (000 $\underline{1}$) and longitudinal cracks (PCs) located on the non-polar surface {1–100}. The longitudinal cracks can be further divided into two subcategories: Type I cracks (PC I) that exist alone on the {1 $\underline{1}$

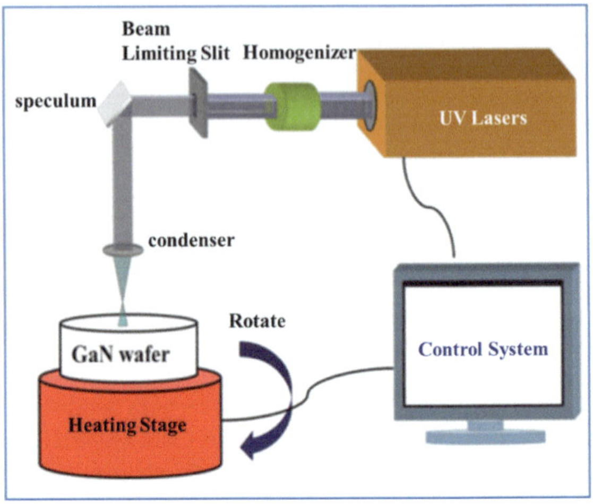

Fig. 3.63 Schematic diagram of the laser lift-off device

00} polar surface and Type II cracks (PC II) that accompany lateral cracks. PC II longitudinal cracks are always located at the forefront of lateral cracks, and their extension originates from the forefront of lateral cracks. By keeping the stress caused by the shock wave generated by laser irradiation below the damage threshold of GaN, the generation of cracks during the LLO process can be effectively avoided. Xu et al. [173] independently developed LLO equipment, achieving laser lift-off separation of 4-inch free-standing gallium nitride and sapphire, as shown in Fig. 3.64.

3.6 Development Trends

Although the GaN single crystal substrate prepared by HVPE growth is currently the main method for commercial production of GaN single crystal substrates, in order to meet the needs of future industrial development, further breakthroughs are still needed in reducing dislocation density and cost. To address these challenges, further integration of HVPE growth technology, ammonothermal method, and flux method to achieve larger size, lower cost, and higher quality GaN single crystal substrates will be an important development trend.

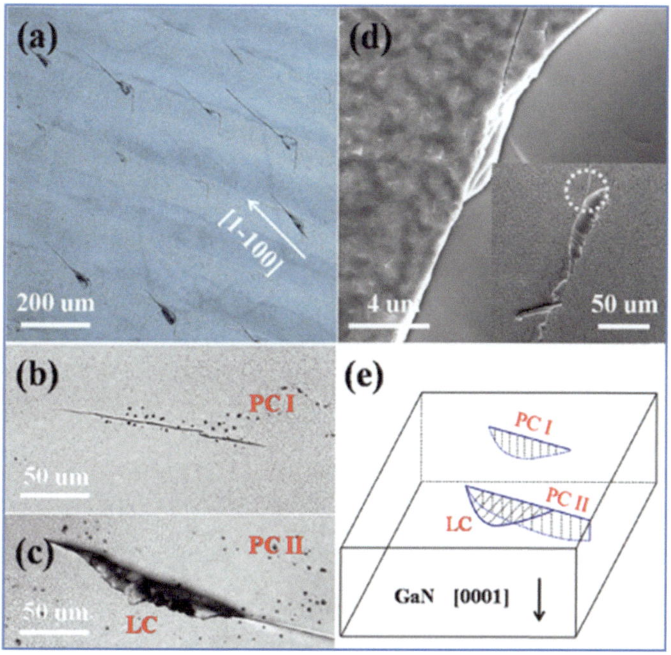

Fig. 3.64 Cracks during the laser lift-off process

References

1. Marinace JC (1960) Epitaxial vapor growth of GE single crystals in a closed-cycle process. IBM J Res Dev 4(3):248–255
2. Knight JR, Effer D, Evans PR (1965) Preparation of high purity gallium arsenide by vapour phase epitaxial growth. Solid-State Electron 8(2):178–180
3. Oldham WG (1965) Vapor growth of gap on GaAs substrates. J Appl Phys 36(9):2887–3000
4. Moest RR, Shupp BR (1962) Preparation of epitaxial GaAs and gap films by vapor phase reaction. J Electrochem Soc 109(11):1061–1065
5. Tietjen JJ, Maruska HP, Clough RB (1969) Preparation and properties of vapor-deposited epitaxial INAS1-XPX using arsine and phosphine. J Electrochem Soc 116(4):492–500
6. Maruska HP, Tietjen JJ (1969) The preparation and properties of vapor-deposited single-crystal-line GaN. Appl Phys Lett 15(10):327–329
7. Yang S, Ding M (1992) Epitaxial growth technology. National Defense Industry Press, Beijing
8. Yu HQ (2014) The impact of V/III ratio on GaN growth by HVPE. AMR 834–836:221–224
9. Xiu X, Zhang R, Li J et al (2003) The effect of extra HCl and nitriding on HVPE GaN growth
10. Dhanaraj G, Byrappa K, Prasad V et al (2010) Springer handbook of crystal growth
11. Seifert W, Fitzl G, Butter EJ (1981) Study on the growth rate in VPE of GaN. JOCG 52(part-P1):257–62
12. Molnar RJ (1998) Semimetals. Chapter 1 hydride vapor phase epitaxial growth of III–V nitrides, vol 57, pp 1–31
13. Maruska HP (1969) The preparation and properties of vapor-deposited single-crystal-line GaN. JAPL 15(10):327–329
14. Han XF, Hur MJ, Lee JH et al (2014) Numerical simulation of the gallium nitride thin film layer grown on 6-inch wafer by commercial multi-wafer hydride vapor phase epitaxy. J Cryst Growth 406:53–58

15. Ban V et al (1972) Mass spectrometric studies of vapor-phase crystal growth. 119(6):761–765
16. Hemmingsson C, Paskova P, Pozina G et al (2006) Growth of bulk GaN in a vertical hydride vapour phase epitaxy reactor. 40(4–6):205–213
17. Makarov Y, Karpov SY, Prokofyev VG et al (1999) Novel approach to simulation of group-III nitrides growth by MOVPE. 4(4)
18. CGSim, software of STR Group, Inc. http://www.str-soft.com/products/CGSim/.
19. Schineller B, Kaeppeler J, Heuken MJK, Aixtron AG (2007) Vertical-HVPE as a production method for free-standing GaN-substrates
20. Karpov SY, Zimina DV, Makarov YN et al (1999) Modeling study of hydride vapor phase epitaxy of GaN 176(1):439–442
21. Cadoret R, Trassoudaine A (2001) Growth of gallium nitride by HVPE. JJOPCM 13(32):6893
22. Agnès T et al (2004) Temperature influence on the growth of gallium nitride by HVPE in a mixed H2/N2 carrier gas—ScienceDirect 260(1–2):7–12
23. Cadoret R (1999) Growth mechanisms of (00.1)GaN substrates in the hydride vapour-phase method: surface diffusion, spiral growth, H_2 and $GaCl_3$ mechanisms. J Cryst Growth 205(1/2):123–135
24. Aujol E, Napierala J, Trassoudaine A (2001) Thermodynamical and kinetic study of the GaN growth by HVPE under nitrogen. J Cryst Growth 222(3):538–548
25. Segal A, Kondratyev A, Karpov S et al (2004) Surface chemistry and transport effects in GaN hydride vapor phase epitaxy 270(3–4):384–395
26. Dam CEC et al (2004) The effect of HVPE reactor geometry on GaN growth rate—experiments versus simulations. J Cryst Growth 271(1–2):192–199
27. Dam C, Bohnen T, Kleijn CR et al (2007) Scaling up a horizontal HVPE reactor 201(22):8878–8883
28. Meng Z. Simulation of HVPE reactor for GaN material preparation. Graduate School of Chinese Academy of Sciences, Shanghai
29. Wang WN, Stepanov SI (2011) Deposition technique for producing high quality compound semiconductor materials
30. Richter E et al (2005) Reactor and growth process optimization for growth of thick GaN layers on sapphire substrates by reviews 2(7):2313–2316
31. Lan WC, Tsai CD et al (2009) The effects of shower head orientation and substrate position on the uniformity of GaN growth in a HVPE reactor 40(4):475–478
32. Wu J, Zhao L, Wen D et al (2009) New design of nozzle structures and its effect on the surface and crystal qualities of thick GaN using a horizontal HVPE reactor 255(11):5926–5931
33. Attolini G, Carrà S, Di Muzio F et al (2000) A vertical reactor for deposition of gallium nitride 66(2–3):213–218
34. Monemar B, Larsson H, Hemmingsson C et al (2005) Growth of thick GaN layers with hydride vapour phase epitaxy 281(1):17–31
35. Safvi SA et al (1997) Effect of reactor geometry and growth parameters on the uniformity and material properties of GaN-sapphire grown by hydride vapor-phase epitaxy 182(3–4):233–240
36. Liu N, Wu J, Li W et al (2014) Highly uniform growth of 2-inch GaN wafers with a multi-wafer HVPE system. J Crystal Growth 388(4):132–136
37. Kisielowski C, Kruger J, Ruvimov S et al (1996) Strain-related phenomena in GaN thin films. Phys Rev B: Condens Matter 54(24):17745–17753
38. Ager JW, Conti G, Romano LT et al (1998) Stress gradients in heteroepitaxial gallium nitride films. In: Ponce FA, Denbaars SP, Meyer BK et al (eds) Nitride semiconductors, pp 769–774
39. Olsen GH, Ettenberg M (1977) Calculated stresses in multilayered heteroepitaxial structures. J Appl Phys 48(6):2543–2547
40. Paskova T, Darakchieva V, Valcheva E et al (2004) Hydride vapor-phase epitaxial GaN thick films for quasi-substrate applications: Strain distribution and wafer bending. J Electron Mater 33(5):389–394
41. Liu JQ, Wang JF, Qiu YX et al (2009) Determination of the tilt and twist angles of curved GaN layers by high-resolution x-ray diffraction. Semicond Sci Technol 24(12)

42. Paskova et al (2007) Effect of sapphire-substrate thickness on the curvature of thick GaN films grown by hydride vapor phase epitaxy. J Appl Phys 102:123507
43. Hiramatsu K, Detchprohm T, Akasaki I (1993) Relaxation mechanism of thermal-stresses in the heterostructure of GaN grown on sapphire by vapor-phase epitaxy. Jpn J Appl Phys Part 1-Regul Pap Short Notes Rev Pap 32(4):1528–1533
44. Li M, Cheng Y, Yu T et al (2019) Critical thickness of GaN film in controllable stress-induced self-separation for preparing native GaN substrates. Mater Des 180:107985
45. Oshima Y, Eri T, Shibata M et al (2002) Fabrication of freestanding GaN wafers by hydride vapor-phase epitaxy with void-assisted separation. Physica Status Solidi A—Appl Res 194(2):554–558
46. Williams AD, Moustakas TD (2007) Formation of large-area freestanding gallium nitride substrates by natural stress-induced separation of GaN and sapphire. J Cryst Growth 300(1):37–41
47. Habel F, Bruckner P, Tsay JD et al (2005) Hydride vapor phase epitaxial growth of thick GaN layers with improved surface flatness. In: Stutzmann M (ed) Physica Status Solidi C—conferences and critical reviews, vol 2, no 7, pp 2313–2316
48. Gogova D, Larsson H, Kasic A et al (2005) High-quality 2 bulk-like free-standing GaN grown by hydride vapour phase epitaxy on a Si-doped metal organic vapour phase epitaxial GaN template with an ultra low dislocation density. Jpn J Appl Phys Part 1—Regul Pap Brief Commun Rev Pap 44(3):1181–1185
49. Monemar B, Larsson H, Hemmingsson C et al (2005) Growth of thick GaN layers with hydride vapour phase epitaxy. J Cryst Growth 281(1):17–31
50. Aida H, Koyama K, Martin D et al (2013) Growth of thick GaN layers by hydride vapor phase epitaxy on sapphire substrate with internally focused laser processing. Appl Phys Exp 6(3)
51. Hu Q, Wei T-B, Duan R-F et al (2009) Characterization of thick GaN films directly grown on wet-etching patterned sapphire by HVPE. Chin Phys Lett 26(9)
52. Paskova T, Valcheva E, Birch J et al (2001) Defect and stress relaxation in HVPE-GaN films using high temperature reactively sputtered AlN buffer. J Cryst Growth 230(3–4):381–386
53. Paskova T, Paskov PP, Darakchieva V et al (2001) Defect reduction in HVPE growth of GaN and related optical spectra. Physica Status Solidi A—Appl Res 183(1):197–203
54. Nikitina IP, Sheglov MP, Melnik YV et al (1997) Residual strains in GaN grown on 6H-SiC. Diamond Relat Mater 6(10):1524–1527
55. Aksyanov IG, Bessolov VN, Zhilyaev YV et al (2008) Chloride vapor-phase epitaxy of gallium nitride on silicon: effect of a silicon carbide interlayer. Tech Phys Lett 34(6):479–482
56. Goto H, Lee WH, Kinomoto J et al (2005) Growth and characterization of HVPE GaN on c-sapphire with CrN buffer layer. In: Woo JC, Hasegawa H, Kwon YS et al (eds) Compound semiconductors, Proceedings, pp 369–372
57. Molnar RJ, Gotz W, Romano LT et al (1997) Growth of gallium nitride by hydride vapor-phase epitaxy. J Cryst Growth 178(1–2):147–156
58. Dam CEC, Grzegorczyk AP, Hageman PR et al (2006) Method for HVPE growth of thick crack-free GaN layers. J Cryst Growth 290(2):473–478
59. Napierala J, Martin D, Grandjean N et al (2006) Stress control in GaN/sapphire templates for the fabrication of crack-free thick layers. J Cryst Growth 289(2):445–449
60. Lee HJ, Lee SW, Goto H et al (2007) Free standing GaN layers with GaN nanorod buffer layer. Physica Status Solidi C—Curr Top Solid State Phys 4(7):2268
61. Nikolaev V, Golovatenko A, Mynbaeva M et al (2014) Effect of nano-column properties on self-separation of thick GaN layers grown by HVPE. In: Eddy CR, Kuball M, Koleske DD et al (eds) Physica Status Solidi C: Curr Top Solid State Phys 11(3–4):502–504
62. Lee M, Mikulik D, Park S (2017) Thick GaN growth via GaN nanodot formation by HVPE. Cryst Eng Comm 19(6):930–935
63. Zhang L, Shao Y, Hao X et al (2011) Improvement of crystal quality HVPE grown GaN on an H_3PO_4 etched template. Cryst Eng Comm 13(15):5001–5004
64. Usui A, Sunakawa H, Sakai A et al (1997) Thick GaN epitaxial growth with low dislocation density by hydride vapor phase epitaxy. Jpn J Appl Phys Part 2-Lett 36(7B):L899–L902

65. Usui A (1998) Bulk GaN crystal with low defect density grown by hydride vapor phase epitaxy. In: Ponce FA, Denbaars SP, Meyer BK et al (eds) Nitride semiconductors, pp 233–244
66. Hennig C, Richter E, Zeimer U et al (2006) Bowing of thick GaN layers grown by HVPE using ELOG. In: Hildebrandt S, Stutzmann M (eds) Physica Status Solidi C—Curr Top Solid State Phys 3(6):1466–1470
67. Paskova T, Valcheva E, Paskov PP et al (2004) HVPE-GaN: comparison of emission properties and microstructure of films grown on different laterally overgrown templates. Diamond Relat Mater 13(4–8):1125–1129
68. Fujikura H, Yoshida T, Shibata M et al (2017) Recent Progress of High-Quality GaN Substrates by HVPE Method. In: Chyi JI, Fujioka H, Morkoc H et al (eds) Gallium nitride materials and devices Xii. Spie-Int Soc Optical Engineering, Bellingham
69. Hiramatsu K, Matsushima H, Shibata T et al (1998) Selective area growth of GaN by MOVPE and HVPE. In: Ponce FA, Denbaars SP, Meyer BK et al (eds) Nitride semiconductors, 257–268
70. Huang H-H, Chao C-L, Chi T-W et al (2009) Strain-reduced GaN thick-film grown by hydride vapor phase epitaxy utilizing dot air-bridged structure. J Cryst Growth 311(10):3029–3032
71. Richter E, Hennig C, Weyers M et al (2005) Reactor and growth process optimization for growth of thick GaN layers on sapphire substrates by HVPE. J Cryst Growth 277(1–4):6–12
72. Richter E, Hennig C, Kissel H et al (2005) Growth optimization for thick crack-free GaN layers on sapphire with HVPE. In: Stutzmann M (ed) Physica Status Solidi C—Conf Crit Rev 2(7):2099–2103
73. Bohyama S, Yoshikawa K, Naoi H et al (2002) High quality GaN grown by raised-pressure HVPE. Phys Status Solidi A-Appl Mat 194(2):528–531
74. Luo W, Wu J, Goldsmith J et al (2012) The growth of high-quality and self-separation GaN thick-films by hydride vapor phase epitaxy. J Cryst Growth 340(1):18–22
75. Valcheva E, Paskova T, Persson POA et al (2002) Misfit defect formation in thick GaN layers grown on sapphire by hydride vapor phase epitaxy. Appl Phys Lett 80(9):1550–1552
76. Wei T-B, Ma P, Duan R-F et al (2007) Columnar structures and stress relaxation in thick GaN films grown on sapphire by HVPE. Chin Phys Lett 24(3):822–824
77. Huang H-H, Chen K-M, Tu L-W et al (2008) A novel technique for growing crack-free GaN thick film by hydride vapor phase epitaxy. Jpn J Appl Phys 47(11):8394–8396
78. Voronenkov VV, Bochkareva NI, Gorbunov RI et al (2013) Two modes of HVPE growth of GaN and related macrodefects. In: Toropov A, Ivanov S (eds) Physica Status Solidi C: Curr Top Solid State Phys 10(3):468–471
79. Tsyuk A, Gorbunov R, Voronenkov V et al (2011) Effect of growth parameters on stress in HVPE GaN films. In: Bardwell JA, Hunter GW, Stokes EB et al (eds) Wide bandgap semiconductor materials and devices, vol 12, pp 73–81
80. Lee H-J, Goto T, Fujii K et al (2010) Novel approach to the fabrication of a strain- and crack-free GaN free-standing template: self-separation assisted by the voids spontaneously formed during the transition in the preferred orientation. J Cryst Growth 312(2):198–201
81. Geng H, Yamaguchi AA, Sunakawa H et al (2011) Residual strain evaluation by cross-sectional micro-reflectance spectroscopy of freestanding GaN grown by hydride vapor phase epitaxy. Jpn J Appl Phys 50(1)
82. Lee M, Mikulik D, Kim J et al (2013) A novel growth method of freestanding GaN using in situ removal of Si substrate in hydride vapor phase epitaxy. Appl Phys Exp 6(12)
83. Jasinski J, Liliental-Weber Z (2002) Extended defects and polarity of hydride vapor phase epitaxy GaN. J Electron Mater 31(5):429–436
84. Richter E, Gruender M, Netzel C et al (2012) Growth of GaN boules via vertical HVPE. J Cryst Growth 350(1):89–92
85. Lee CK, Chen YB, Chang SC et al (2002) Micro-photoluminescence from V-shape inverted pyramid in HVPE grown GaN film. In: Wehrspohn RB, Marz R, Noda S et al (eds) Materials and devices for optoelectronics and microphotonics, pp 15–20
86. Zhang YM, Wang JF, Zheng SN et al (2019) Optical and electrical characterizations of the V-shaped pits in Fe-doped bulk GaN. Appl Phys Express 12(7):4

87. Zhang Y, Wang J, Su X et al (2019) Investigation of pits in Ge-doped GaN grown by HVPE. Jpn J Appl Phys 58(12)
88. Wei TB, Duan RF, Wang JX et al (2007) Characterization of free-standing GaN substrate grown through hydride vapor phase epitaxy with a TiN interlayer. Appl Surf Sci 253(18):7423–7428
89. Wu J, Zhao L, Wen D et al (2009) New design of nozzle structures and its effect on the surface and crystal qualities of thick GaN using a horizontal HVPE reactor. Appl Surf Sci 255(11):5926–5931
90. Richter E, Zeimer U, Brunner F et al (2010) Boule-like growth of GaN by HVPE. Physica Status Solidi C: Curr Top Solid State Phys 7(1):28–31
91. Lee W, Lee HJ, Park SH et al (2012) Cross sectional CL study of the growth and annihilation of pit type defects in HVPE grown (0001) thick GaN. J Cryst Growth 351(1):83–87
92. Srikant V, Speck JS, Clarke DR (1997) Mosaic structure in epitaxial thin films having large lattice mismatch. J Appl Phys 82(9):4286–4295
93. Chierchia R, Bottcher T, Heinke H et al (2003) Microstructure of heteroepitaxial GaN revealed by x-ray diffraction. J Appl Phys 93(11):8918–8925
94. Liu JQ, Qiu YX, Wang JF et al (2009) High-resolution x-ray diffraction studies of highly curved GaN layers prepared by hydride vapor phase epitaxy 7518:75180G
95. Weyher JL, Lazar S, Macht L et al (2007) Orthodox etching of HVPE-grown GaN. J Cryst Growth 305(2):384–392
96. Usami S, Ando Y, Tanaka A et al (2018) Correlation between dislocations and leakage current of p-n diodes on a free-standing GaN substrate. Appl Phys Lett 112(18)
97. Kapolnek D, Wu XH, Heying B et al (1995) Structural evolution in epitaxial metalorganic chemical-vapor-deposition grown GaN films on sapphire. Appl Phys Lett 67(11):1541–1543
98. Amano H, Sawaki N, Akasaki I et al (1986) Metalorganic vapor-phase epitaxial-growth of a high-quality GaN film using an AIN buffer layer. Appl Phys Lett 48(5):353–355
99. Nakamura S, Senoh M, Nagahama S et al (1998) Continuous-wave operation of InGaN/GaN/AlGaN-based laser diodes grown on GaN substrates. Appl Phys Lett 72(16):2014–2016
100. Molnar RJ, Maki P, Aggarwal R et al (1996) Gallium nitride thick films grown by hydride vapor phase epitaxy. MRS Proc 423:221–226
101. Tavernier PR, Etzkorn EV, Wang Y et al (2000) Two-step growth of high-quality GaN by hydride vapor-phase epitaxy. Appl Phys Lett 77(12):1804–1806
102. Kim MD, Kim TW (2005) Improvement of the crystallinity of GaN epitaxial films grown on sapphire substrates due to the use of AlN quantum dot buffer layers. J Mater Sci 40(20):5533–5535
103. Sakai A, Sunakawa H, Usui A (1997) Defect structure in selectively grown GaN films with low threading dislocation density. Appl Phys Lett 71(16):2259–2261
104. Lipski F, Wunderer T, Schwaiger S et al (2010) Fabrication of freestanding 2-GaN wafers by hydride vapour phase epitaxy and self-separation during cooldown. Phys Status Solidi A—Appl Mat 207(6):1287–1291
105. Motoki K, Okahisa T, Matsumoto N et al (2001) Preparation of large freestanding GaN substrates by hydride vapor phase epitaxy using GaAs as a starting substrate. Jpn J Appl Phys Part 2—Lett Exp Lett 40(2B):L140–L3
106. Motoki K, Okahisa T, Hirota R et al (2007) Dislocation reduction in GaN crystal by advanced-DEEP. J Cryst Growth 305(2):377–383
107. Oshima Y, Eri T, Shibata M et al (2003) Preparation of freestanding GaN wafers by hydride vapor phase epitaxy with void-assisted separation. Jpn J Appl Phys Part 2-Lett 42(1A-B): L1–L3
108. Geng H, Sunakawa H, Sumi N et al (2012) Growth and strain characterization of high quality GaN crystal by HVPE. J Cryst Growth 350(1):44–49
109. Fujito K, Kubo S, Nagaoka H et al (2009) Bulk GaN crystals grown by HVPE. J Cryst Growth 311(10):3011–3014
110. Gu H, Ren G, Zhou T et al (2016) Study of optical properties of bulk GaN crystals grown by HVPE. J Alloy Compd 674:218–222

111. Maruska HP, Tietjen JJ (1969) Preparation and properties of vapor-deposited single-crystalline GaN. Appl Phys Lett 15(10):327–329
112. Fujikura H, Konno T, Yoshida T et al (2017) Hydride-vapor-phase epitaxial growth of highly pure GaN layers with smooth as-grown surfaces on freestanding GaN substrates. Jpn J Appl Phys 56(8):9
113. Xu K, Wang JF, Ren GQ (2015) Progress in bulk GaN growth. Chin Phys B 24(6):16
114. Gu H, Ren G, Zhou T et al (2016) The electrical properties of bulk GaN crystals grown by HVPE. J Cryst Growth 436:76–81
115. Elsner J, Jones R, Heggie MI et al (1998) Deep acceptors trapped at threading-edge dislocations in GaN. Phys Rev B 58(19):12571–12574
116. Kang T-S, Ren F, Gila BP et al (2015) Investigation of traps in AlGaN/GaN high electron mobility transistors by sub-bandgap optical pumping. J Vac Sci Technol, B 33(6):061202
117. Shalish I, Kronik L, Segal C et al (1999) Yellow luminescence and related deep levels in unintentionally doped GaN films. Phys Rev B 59(15):9748–9751
118. Nakamura S, Mukai T, Senoh M (1992) SI-doped and Ge-doped GaN films grown with GaN buffer layers. J J Appl Phys Part 1—Regul Pap Short Notes Rev Pap 31(9A):2883–2888
119. Fomin AV, Nikolaev AE, Nikitina IP et al (2001) Properties of Si-doped GaN layers grown by HVPE. Physica Status Solidi A—Appl Res 188(1):433–437
120. Usui A, Sunakawa H, Kuroda N et al (1998) Recent progress in epitaxial lateral overgrowth technique for growing bulk GaN by HVPE. Ohmsha Ltd., Tokyo
121. Oshima Y, Yoshida T, Eri T et al (2006) Thermal and electrical properties of high-quality freestanding GaN wafers with high carrier concentration. Jpn J Appl Phys Part 1—Regul Pap Brief Commun Rev Pap 45(10A):7685–7687
122. Richter E, Hennig C, Zeimer U et al (2006) N-type doping of HVPE-grown GaN using dichlorosilane. Physica Status Solidi (A) 203(7):1658–1662
123. Hofmann P, Roder C, Habel F et al (2016) Silicon doping of HVPE GaN bulk-crystals avoiding tensile strain generation. J Phys D-Appl Phys 49(7):7
124. Iwinska M, Sochacki T, Amilusik M et al (2016) Homoepitaxial growth of HVPE-GaN doped with Si. J Cryst Growth 456:91–96
125. Park HJ, Kim HY, Bae JY et al (2012) Control of the free carrier concentrations in a Si-doped freestanding GaN grown by hydride vapor phase epitaxy. J Cryst Growth 350(1):85–88
126. Dadgar A, Veit P, Schulze F et al (2007) MOVPE growth of GaN on Si—substrates and strain. Thin Solid Films 515(10):4356–4361
127. Romanov AE, Speck JS (2003) Stress relaxation in mismatched layers due to threading dislocation inclination. Appl Phys Lett 83(13):2569–2571
128. Cantu P, Wu F, Waltereit P et al (2005) Role of inclined threading dislocations in stress relaxation in mismatched layers. J Appl Phys 97(10)
129. Fritze S, Dadgar A, Witte H et al (2012) High Si and Ge n-type doping of GaN doping—limits and impact on stress. Appl Phys Lett 100(12):4
130. Markurt T, Lymperakis L, Neugebauer J et al (2013) Blocking growth by an electrically active subsurface layer: the effect of Si as an antisurfactant in the growth of GaN. Phys Rev Lett 110(3):5
131. Nenstiel C, Bugler M, Callsen G et al (2015) Germanium—the superior dopant in n-type GaN. Physica Status Solidi-Rapid Res Lett 9(12):716–721
132. Gotz W, Kern RS, Chen CH et al (1999) Hall-effect characterization of III-V nitride semiconductors for high efficiency light emitting diodes. Mater Sci Eng B-Solid State Mater Adv Technol 59(1–3):211–217
133. Gotz W, Johnson NM, Chen C et al (1996) Activation energies of Si donors in GaN. Appl Phys Lett 68(22):3144–3146
134. Oshima Y, Yoshida T, Watanabe K et al (2010) Properties of Ge-doped, high-quality bulk GaN crystals fabricated by hydride vapor phase epitaxy. J Cryst Growth 312(24):3569–3573
135. Fritze S, Dadgar A, Witte H et al (2012) High Si and Ge n-type doping of GaN doping—limits and impact on stress. Appl Phys Lett 100(12)

136. Wieneke M, Witte H, Lange K et al (2013) Ge as a surfactant in metal-organic vapor phase epitaxy growth of a-plane GaN exceeding carrier concentrations of 10(20) cm(-3). Appl Phys Lett 103(1)
137. Hofmann P, Krupinski M, Habel F et al (2016) Novel approach for n-type doping of HVPE gallium nitride with germanium. J Cryst Growth 450:61–65
138. Iwinska M, Takekawa N, Ivanov VY et al (2017) Crystal growth of HVPE-GaN doped with germanium. J Cryst Growth 480:102–107
139. Vaudo RP, Xu XP, Salant A et al (2003) Characteristics of semi-insulating, Fe-doped GaN substrates. Physica Status Solidi A—Appl Res 200(1):18–21
140. Richter E, Gridneva E, Weyers M et al (2016) Fe-doping in hydride vapor-phase epitaxy for semi-insulating gallium nitride. J Cryst Growth 456:97–100
141. Fujikura H, Konno T, Yoshida T et al (2017) Hydride-vapor-phase epitaxial growth of highly pure GaN layers with smooth as-grown surfaces on freestanding GaN substrates. Jpn J Appl Phys 56(8)
142. Kubota M, Onuma T, Ishihara Y et al (2009) Thermal stability of semi-insulating property of Fe-doped GaN bulk films studied by photoluminescence and monoenergetic positron annihilation techniques. J Appl Phys 105(8):9
143. Gladkov P, Humlicek J, Hulicius E et al (2010) Effect of Fe doping on optical properties of freestanding semi-insulating HVPE GaN:Fe. J Cryst Growth 312(8):1205–1209
144. Fang ZQ, Claflin B, Look DC et al (2008) Deep centers in semi-insulating current topics in solid state physics Fe-doped native GaN substrates grown by hydride vapour phase epitaxy. Physica Status Solidi C—Curr Top Solid State Phys 5(6):1508-+
145. Jarasiunas K, Kadys A, Aleksiejunas R et al (2009) Optical nonlinearities and carrier dynamics in semi-insulating crystals. In: Krotkus A, Valusis G, Adomavicius R (eds) Physica Status Solidi C—Curr Top Solid State Phys 6(12):2846–2848
146. Fang Y, Wu X, Yang J et al (2015) Effect of Fe-doping on nonlinear optical responses and carrier trapping dynamics in GaN single crystals. Appl Phys Lett 107(5)
147. Dalpian GM, Da Silva JLF, Wei S-H (2009) Ferrimagnetic Fe-doped GaN: an unusual magnetic phase in dilute magnetic semiconductors. Phys Rev B 79(24)
148. Theodoropoulou N, Hebard AF, Chu SNG et al (2002) Use of ion implantation to facilitate the discovery and characterization of ferromagnetic semiconductors. J Appl Phys 91(10):7499–7501
149. Akinaga H, Nemeth S, De Boeck J et al (2000) Growth and characterization of low-temperature grown GaN with high Fe doping. Appl Phys Lett 77(26):4377–4379
150. Zhang M, Cai D, Zhang Y et al (2017) Investigation of the properties and formation process of a peculiar V-pit in HVPE-grown GaN film. Mater Lett 198:12–15
151. Wei TB, Duan RF, Wang JX et al (2008) Hillocks and hexagonal pits in a thick film grown by HVPE. Microelectron J 39(12):1556–1559
152. Montes Bajo M, Hodges C, Uren MJ et al (2012) On the link between electroluminescence, gate current leakage, and surface defects in AlGaN/GaN high electron mobility transistors upon off-state stress. Appl Phys Lett 101(3):4
153. Chen Y, Takeuchi T, Amano H et al (1998) Pit formation in GaInN quantum wells. Appl Phys Lett 72(6):710–712
154. Lucznik B, Pastuszka B, Weyher JL et al (2009) Bulk GaN crystals and wafers grown by HVPE without intentional doping. Physica Status Solidi C: Curr Top Solid State Phys 6(2):S297–S300
155. Voronenkov V, Bochkareva N, Gorbunov R et al (2013) Nature of V-shaped defects in GaN. Jpn J Appl Phys 52(8)
156. Zhang Y, Wang J, Zheng S et al (2019) Optical and electrical characterizations of the V-shaped pits in Fe-doped bulk GaN. Appl Phys Exp 12(7)
157. Puzyrev YS, Schrimpf RD, Fleetwood DM et al (2015) Role of Fe impurity complexes in the degradation of GaN/AlGaN high-electron-mobility transistors. Appl Phys Lett 106(5):4
158. Axelsson O, Billstrom N, Rorsman N et al (2016) Impact of trapping effects on the recovery time of GaN based low noise amplifiers. IEEE Microwave Wirel Compon Lett 26(1):31–33

159. Oshimura Y, Takeda K, Sugiyama T et al (2010) AlGaN/GaN HFETs on Fe-doped GaN substrates. Physica Status Solidi C: Curr Top Solid State Phys 7(7–8)
160. Williams AD (2007) Moustakas TDJJOCG. Formation of large-area freestanding gallium nitride substrates by natural stress-induced separation of GaN and sapphire 300(1):37–41
161. Li M, Cheng Y, Yu T et al (2019) Critical thickness of GaN film in controllable stress-induced self-separation for preparing native GaN substrates. 180:107985
162. Yoshida T, Oshima Y, Eri T et al (2008) Preparation of 3 inch freestanding GaN substrates by hydride vapor phase epitaxy with void-assisted separation. 205(5):1053–1055
163. Hennig C, Richter E, Weyers M et al (2007) Self-separation of thick two inch GaN layers grown by HVPE on sapphire using epitaxial lateral overgrowth with masks containing tungsten 4(7):2638–2641
164. Lee HJ, Lee SW, Goto H et al (2007) Self-separated freestanding GaN using a NH_4Cl interlayer 91(19):L217
165. Sato T, Okano S, Goto T et al (2013) Nearly 4-inch-diameter free-standing GaN wafer fabricated by hydride vapor phase epitaxy with pit-inducing buffer layer. 52:279–287
166. Zhang L, Dai Y, Wu Y et al (2014) Epitaxial growth of a self-separated GaN crystal by using a novel high temperature annealing porous template 16(38):9063–9068
167. Nikolaev V, Golovatenko A, Mynbaeva M et al (2014) Effect of nano-column properties on self-separation of thick GaN layers grown by HVPE 11(3–4):502–504
168. Usui A, Ichihashi T, Kobayashi K et al (2015) Role of TiN film in the fabrication of freestanding GaN wafers using hydride vapor phase epitaxy with void-assisted separation 194(2):572–575
169. Kunook C et al (2010) Transferable GaN layers grown on ZnO-coated graphene layers for optoelectronic devices
170. Kim J, Bayram C et al (2014) Principle of direct van der Waals epitaxy of single-crystalline films on epitaxial graphene 5(1):4836
171. Wong WS et al (1998) Damage-free separation of GaN thin films from sapphire substrates
172. Su XJ, Xu K, Xu Y et al (2013) Shock-induced brittle cracking in HVPE-GaN processed by laser lift-off techniques
173. Xu K, Wang JF, Ren GQJCPB (2015) Progress in bulk GaN growth 24(6):066105

Chapter 4
Ammonothermal Method

4.1 History

At present, the main methods of growing Gallium Nitride (GaN) single crystal substrates include High-nitrogen-pressure solution growth (HNPSG), Na-flux method (Na Flux), Hydride vapor phase epitaxy (HVPE) and Ammonothermal method. HNPSG has stringent requirements for equipment (pressure greater than 1 Gpa, temperature greater than 1500 °C, making the production of large-sized crystals and industrialization challenging. The remaining three methods have their own advantages and disadvantages: HVPE has a rapid growth rate and is easy to obtain large-size crystal, which is the most promising method for commercial applications. However, the disadvantages of HVPE are obvious: high cost, high crystal defect density (> 10^5 cm^{-2}), small radius of curvature and environmental pollution. The Na-Flux Method is susceptible to polycrystal formation due to a high density of spontaneous nucleation, making the growth of thick crystals challenging. The GaN ammonothermal growth technology is analogous to the hydrothermal growth technology of quartz and ZnO crystals, and has many advantages: (1) It operates under conditions close to thermodynamic equilibrium, resulting in high crystal quality; (2) It can simultaneously grow on hundreds or thousands of seed crystals, facilitating mass production; (3) Free-standing GaN single crystals can be directly obtained without the need to strip the growth substrate; (4) The growth equipment is straightforward, and the growth process functions as a closed system. No additional supply of growth raw materials is necessary, resulting in low costs; (5) The hydrothermal method boasts a nearly 100-year history of producing a variety of crystals. The accumulated experimental experience, research methods and abundant theoretical knowledge can directly guide GaN single crystal grown by ammonothermal method. Although the ammonothermal technology has disadvantages, such as high growth pressure and low growth rate, recent increases in international investment in ammonothermal research have led to progressive advancements in the understanding of the physical and chemical processes within the ammonothermal system, as well as in-depth research

K. Xu et al., *GaN Single Crystal Growth and Application*, Wide Bandgap Semiconductors, https://doi.org/10.1007/978-981-96-7572-2_4

into related issues. The ammonothermal growth technology of GaN is progressively advancing towards industrialization.

The early ammonothermal method was primarily employed for material synthesis, especially for the synthesis of binary and ternary transition metal nitrides. Transition metal nitrides are usually high melting point compounds that are easily decomposed at high temperature and difficult to synthesize. Ammonothermal synthesis is an effective method for preparing new nitride material. An additional advantage of this method is that it uses chemical reaction in supercritical ammonia to lower the reaction barrier and can prepare nitride at relatively low temperature. In the 1990s, AlN and GaN polycrystals were prepared using the ammonothermal synthesis, which laid the foundation for the ammonothermal growth of GaN single crystal [1–3].

Although the ammonothermal synthesis of GaN was achieved in 1995, it was not until 2000 that millimeter-sized GaN crystals were grown using the ammonothermal technology: In 2001, Ketchum and Kolis conducted the ammonothermal synthesis of GaN in an alkaline mineralizer ammonothermal environment of 400 °C and 240 MPa. It took ten days to grow a GaN crystal of $0.5 \times 0.2 \times 0.1$ mm^3 [4]. In 2002, Purdy et al. studied GaN recrystallization in an acidic mineralizer ammonothermal environment and grew cone-shaped GaN crystal with a width of 0.1 mm [5]. In 2003, Callahan et al. utilized GaN polycrystals and alkaline mineralizer to facilitate mass transfer growth on HVPE-GaN seed crystals [6]. In 2004, Yoshikawa et al. used NH_4Cl as mineralizer and Pt as lining to achieve needle-like GaN crystal growth at a pressure of 135 MPa [7]; In 2006, Wang and Callahan reported the growth of GaN crystal with a growth rate of 50 μm/day and a size of $10 \times 10 \times 1$ mm^3 [8]; In 2006, Kagamitani et al. used NH_4Cl as mineralizer to achieve the ammonothermal growth with a maximum growth rate of 27.5 μm/day on the N surface [9]. In recent years, the ammonothermal method has developed rapidly, and made great progress. The size of GaN crystal growth has quickly expanded to 2–4 inches, and the dislocation density has been reduced to the order of 10^3 cm^{-2}. Concurrently, several companies have emerged globally for the ammonothermal growth of GaN single crystals, including Poland's AMMONO, Japan's Mitsubishi Chemical, and the United States' SixPoint and Soraa. The ammonothermal growth of GaN is progressing toward industrialization.

4.2 Principle of Growth

Ammonothermal method is a technique of growing crystal from supercritical ammonia at high temperature and high pressure (400–750 °C, 1000–6000 bar). This technique is similar to the hydrothermal method of growing quartz. The solvent is replaced from supercritical water in hydrothermal method to supercritical ammonia, facilitating the growth of nitrides (as opposed to oxides). The crystallization process occurs in an autoclave. The autoclave is constructed from specialized steel that is resistant to high temperatures, high pressures, as well as acids and alkalis. The autoclave is divided into a dissolution zone (the raw material zone), where GaN or Ga raw material are placed and a crystallization zone, where GaN seed crystal are suspended.

The autoclave is filled with ammonia and mineralizer that assists in dissolution of the raw material. Due to the temperature difference between the crystallization zone and the dissolution zone, convection occurs. Convection brings saturated solution of the dissolution zone to the crystallization zone, where a supersaturated solution is formed because of reduced solubility, thereby precipitating GaN. The solution that has already precipitated solutes flows back into the dissolution zone to continue dissolving the raw material. The seed crystal can continuously grow by repeating this cycle.

The working principle of a typical ammonothermal crystal growth equipment is shown in Fig. 4.1. During the growth process, the autoclave is positioned vertically, and is divided into a raw material dissolution zone and a growth zone through baffles with holes. Normally, the solubility temperature coefficient of GaN under the function of acidic mineralizer is positive. The raw materials are placed in the lower part of the autoclave, and the seed crystal is placed in the upper part of the autoclave (as shown in Fig. 4.1a). Under the influence of alkaline mineralizer, the solubility temperature coefficient of GaN is negative (the solubility of GaN decreases as the temperature increases). The raw materials are placed in the upper part of the autoclave, and the seed crystal is placed in the lower part of the autoclave (as shown in the Fig. 4.1b). Taking the positive solubility temperature coefficient as an example, the autoclave establishes two temperature zones: the upper and lower temperature zones, through temperature regulation. The temperature T_1 of the lower part exceeds the temperature T_2 of the upper part, resulting in strong convection of the solution. The solution in the raw material zone exhibits higher solubility. Through convection, GaN is transported to the growth zone, which has lower solubility. The solution reaches supersaturation, causing GaN to crystallize on the seed crystal, thereby achieving the growth of GaN single crystals. In order to achieve uniform and rapid growth of all crystal in the autoclave, it is necessary to accurately control the flow field in the autoclave. The main control methods of the ammonothermal growth system include temperature, temperature gradient, baffle structure and position, filling degree, and mineralizer concentration and ratio, etc.

4.3 Processes and Growth Equipment

4.3.1 Ammonothermal Growth Process

The ammonothermal growth process is significantly more complex than the hydrothermal process, necessitating a redesign of the associated growth equipment. This complexity arises primarily from the following reasons: Firstly, the solvent in the ammonothermal method is ammonia, which is an irritating gas at room temperature and pressure and cannot be poured directly into the high-pressure autoclave like water in the hydrothermal method. The solvent must be added in a closed system throughout the entire process. Secondly, the mineralizer, especially the alkaline mineralizer,

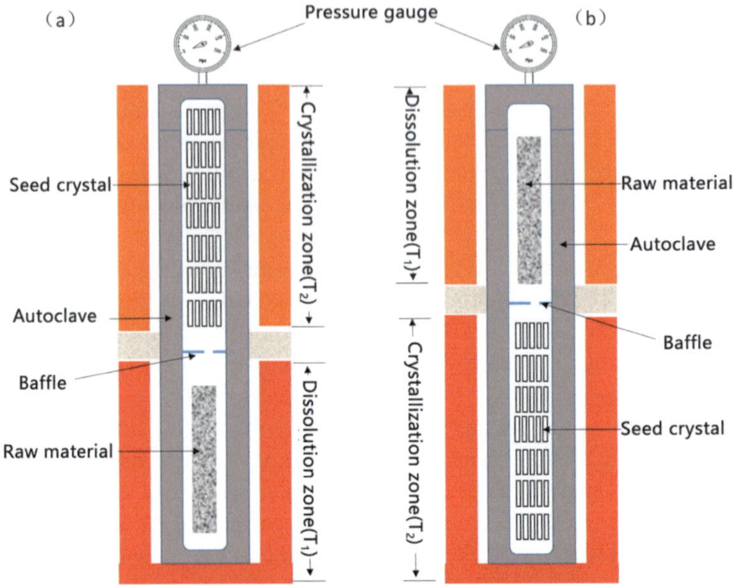

Fig. 4.1 Schematic diagram of GaN grown by ammonothermal. **a** Schematic diagram of growth under positive solubility temperature coefficient. **b** Schematic diagram of growth under negative solubility temperature coefficient

reacts with water and oxygen. Given that the crystal being grown is gallium nitride, it is essential to avoid the introduction of oxygen and oxygen-containing elements in water, necessitating a redesign of the raw material loading process. Thirdly, the growth temperature and pressure in the ammonothermal method are significantly higher than those in the hydrothermal method. Traditional materials and mechanical strengths of hydrothermal high-pressure autoclaves cannot meet the requirements of the ammonothermal growth process, necessitating the reselection and redesign of high-pressure autoclaves with enhanced safety and security requirements compared to those of the hydrothermal method. The main process flow of the ammonothermal method is illustrated in Fig. 4.2. Prior to loading the autoclave, the high-pressure autoclave undergoes dewatering and deoxygenation. The mineralizer, GaN polycrystalline or Ga raw materials, and seed crystal racks are placed in a water-insulated and oxygen-insulated system. Subsequently, these materials are loaded into the high-pressure autoclave, followed by the injection of a predetermined amount of ammonia into the autoclave within the closed system. Then seal the autoclave and heat it to keep a temperature difference between the upper and lower zones. Set up the growth cycle at last.

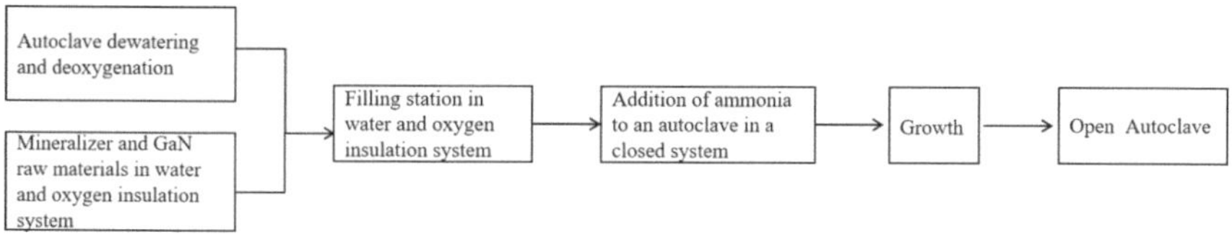

Fig. 4.2 Ammonothermal growth process

4.3.2 Ammonothermal Growth Equipment

Ammonothermal process is complex, the growth temperature and pressure are high, so the requirement of ammonothermal growth equipment is also high. The main equipments of ammonothermal method include autoclave, liquid ammonia filling device, temperature and pressure control device, liner, water and oxygen insulation system, etc.

1. Autoclave

The high-pressure autoclave is a critical component for ammonia thermal crystal growth, and the quality of crystal growth is directly correlated with its performance. Autoclaves utilized in the ammonothermal growth process typically operate at temperatures exceeding 400 °C and pressures ranging from 100 to 300 MPa, or even higher. Traditional hydrothermal autoclave materials struggle to meet the rigorous requirements of ammonia thermal growth, particularly in terms of creep and enduring strength, when exposed to temperatures above 400 °C. Consequently, these materials are unsuitable for ammonothermal growth, necessitating a reevaluation of material design for ammonothermal autoclaves. An ideal ammonothermal autoclave should possess the following characteristics: 1. Strong corrosion resistance to both acids and alkalis. 2. A reliable and straightforward sealing structure for the autoclave body that facilitates easy assembly and disassembly. 3. An appropriate length-to-diameter ratio to maintain the required growth temperature gradient. 4. Excellent mechanical properties at high temperatures. 5. Capable of withstanding prolonged high-pressure and high-temperature conditions to ensure continuous operation over extended periods.

Due to the late development of the ammonothermal process, material requirements and design criteria have not been established for the autoclaves of ammonothermal process. When selecting suitable materials for high-pressure autoclaves, the primary considerations include the experimental temperature and pressure conditions, as well as corrosion resistance within the designated pressure–temperature range of a given solvent or solution. If the reaction occurs directly within the autoclave, corrosion resistance is the primary criterion for selecting high-pressure autoclave materials. In addition to corrosion resistance, key characteristics of high-pressure autoclave materials include their creep fracture strength, ultimate tensile strength, and yield strength. Given the temperature and pressure requirements commonly encountered in the ammonothermal process, materials such as nickel-based high-temperature alloys are often chosen for the fabrication of high-pressure autoclaves. These alloys can withstand complex stresses and demonstrate long-term reliability in oxidative and gas-corrosive atmospheres at temperatures ranging from 600 to 1000 °C, such as Rene 41 [4]. Inconel alloy 625, Alloy 718 and other types of alloys [10, 11].

The sealing structure of the high-pressure autoclave is a critical component, and the safe and stable operation of the autoclave largely dependent on the integrity of the sealing structure. The sealing mechanisms of high-pressure autoclaves can be broadly categorized into three types: forced sealing, semi-self-tightening sealing, and self-tightening sealing. Since forced seals and semi-self-tightening seals are not adequate for autoclaves with pressures over 100 MPa, ammonothermal seal usually adopt self-tightening seal structure, especially the modified Bridgman structure. This structure involves the threaded connection of the top cover with the cylinder. The internal pressure is transmitted to the sealing ring, which is supported at the lower end of the top cover through a sealing plug. The direction of the internal pressure opposes the pre-tightening force exerted on the top cover. When the internal pressure reaches a certain value, the sealing plug floats, causing the sealing ring to expand under pressure, thereby achieving the seal. In this way, as the internal pressure increases, the pressure on the sealing ring per unit area also increases, ensuring a more reliable seal.

2. Temperature and pressure control devices

Ammonothermal GaN bulk single crystal growth is generally used a temperature difference method, where the autoclave through the baffle is divided into the growth zone and the dissolution zone. The crystal growth needs to be in the dissolution zone and the growth zone to establish a constant and stable temperature difference. Furthermore, to achieve uniform crystal growth, it is essential to maintain consistent temperatures in the respective areas of the growth and dissolution zones, necessitating a step-type temperature difference near the baffle. Besides the internal design of the autoclave, the design of the heating unit is a critical factor in establishing a constant and stable step-type temperature difference.

The heating tape fixed on the autoclave is used for heating, and the power of the heating tape can be adjusted according to the needs. To facilitate temperature elevation and control, the heating systems for the growth zone and dissolution zone are independently controlled in two groups. The outside of the heating system is insulated with a heat preservation furnace made of heat preservation material. To facilitate control of the temperature difference between the culture area and the growth area, a baffle made of heat-insulating material is inserted, preventing hot air from convecting from the culture area to the growth area and thereby ensuring a constant and stable step-type temperature difference between the two areas. The temperature is typically controlled by a combination of a computer and a silicon-controlled rectifier system, which prepares a temperature program for autoclave operation based on process conditions and inputs the growth program into the computer. During crystal growth, the designed temperature value is compared with the signal transmitted by the thermocouple through the computer, and the power input is continuously adjusted by the SCR to stabilize the autoclave temperature at the set temperature.

Pressure detection in ammonothermal autoclaves is commonly carried out using either pressure gauges or pressure sensors. Pressure gauges typically utilize a spring tube structure, and the selection of a 250 MPa or 400 MPa pressure gauge is based on the growth pressure. The pressure inside the autoclave can be easily read from the dial, providing an intuitive indication. Pressure sensors, operating on the resistance strain principle, directly sense the measured pressure using a thick-walled cylindrical strain tube equipped with resistance strain gauges. Radial expansion of the strain tube leads to an increase in the resistance of the strain gauges, resulting in an unbalanced voltage (current) output across the diagonal ends of the bridge circuit. The magnitude of this voltage (current) is proportionate to the pressure applied to the sensor. Pressure sensors serve not only for pressure detection but also for pressure control, incorporating features such as alarms, heating cessation, and accident prevention. The pressure sensor allows for the pre-setting of a pressure deviation value. If the signal from the sensor exceeds the preset warning deviation value, the receiver can take control of the power supply, halting heating to prevent accidents.

3. Other equipment

In addition to the autoclave, temperature and pressure control system, ammonia crystal growth equipment and liquid ammonia filling system and lining, etc. The GaN ammonia single crystal growth process is more complex than the artificial crystal hydrothermal growth process, primarily due to the use of ammonia as the solvent. Ammonia, at room temperature, is a gas with a strong pungent odor, highly soluble in water, and does not behave like water in an open environment. In addition, under specific temperature and solvent concentration conditions, the pressure within the autoclave is dependent on the filling degree. If the filling degree is too low, the required growth pressure cannot be achieved at the set growth temperature; conversely, if the filling degree is too high, the pressure increases rapidly, potentially leading to accidents. Therefore, the ammonothermal method for accurately filling liquid ammonia has stringent requirements. The precise filling of liquid ammonia can be achieved through pressurized filling or cooling filling methods. In the pressurized filling system, ammonia is injected into the vessel through the cylinder. However, the fouled oil in the cylinder mixes in the vessel and contaminates the system. Another better method is to operate at low temperatures, which is suitable for small autoclaves. The autoclave is evacuated to vacuum and cooled to below the boiling point of NH_3. A mass flow controller is then used to fill the autoclave with high purity ammonia. To obtain GaN single crystals with low impurity content, it is essential to incorporate a lining made of high-purity noble metals that are resistant to corrosion by supercritical ammonia and mineralizers. In particular, acidic mineralizers corrode commonly used high-temperature alloys, so in order to protect the autoclave and reduce the doping of the wall with impurities, it is necessary to use an inert metal lining. Autoclaves used for acidic ammonia thermal growth processes are usually equipped with liners made of chemically inert platinum (Pt) to prevent corrosion of the autoclave and doping of impurities into the growing crystals, baffles, and seed crystal holders are made of Pt [12].

4.4 Ammonothermal Chemistry and Solubility Control

The type and concentration of solvents determine the ammonothermal process and key parameters, such as the solubility of raw materials, phase, kinetics of single crystal growth and growth mechanism. The thermodynamic conditions for the growth of ammonothermal crystals can be expressed as:

$$\Delta G = RT\ln\frac{Q}{K} = RT\ln\frac{C_0}{C} < 0 \tag{4.1}$$

where ΔG is the change in Gibbs free energy, R is the ideal gas constant, T is the temperature, C_0 is the equilibrium concentration of the solution, and C is the actual concentration of the solution. Equation (4.1) indicates that solubility is a key parameter for the growth of GaN crystals by ammonothermal method. At present, there is no theory that can predict the solubility of GaN in actual solutions. However, many solubility-related questions can be understood through physical and chemical principles or laws. The solubility of GaN can be understood using the principle of similarity and solubility, which means that polar materials with higher dielectric constant (ε) are easily soluble in polar solvents. The dielectric constant of NH_3 is much lower than that of water, making it a solvent with weaker polarity. Therefore, the solubility of GaN in ammonia is very low, preventing a meaningful growth rate in pure ammonia. The ammonothermal growth of GaN requires mineralizers to improve the solubility of GaN, and the solubility of GaN in solution is a key link for successful crystal growth. The primary difference in the growth process arises from the use of different mineralizers. Mineralizers often cause significant differences in many important growth parameters and physicochemical processes during the growth process, which also determine the growth process, equipment, and factory configuration of the ammonothermal growth system.

At present, the selection of ammonothermal growth mineralizers internationally is mainly divided into two categories: alkaline mineralizers and acidic mineralizers. Alkaline mineralizers mainly include $NaNH_2$, KNH_2, $LiNH_2$, KN_3, etc.; Acidic mineralizers mainly include NH_4Cl, NH_4F, NH_4I, etc. The following will briefly explain the ammonothermal chemistry in alkaline mineralizer and acidic mineralizer systems.

In an alkaline mineralizer system, GaN raw materials first react with the mineralizer to form intermediate compounds. Under appropriate temperature control, the intermediate compounds are transported by convection to the seed crystal growth zone, and the solution reaches supersaturation. Since GaN does not require nucleation energy for growth on the seed crystal, it has lower energy than spontaneous nucleation, allowing it to preferentially grow on the seed crystal. Therefore, the chemical reactions that occur in the dissolution zone and the growth zone are reciprocal reactions. Taking KNH_2 mineralizer as an example, the reactions that occur in the dissolution zone are as follows:

$$KNH_2 + GaN + 2NH_3 \rightarrow KGa(NH_2)_4 \tag{4.2}$$

The reactions that occur in the growth zone are as follows:

$$KGa(NH_2)_4 \rightarrow KNH_2 + GaN + 2NH_3 \quad (4.3)$$

For $NaNH_2$ mineralizer, the intermediate compound formed is $Na_2Ga(NH_2)_5$, with the following chemical reaction formula:

$$GaN + 2NaNH_2 + 2NH_3 \rightleftharpoons Na_2Ga(NH_2)_5 \quad (4.4)$$

In addition to using GaN polycrystalline as raw materials, Ga metal can also be used as raw materials, and the chemical reactions that occur are (using KNH_2 as an example) [13]:

$$KNH_2 + Ga + NH_3 \rightarrow KNH_2 + GaN + 1.5H_2 \quad (4.5)$$

In acidic mineralizer systems, the main mineralizers added include NH_4Cl, NH_4F, NH_4I, etc. There are many types of intermediate compounds formed, such as $[Ga(NH_3)_6]X_3 \cdot NH_3$ (X = Br, I), $[Ga(NH_3)_5Cl]Cl_2$, $Ga(NH_3)_3F_3$, etc. The intermediate compounds formed by the acidic mineralizer system are shown in Table 4.1. In the ammonothermal system, the coordination structure of intermediate compounds may change, which mainly depends on concentration, temperature, and process pressure [14]. Similar to the growth principle in alkaline mineralizer systems, under controlled temperature conditions, these intermediate compounds are transported via convection to the seed crystal growth zone, where the solution reaches supersaturation and GaN crystal growth occurs on the seed crystal.

As described in the chemical reaction process above, the growth of GaN crystals involves the transformation of GaN polycrystals (using GaN polycrystalline raw materials as an example) into GaN single crystals through intermediate compounds. The key to crystal growth lies in the control of solubility in the dissolution zone and growth zone. In the ammonothermal system, the solubility can be adjusted by

Table 4.1 Intermediate compounds formed by acidic mineralizers

Number of coordinated NH_3 ligands	Complex unit	Halide investigated
1	${}^{0}[Ga(NH_3)X_3]$	Br
1	${}^{1}_{\infty}[Ga(NH_3)XX_{4/2}]$	F
2	${}^{2}_{\infty}[Ga(NH_3)_2X_2X_{2/2}]$	F
3	${}^{0}[Ga(NH_3)_2X_4]^-$, ${}^{0}[Ga(NH_3)_4X_2]^+$	F
5	${}^{0}[Ga(NH_3)_5X]^{2+}$	Cl
6	${}^{0}[Ga(NH_3)_6]^{3+}$	Br, I

controlling temperature. Therefore, understanding the solubility temperature characteristics of GaN under the action of mineralizers is very important. Moreover, the dissolution characteristics of GaN under the action of mineralizers vary with temperature, resulting in significant differences in many important growth parameters and physicochemical processes during the growth process. This determines the growth process and equipment configuration of the ammonothermal growth system. If the solubility temperature coefficient of gallium nitride is positive, it is necessary to place the GaN precursor in the high-temperature zone at the lower part of the high-pressure vessel and the seed crystal in the low-temperature zone at the upper part of the high-pressure vessel. When the solubility temperature coefficient of gallium nitride is negative, the seed crystal needs to be suspended in the high-temperature zone at the bottom of the high-pressure vessel, and the GaN precursor needs to be placed in the low-temperature zone at the top of the high-pressure vessel.

Due to the sensitivity of GaN solubility to intermediate compound reactions, ammonia decomposition, temperature and pressure changes, it is difficult to provide relevant physical parameters for solubility calculation. Currently, the solubility of GaN in various mineralizers ammonothermal systems is mainly measured experimentally. Ehrentraut et al. found that the solubility of GaN increased with increasing temperature in the temperature range of 250–550 °C using NH_4Cl as a mineralizer [15]. Yoshida et al. found that the solubility temperature coefficient of GaN was negative when the temperature exceeded 650 °C, indicating that the solubility of GaN decreased as the temperature rose [16]. The ammonothermal system using $NaNH_2$ as the mineralizer also showed a similar trend of solubility with temperature variation, with a positive temperature coefficient for GaN solubility below 600 °C and a negative temperature coefficient for GaN solubility above 600 °C [17]. Ehrentraut et al. summarized the temperature dependent solubility of GaN under different mineralizers in different study groups in Fig. 4.3a [18]. Tomida et al. investigated the temperature dependent solubility of GaN in acidic mineralizers (as shown in Fig. 4.3b) and found that NH_4I has better solubility temperature characteristics compared to NH_4Cl and NH_4Br [19]. When the acidic mineralizer mentioned above is below 600 °C, the solubility of GaN increases with temperature, while the solubility temperature coefficient of NH_4F is negative in the range of 550–650 °C [20].

In addition to temperature, methods for regulating solubility include pressure, concentration of mineralizers, and ratio of mineralizers. Both acidic and alkaline mineralizers exhibit an increasing solubility of gallium nitride with increasing pressure [20, 21]. Tomida et al. found that the slope of the solubility curve changes with the ratio of mixed mineralizers, so the solubility can be controlled by adjusting the ratio of mineralizers [22].

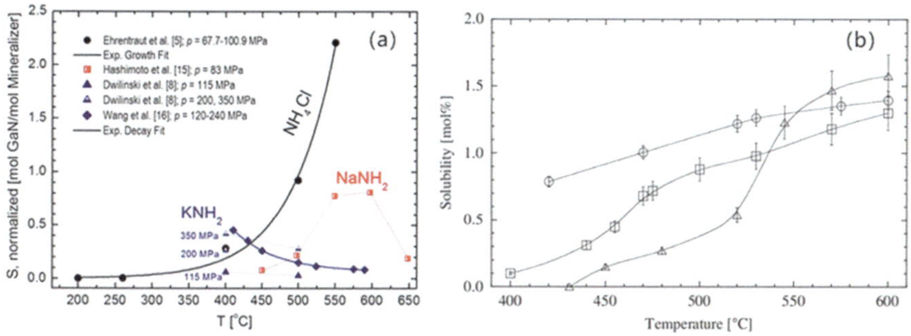

Fig. 4.3 **a** Solubility temperature characteristics of main mineralizers. **b** Solubility temperature characteristics of acidic mineralizers

4.5 Crystal Growth Progress

Since the acidic mineralizer and basic mineralizer systems have obvious differences in equipment requirements, growth processes, raw material filling methods, etc., the ammonothermal growth of GaN is divided into basic mineralizer ammonothermal growth and acid mineralizer ammonothermal growth according to the choice of mineralizer. The progress of these two growth systems will be described below.

4.5.1 Progress in Basic Mineralizer Ammonothermal Crystal Growth of GaN

In 1995, Polish scientist R. Dwilinski et al. first added $LiNH_2$ and KNH_2 mineralizers to the ammonothermal system to grow GaN powders. In the decade since then, there have been relatively few reports on the growth of GaN with basic mineralizers, and only a few articles are reported of synthesis of GaN polycrystalline powder and temperature difference growth. The size of the grown GaN crystals is small [1]. Until 2005, Hashimoto et al. used $NaNH_2$ and NaI as mineralizers and HVPE-grown GaN as seed crystals for temperature difference growth. The growth size exceeded 1 inch, but the crystal quality was poor. The X-ray rocking curve measured from the Ga-face and N-face showed full-width at half-maximum (FWHM) of the (0002) reflection exceeds 1500 arcsec [23]. In 2006, Wang and Callahan et al. continued to report the thermochemical thermodynamics of alkaline mineralizer ammonothermal and the results using HVPE-GaN as seed crystal. The X-ray rocking curve measured from the Ga-face showed full-width at half-maximum (FWHM) of 535 arcsec and 751 arcsec for (002) and (102) reflections, respectively. The X-ray rocking curve measured from the N-face showed full-width at half-maximum (FWHM) values of 859 arcsec and 961 arcsec for the (002) and (102) reflections, respectively, with the c-plane growth rate reaching 50 μm/day [24]. T. Hashimoto et al. used $NaNH_2$ as mineralizers to conduct a long-term (82 days) ammonothermal growth verification. The growth thickness exceeded 5 mm, and the growth rate on the c surface reached 55 μm/day [25].

In 2008, the ammonothermal growth of alkaline mineralizers achieved milestone progress. Dwilinski et al. from Poland used KNH_2 as mineralizers to achieve high-quality growth of GaN [26]. The X-ray rocking curve measured showed full-width at half-maximum (FWHM) of 17 arcsec for (002) reflection and the curvature radius of GaN reached over 100 m (as shown in Fig. 4.4a, b), with a corrosion pit density of 5×10^3 cm^{-2}. The above technical indicators are significantly higher than the indicators of crystals grown by the mainstream growth technology HVPE, which has also sparked a debate on which technology is the final choice for future GaN single crystal substrate mass production between ammonothermal and HVPE on the Compound Semiconductor [27, 28]. Subsequently, Polish researchers quickly expanded the size of the *c*-sided GaN to 2-inch, with a growth thickness reaching the centimeter level (Fig. 4.4d) [29]. In 2018, the growth size of the *c* surface of GaN approached 3 inches (Fig. 4.4c) [30].

In addition to the Polish research group, other international research groups and companies have also made rapid progress in the growth of basic mineralizers since 2008. UCSB has improved the experimental device of ammonothermal growth GaN, added a silver capsule to the high-pressure reactor, and has yielded an ultrahigh purity (UHP) growth environment, significantly reducing the impurities introduced by the

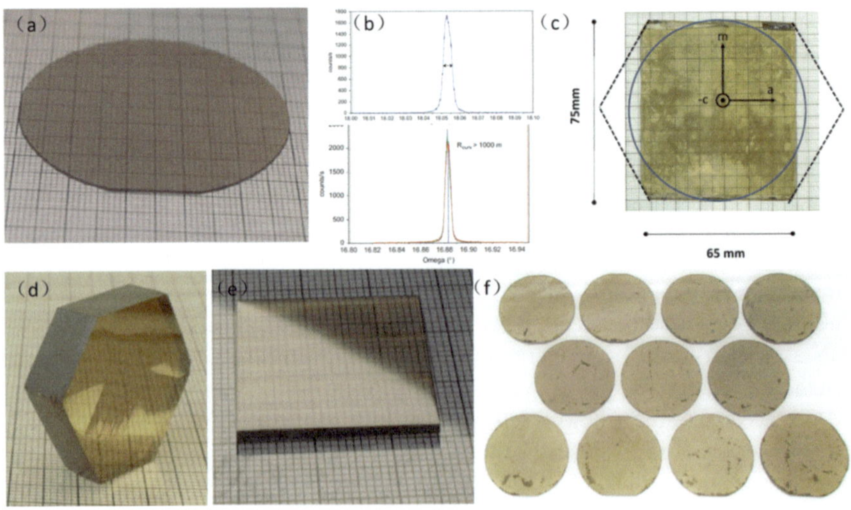

Fig. 4.4 Ammonothermal GaN crystal growth of alkaline mineralizer: **a** 2-inch GaN substrate from Ammono, Poland. **b** (002) X-ray rocking curve with a half-height at width of 17 arcsec and a curvature radius greater than 1000 m (Ammono Corporation). **c** GaN crystals of approximately 3-inch. **d** The thick 1-in AMMONO-GaN crystal. **e** A non-polar m-face GaN substrate cut from a thick crystal. **f** Mass production of alkaline ammonothermal GaN crystals, crystals grown in a single furnace by SixPoint, USA

high-pressure reactor material [31]. SixPoint achieved mass production of 2-inch GaN using HVPE grown GaN as seed crystals (Fig. 4.4f). The 2-inch GaN wafers cut from bulk single crystals had full-width at half-maximum (FWHM) of 25 arcsec for (002) reflection [32].

In the preparation of non-polar and semi-polar GaN, Ammono company obtained non-polar and semi-polar GaN single crystal substrates by cutting long and thick crystals (Fig. 4.4e) [33]. Based on an analysis of the growth habits of different crystal planes, the research team from the Suzhou Institute of Nanotechnology, Chinese Academy of Sciences, employed HVPE-GaN crystals with various orientations as seed crystals to directly grow non-polar and semi-polar GaN crystals. The growth rates of different crystal planes were obtained using cathodoluminscence. Based on the growth rates, a kinetic Wulff's plot for the growth of GaN by ammonothermal method was constructed, and the evolution of the growth equilibrium shape of each seed crystal orientation was predicted based on the kinetic Wulff's plot [34].

Beyond crystal size and dislocations, the electrical and optical properties are critical to the performance of crystals in device applications. Impurities and native point defects are the primary factors influencing the electrical and optical properties of crystals. Regarding unintentional doping, the ammonothermal process operates as a closed-system growth method, meaning that all materials exposed to the ammonothermal environment may become incorporated into the crystal. There are four main sources of impurities in the ammonothermal system: (1) impurities in raw materials (such as ammonia, Ga source materials, GaN crystals, and mineralizers); (2) atmospheric or organic pollutants that are not completely removed during the loading process of raw materials; (3) impurities in structural materials (such as high-pressure vessel walls and seed crystal racks); (4) incorporation of necessary materials for ammonothermal growth, such as elements from the mineralizer itself into the crystals.

Suihkonen et al. summarized the impurity elements in the ammonothermal growth system of alkaline and acidic mineralizers in Fig. 4.5 and Table 4.2 [35]. The two main impurities commonly found in acidic and alkaline ammonothermal GaN crystals are hydrogen (H) and oxygen (O). Hydrogen mainly combines in atomic form as interstitial or is related to defects, such as vacancies. Usually, hydrogen passivates p-type dopants in GaN, such as H-Mg complexes. H can move under high temperature or high current, leading to passivation of p-type GaN layers. In the ammonothermal system, the content of H measured by SIMS ranges from 10^{17} to 10^{20} cm^{-3}. A strong correlation exists between the hydrogen content and the concentration of Ga-H composite vacancies. The doping of oxygen has three main effects on GaN crystals. One reason is that oxygen acts as a shallow n-type donor, leading to an increase in the concentration of free electrons. These free electrons can generate sub band gaps through phonon assisted mechanisms, leading to visible light absorption and causing the crystal to appear colored [36]. The second is that the doping of oxygen in the lattice and the generation of free electrons also increase the lattice constant [37–40], and lattice mismatch may lead to a large accumulation of strain in thick crystals, resulting in cracks. Thirdly, oxygen impurities can reduce the thermal conductivity of GaN. Compared with the sample with oxygen content of 4×10^{16} cm^{-3}, the thermal conductivity of the sample with oxygen content of 2.6×10^{18} cm^{-3} and 1.1×10^{20} cm^{-3} is reduced by 10% and 32–42%, respectively [41]. The oxygen content of crystals grown in the early stage of ammonothermal is in the range of 10^{20} cm^{-3}. With the improvement of the growth process, the reported oxygen content is generally in the range of 10^{17}–10^{19} cm^{-3}. In addition to impurities, Ga vacancies and their H and O composite vacancies have important effects on the optical and electrical properties of GaN crystals. Suihkonen et al. detected the composite vacancies of Ga vacancies with H and O through positron annihilation spectroscopy [42]. The influence and formation mechanism of related vacancies on crystal properties are currently being studied.

By controlling impurities and native point defects, ammonothermal crystals can meet the requirements of various devices for electrical resistivity. Zajac et al. used O-doped carriers with a concentration of 10^{19} cm^{-3} and a resistivity of 10^{-3} Ω cm; Doping different contents of Mg can achieve the growth of p-type conductive and semi insulating single crystals, respectively; The use of Mn doping can achieve a resistivity of 10^{12} Ω cm (as shown in Fig. 4.6) [30].

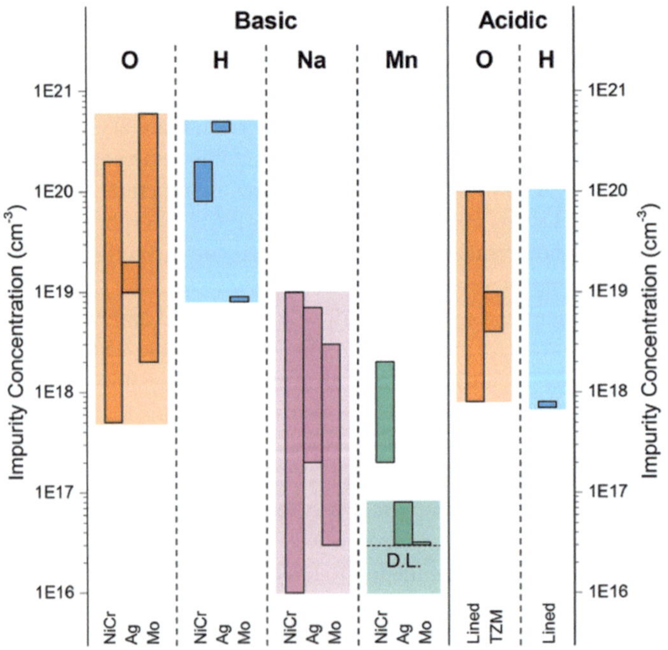

Fig. 4.5 Main impurity elements and contents in ammonothermal growth system of acidic mineralizer and basic mineralizer

Table 4.2 Impurity elements (excluding elements in Fig. 4.4) and their contents in the ammonothermal growth system of basic mineralizer and acidic mineralizer measured by SIMS

		Impurity Concentration (cm^{-3})			
		Below D.L	High-10^{16} to low-10^{17}	$\approx$ Low-10^{18}	
Basic	NiCr	K, Li, Ti, Co, Cr, Ni	C, Mg, Fe, Mo	Al, Si	
	Ag	Mo	C, Mg, Fe, Mn, Al, Si		
	Mo	Fe, Mn, Mo	C, Mg, Si		
Acdic	Lined	Transition metals, alkaline metals			
	TZM	Cr, Mo, Zr, Nb, Cl, Ti	C, Fe		

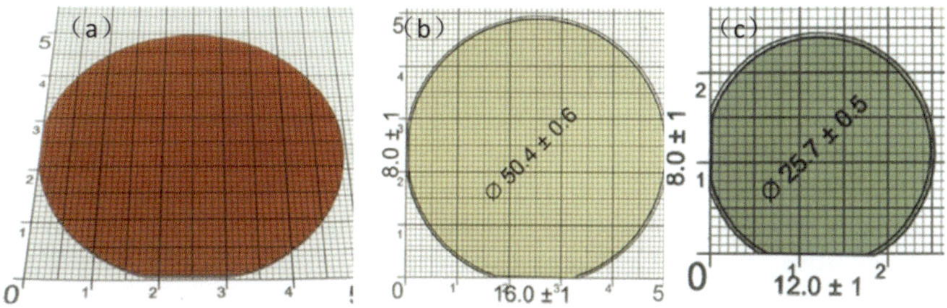

Fig. 4.6 Electrical properties of GaN single crystal. **a** 2-inch Mn-doped semi-insulating GaN single crystal substrate. **b** 2-inch Mg-doped semi-insulating GaN single crystal substrate. **c** 1-inch Mg-doped P-type GaN single crystal substrate

4.5.2 *Progress in Ammonothermal GaN Crystal Growth with Acid Mineralizers*

Acidic mineralizers have been used in the synthesis of GaN for a long time. In 1999, Purdy described the synthesis of bulk *c*-GaN by ammonothermal reactions of gallium metal and compounds under acidic (NH_4X; X = Cl, Br, I) conditions [43]. By 2002, the crystals exhibited a tapered shape, becoming wider as they grew, with widths reaching up to 0.1 mm [5]. A. Yoshikawa et al. demonstrated that needle-shaped and hexagonal GaN crystals could be grown using the ammonothermal method with a Ga metal precursor, supercritical ammonia solution, and NH_4Cl mineralizer, with the inner wall of the autoclave coated with platinum (Pt). Reaction from Ga metal could be carried out at a temperature of 500 °C and pressure of 135 MPa [7]. In 2007, Y. Kagamitani et al. reported that single-crystalline GaN films of $\leq$ 110 mm in thickness have been fabricated on hydride vapor phase-grown (0001) GaN substrates by employing a relatively low-pressure ($\leq$ 170 MPa) ammonothermal growth method with NH_4Cl as a mineralizer. The maximum growth speed on the (000 $\overline{1}$) face of 27.5 mm/day was achieved for the film grown at a supersaturation from a combined Ga/GaN precursor [9]. In 2008, Ehrentraut et al. first reported that 2-inch GaN crystal grown by an acidic ammonothermal approach [44]. They studied the self-nucleation of GaN crystals and found that the yield increased with increasing temperature and acidity of the mineralizer as shown in Fig. 4.7. The cubic phase of GaN (JCPDS-International Centre for Diffraction Data card no. 88-2364) forms preferentially at lower temperatures and in environments with more acidic mineralizers. Ehrentraut et al. reported thar GaN grown by the acidic ammonothermal technique and nucleated on HVPE seed crystals: a 1 cm^2 large crystal, a 3 cm diameter large crystal, and a 2-inch GaN crystal (as shown in Fig. 4.8) [45]

In 2013, Soraa developed a novel ammonothermal approach. Soraa's approach, known as scalable compact rapid ammonothermal (SCoRA), utilizes internal heating to circumvent the material property limitations of conventional ammonothermal reactors. The SCoRA reactor has demonstrated capability for temperatures and pressures as high as 750 °C and 600 MPa. The configuration shown in Fig. 4.9a is appropriate for a mineralizer with a positive solubility coefficient, e.g., NH_4Cl. Examples of as-grown SCoRA crystals are shown in Fig. 4.9b–d [46]. Dislocation densities in *c*-GaN material are about 10^5 cm^{-2} (as shown in Fig. 4.9). In a late study, they used an acidic mineralizer with a capsule wall material selected to resist corrosion, they found that using native SCoRA seeds instead of HVPE seeds produced boules with significantly better crystal quality (as shown in Fig. 4.10), the absorption spectrum of high conductivity SCoRA GaN is comparable to that of similarly conductive HVPE

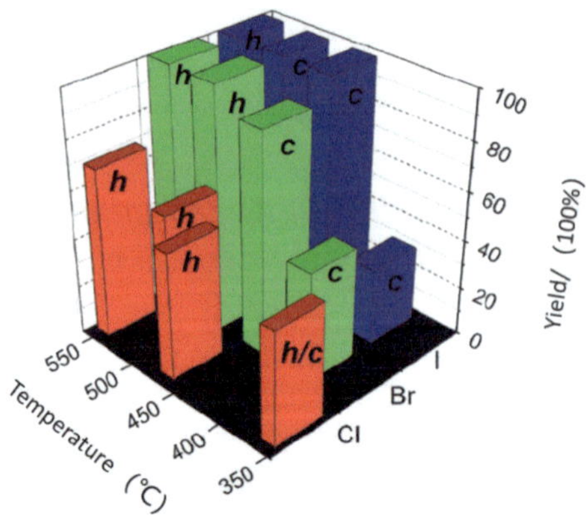

Fig. 4.7 Yield of GaN over temperature for three different acidic mineralizers

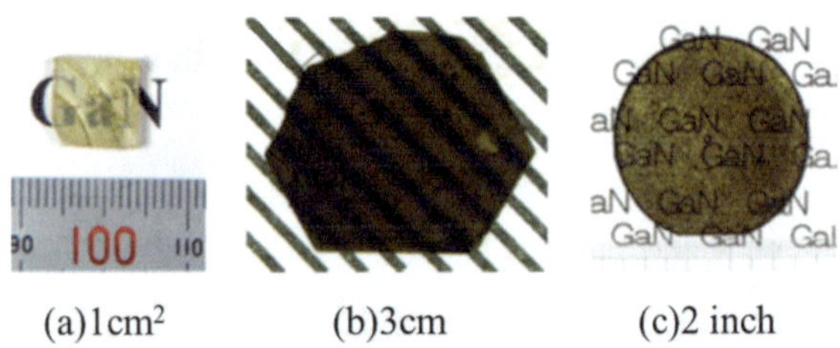

Fig. 4.8 1 cm^2, 3 cm, and 2 inch GaN crystals grown by acid mineralizer

substrates, radii of curvature of 46 m in the [100] direction and 236 m in the [110] direction are obtained. The (201) rocking curve has a FWHM of 28 arcsec [47, 48]

In recent years, Mitsubishi Chemical of Japan has made rapid progress in the ammonothermal growth of GaN with acid mineralizers. They proposed an original ammonothermal method known as SCAAT™ which enables obtaining strain-free, high quality and large size bulk GaN crystals. In 2015, They showed *m*-plane as-grown bulk GaN crystal which was grown on *m*-plane (10–10) seed of rectangular plate shape cut from SCAAT™ GaN crystal with a growth rate of several hundred micrometres per day. The etch pit density (EPD) was estimated at 2.5×10^2 cm^{-2}. The full width at half maximum (FWHM) on (10–10) plane was 13.3 arcsec and (10–12) plane was 8.8 arcsec. The EPD of 2 inch size *c*-plane GaN was 1×10^2 ~ 1×10^4 cm^{-2}, the FWHM on (0004) plane was 10.0 arcsec and (10–12) plane was 16.7 arcsec (as shown in Fig. 4.11) [49]. In 2020, Mikawa et al. presented a nearly 4-inch size *c*-plane GaN grown by SCAAT™, off-angle distributions are still small; $\pm$ 0.006° along *a*-axis and *m*-axis both, even the crystal size was enlarged. One of the advantages of SCAAT™ is that the crystals are grown under almost equilibrium conditions, which is beneficial for large-size crystal growth. As a result, uniform growth was achieved regardless of the position in the crystal; center or edge [50].

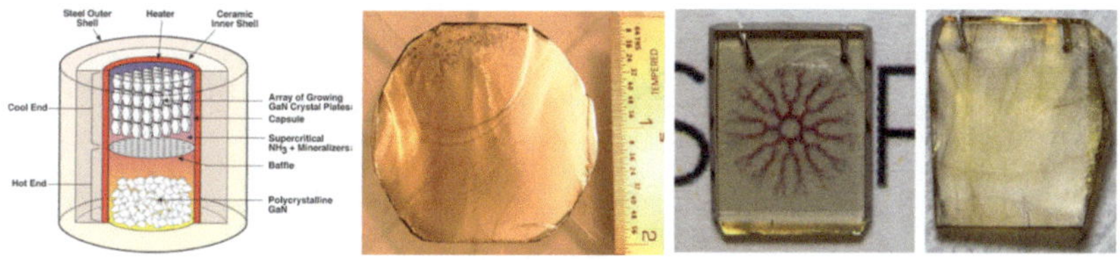

(a) (Color online) Schematic of internally-heated high pressure SCoRA apparatus

(b)2-inch c-plane with macro-steps

(c)7mm×5 mm m-plane

(d)7mm× 5 mm $(10\bar{1}\bar{1})$ semipolar

Fig. 4.9 **a** Schematic of internally-heated high pressure SCoRA apparatus. **b** photographs of SCoRA-grown bulk GaN crystals: 2-inch *c*-plane with macro-steps. **c** 7 mm × 5 mm *m*-plane. **d** 7 mm × 5 mm $(10\,\overline{11})$ semipolar, lightly etched (Color online)

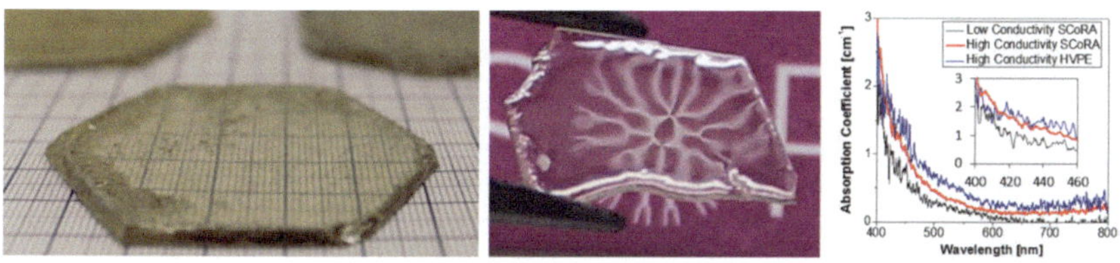

(a) as-grown SCoRA boules

(b) Double side polished SCoRA crystal

(c) Optical absorption coefficients of SCoRA and HVPE GaN as a function of wavelength.

Fig. 4.10 **a** Photographs of as-grown SCoRA boules. **b** Doubleside polished SCoRA crystal. **c** Optical absorption coefficients of SCoRA and HVPE GaN as a function of wavelength

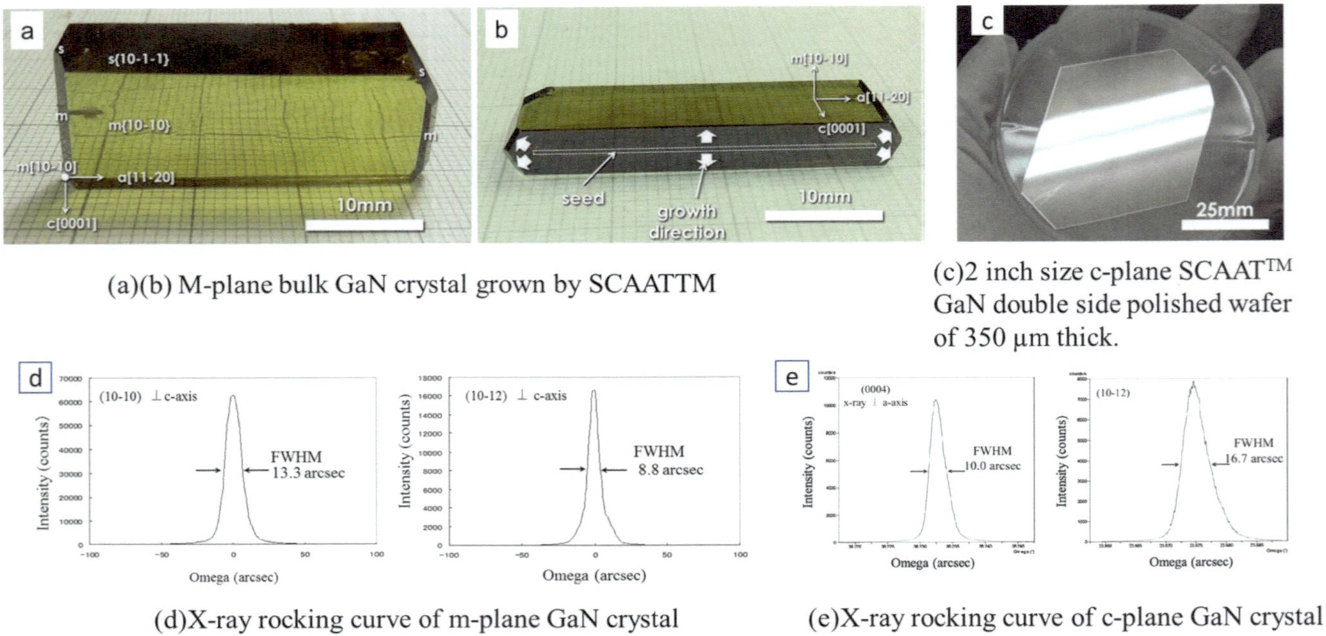

Fig. 4.11 **a**, **b** *m*-plane bulk GaN crystal grown by SCAAT™. **c** 2-inch size *c*-plane SCAAT™ GaN double side polished wafer of 350 μm thick. **d** X-ray rocking curve of *m*-plane GaN crystal. **e** X-ray rocking curve of *c*-plane GaN crystal

4.6 Development Trend

Ammonothermal method is currently the most feasible technique for growing large bulk GaN crystal. Regardless of the acidic or alkaline ammonothermal growth system, the crystal quality and size have been significantly improved. Compared with HVPE growth technique, the ammonothermal growth technique has the following advantages: (1) the crystal quality is high, and the dislocation density is about two orders of magnitude lower than HVPE-GaN; (2) the crystal lattice exhibits minimal warpage, and the radius of curvature reaches several hundred meters, which is advantageous for controlling the substrate bevel angle. (3) the ammonothermal growth technique is similar to the hydrothermal growth technique for growing quartz and ZnO, and is easy to mass produce and low cost; (4) high-quality non-polar and semi-polar GaN substrates can be directly grown, and high-quality non-polar and semi-polar GaN substrates can also be obtained by cutting from thick crystal.

At present, the industrialization of the ammonothermal growth technique still needs to solve the following problems: (1) the manufacturing of large-diameter autoclaves to meet the growth of GaN of 4–6 inches or larger; (2) control of impurities and intrinsic point defects in ammonothermal GaN to meet the optical and electrical properties of the substrates required by various devices; (3) high-quality ammonothermal GaN single crystal are generally grown from high-quality seed crystal cultured by ammonothermal. The number of large-sized and high-quality GaN seed crystal is relatively less. To realize the industrialization of the ammonothermal growth technique, a certain amount of high-quality large-sized GaN seed crystal are needed.

In summary, the ammonothermal growth technique has proven to offer significant advantages in terms of crystallization quality. In recent years, ammonothermal growth has seen rapid progress in terms of crystal size, point defects, and electrical control. It is expected that the ammonothermal method will gradually dominate the industrialization of GaN single crystal substrates.

References

1. Dwilinski R, Wysmolek A, Baranowski J et al (1995) GaN synthesis by ammonothermnal method. Acta Pysica Polonica 88(5):833–836
2. Peters D (1990) Ammonothermal synthesis of aluminum nitride. J Cryst Growth 104(2):411–418
3. Dwilinsk R, Baranowski JM, Kaminska M et al (1996) On GaN crystallization by ammonothermal method. Acta Pysica Polonica 90(4):763–766
4. Ketchum DR, Kolis JW (2001) Crystal growth of gallium nitride in superitical ammonia. J Cryst Growth 222(3):431–434
5. Purdy AP, Jouet RJ, George CF (2002) Ammonothermal recrystallization of gallium nitride with acidic mineralizers. Cryst Growth Des 2(2):141–145
6. Callahan MJ, Wang BG, Bouthillette LO et al (2003) Growth of GaN crystals under ammonothermal conditions. Mater Res Soc Symp Proc 798:263–268

7. Yoshikawa A, Ohshima E, Fukuda T et al (2004) Crystal growth of GaN by ammonothermal method. J Cryst Growth 260(1–2):67–72
8. Wang BG, Callahan MJ, Rakes KD et al (2006) Ammonothermal growth of GaN crystals in alkaline solutions. J Cryst Growth 287:376–380
9. Kagamitani Y, Ehrentraut D , Yoshikawa A et al (2006) Ammonothermal epitaxy of thick GaN film using NH_4Cl mineralizer. Jpn J Appl Phys 45(5A):4018–4020
10. Hertweck B, Steigerwald TG, Alt NSA et al (2014) Different corrosion behavior of autoclaves made of nickel base alloy 718 in ammonobasic and ammonoacidic environments. J Supercrit Fluids 95:158–166
11. Hertweck B, Steigerwald TG, Alt NSA et al (2014) Corrosive degeneration of autoclaves for the ammonothermal synthesis: experimental approach and first results. Chem Eng Technol 37:1903–1906
12. Ehrentraut D, Kagamitani Y, Fukuda T et al (2008) Reviewing recent developments in the acid ammonothermal crystal growth of gallium nitride. J Cryt Growth 310(17):3902–3906
13. Wang BG, Callahan MJ (2006) Ammonothermal synthesis of III-Nitride crystals. Cryst Growth Des 6(6):1227–1246
14. Zhang SY, Hintze F, Schnick W et al (2013) Intermediates in ammonothermal GaN crystal growth under ammonoacidic conditions. Eur J Inorg Chem 2013(31):5387–5399
15. Ehrentraut D, Kagamitani Y, Yokoyama C et al (2008) Physico-chemical features of the acid ammonothermal growth of GaN. J Cryst Growth 310(5):891–895
16. Yoshida K, Aoki K, Fukuda T (2014) High-temperature acidic ammonothermal method for GaN crystal growth. J Cryst Growth 393(1):93–97
17. Hashimoto T, Saito M, Fujito K et al (2007) Seeded growth of GaN by the basic ammonothermal method. J Cryst Growth 305(2):311–316
18. Ehrentraut D, Fukuda T (2010) Ammonothermal crystal growth of gallium nitride: a brief discussion of critical issues. J Cryst Growth 312(18):2514–2518
19. Tomida D, Kuribayashi T, Suzuki K et al (2011) Effect of halogen species of acidic mineralizer on solubility of GaN in supercritical ammonia. J Cryst Growth 325(1):52–54
20. Bao QX, Saito M, Hazu K et al (2013) Ammonothermal crystal growth of GaN using an NH_4F mineralizer. Cryst Growth Des 13(10):4158–4161
21. Dwilinski R, Doradzinski R, Garczynski J et al (2008) Excellent crystallinity of truly bulk ammonothermal GaN. J Cryst Growth 310(17):3911–3916
22. Tomida D, Kuroda K, Nakamura K et al (2018) Temperature dependent control of the solubility of gallium nitride in supercritical ammonia using mixed mineralizer. Chem Cent J 12(127):1–6
23. Hashimoto T, Fujito K, Saito M et al (2005) Ammonothermal growth of GaN on an over-l-inch seed crystal. Jpn J Appl Phys 44(52):L1570–L1572
24. Wang BG, Callahan MJ, Rakes KD et al (2006) Ammonothermal growth of GaN crystals in alkaline solutions. J Cryst Growth 287(2):376–380
25. Hashimoto T, Wu F, Speck JS et al (2007) Growth of bulk GaN crystals by the basic ammonothermal method. Jpn J Appl Phys 310(17):L889–L891
26. Dwilinski R, Doradzinski R, Garczynski J et al (2008) Excellent crystallinity of truly bulk ammonothermal GaN. J Cryst Growth 310(17):3911–16
27. Bulk GaN (2010) Ammonothermal trumps HVPE. Compound Semicond 2:12-16
28. Kyma Responds to CS Article (2010) Bulk GaN: ammonothermal trumps HVPE. Compound Semicond 3:93
29. Dwilinski R, Doradzinski R, Garczynski J et al (2010) Recent achievements in AMMONO-bulk method. J Cryst Growth 312(18):2499–2502
30. Zajac M, Kucharski R, Grabianska K et al (2018) Basc ammonothermal growth of gallium nitride-state of the art, challenges, perspectives. Progr Cryst Growth Character Mater 64(3):63–74
31. Pimputkar S, Kawabata S, Speck JS et al (2014) Improved growth rates and purity of basic ammonothermal GaN. J Cryst Growth 403(1):7–17
32. Hashimoto T, Letts ER, Key D et al (2019) Two inch GaN substrates fabricated by the near equilibrium ammonothermal (NEAT) method. Jpn J Appl Phys 58(C):SC1005

33. Kucharski R, Zajac M, Doradzinski R et al (2012) Non-polar and semi-polar ammonothermal GaN substrates. Semicond Sci Technol 27(2):024007
34. Li TK, Ren GQ, Su XJ et al (2019) Growth behavior of ammonothermal GaN crystals grown on non-polar and semi-polar HVPE GaN seeds. CrystEngComm 21(33):4665–4830
35. Suihkonen S, Pimputkar S, Sintonen S et al (2017) Defects in single crystalline ammonothermal gallium nitride. Adv Electron Mater 3(6):1600496
36. Pimputkar S, Suihkonen S, Imade M et al (2015) Free electron concentration dependent sub-bandgap optical absorption characterization of bulk GaN crystals. J Cryst Growth 432(15):49–53
37. Sintonen S, Kivisaari P, Pimputkar S et al (2016) Incorporation and effects of impurities in different growth zones within basic ammonothermal GaN. J Cryst Growth 456(15):43–50
38. Krysko M, Sarzynski M, Domagala J et al (2005) The influence of lattice parameter variation on microstructure of GaN single crystals. J Alloy Compd 401(1–2):261–264
39. Walle CVD (2003) Effects of impurities on the lattice parameters of GaN. Phys Rev B 68(16):165209
40. Darakchieva V, Monemar B, Usui A (2007) On the lattice parameters of GaN. Appl Phys Lett 91(3):31911
41. Jezowski A, Churiukova O, Mucha J et al (2015) Thermal conductivity of heavily doped bulk crystals GaN: O. Free carriers contribution. Mater Res Exp 2(8):85902
42. Tuomisto F, Makkonen I (2013) Defect identification in semiconductors with positron annihilation: experiment and theory. Rev Mod Phys 85(4):1583–1631
43. Purdy AP (1999) Ammonothermal synthesis of cubic gallium nitride. Chem Mater 11(7):1648–1651
44. Ehrentraut D, Kagamitani K, Fukuda T et al (2008) Reviewing recent developments in the acid ammonothermal crystal growth of gallium nitride. J Cryst Growth 310(17):3902–3906
45. Ehrentraut D, Fukuda T (2010) The ammonothermal crystal growth of gallium nitride. A technique on the up rise. Proc IEEE 98(7):1816–1323
46. Ehrentraut D, Pakalapati RT, Kamber DS et al (2013) High quality, low cost ammonothermal bulk GaN substrates. Jpn J Appl Phys 52(85):08JA01
47. Jiang W, Ehrentraut D, Kamber DS et al (2014) Ammonothermal bulk GaN substrates for LED. Proc SPIE 9003:900313
48. Jang WK, Ehrentraut D, Cook J et al (2015) Transparent, conductive bulk GaN by high temperature ammonothermal growth. Physica Status Solidi (B) 252(5):1069–1074
49. Mikawa Y, Ishinabe T, Kawabata S et al (2015) Ammonothermal growth of polar and non-polar bulk GaN crystal. Proc SPIE 9363:936302
50. Mikawa Y, Ishinabe T, Kagamitani Y et al (2020) Recent progress of large size and low dislocation bulk GaN growth. Proc SPIE 11280:1128002

Chapter 5
Na Flux Method

5.1 Development History

In order to obtain a higher growth rate and higher crystal quality of Na-flux GaN, the Na-flux method has undergone a series of technological innovations, as shown Table 5.1. The real development of the Na-flux method of growing GaN began in 2003 when the Na-flux method was combined with the LPE method.

The research on the growth of GaN single crystal by Na Flux method started in 1997. The main research teams include: the research team led by Mori Y. (Osaka University, Japan) and its cooperation unit is Ricoh (Japan), Toyota Central R&D Lab, Sanken Electric Co., etc.; a research team led by Yamane (Tohoku University, Japan) and its cooperation unit is Cornell University (USA), etc.; Nagoya University (Japan), and its cooperative units include the NAGAKI (NGK, Japan); US Air Force/ Naval Research Lab, etc. Among them, the research group of Mori Y from Osaka University in Japan conducted in-depth research on the size of single crystal, crystal quality, crystal growth equipment enlargement, single crystal growth rate and other aspects of the growth of GaN single crystal by Na Flux method, and the growth of LPE combined with the Na-flux method is currently a relatively advanced research in this field. In 2008, Mori's research group used a self-designed and developed Na-flux growth equipment to obtain 3 mm thick, 2-inch GaN single crystal for the first time (as shown in Fig. 5.1a). In 2011, they cooperated with Toyota central R&D Lab/ Sanken Electric co., successfully trial-produced the first 4-inch GaN single crystal grown by Na Flux method (Fig. 5.1c). In addition, they used a spontaneous nucleation single crystal grown by Na flux method as the seed crystal, combined with sapphire mask technology and melt stirring technology, successfully obtained centimeter-level GaN single crystal (Fig. 5.2a–c). Up to now, polar (c-plane) and semi-polar (m-/a-plane) GaN bulk single crystals can be obtained by the Na Flux method, and the dislocation density can be reduced to the order of $10^2/cm^2$.

In order to further obtain high-quality, large-size GaN single crystals, Mori et al. introduced the Necking technique epitaxial combined the melt stirring technology

K. Xu et al., *GaN Single Crystal Growth and Application*, Wide Bandgap Semiconductors, https://doi.org/10.1007/978-981-96-7572-2_5

Table 5.1 Overview of Na Flux method technical improvements

Time	Technological innovation	Research team	Results
1997	$NaN_3 + Ga$	Yamane H. Tohoku University	2 mm, Obtained GaN for the first time
2000	Na + Ga	Yamane H. Tohoku University	Size up to 3 mm
2002	Add Ca, Li	Mori Y. Osaka University	Increased productivity
2003	Combined with LPE	Mori Y. Osaka University	Large size single crystals can be grown
2006	Temperature gradient	Mori Y. Osaka University	Growth rate up to 10 μm/h
2008	Add C source	Mori Y. Osaka University	Inhibit polycrystalline, increase yield
2008	Mixing technology	Mori Y. Osaka University	Further increase the growth rate
2011	Spontaneous nucleation of GaN as seed	Mori Y. Osaka University and Ricoh	8–10 mm Single crystal, growth rate 33 μm/h
2012	Necking technique	Mori Y. Osaka University	Further improve the crystal quality, and can grow large-size GaN single crystals with better uniformity
2014	Two-step method	Mori Y. Osaka University	2 inch GaN, Dislocation density in 75% area is $10^2/cm^2$
2015	Na flux-dipping	Mori Y. Osaka University	Growing thick, low dislocation density GaN
2015	Introduce methane	Mori Y. Osaka University	C-direction growth rate up to 60 μm/h
2017	Tiling technique	Mori Y. Osaka University	111 mm in diameter
2020	FFC-LPE	NIMS	No Inclusion

[3] to obtain large-scale high-quality GaN single crystal, the above technology is expected to obtain a larger size and high-quality GaN single crystal (as shown in Fig. 5.3).

In 2015, Mori et al. introduced the Na flux dipping method [6]. In the process of using this seed crystal dipping technology to grow, it prevents the GaN substrate and the regrowth layer from being re-dissolved. The generation of interfacial dislocations was avoided and a thick and low-density GaN crystal was successfully prepared.

In order to further increase the single crystal growth rate, in 2015, Mori et al. introduced methane gas as a carbon source, obtained a GaN single crystal with c-direction growth rate up to 60 μm/h, as shown in Fig. 5.4. Moreover, the single crystal

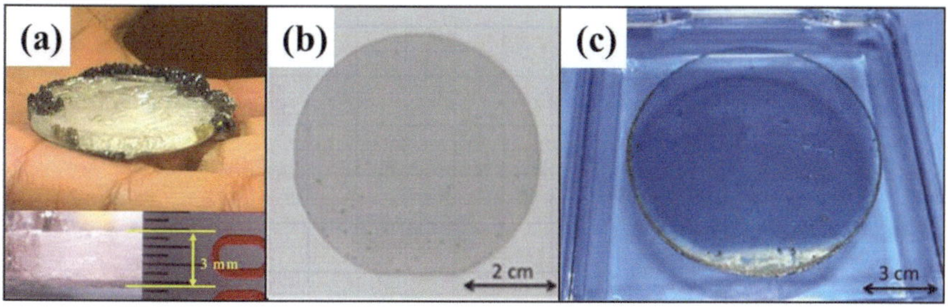

Fig. 5.1 **a** For the first time using the Na Flux method, a GaN single crystal of 3 mm thick 2 inch was obtained [1]; Using HVPE growth GaN single crystal as seed crystal, Na Flux method growth **b** 2 inch and **c** 4 inch single crystal of GaN [2]

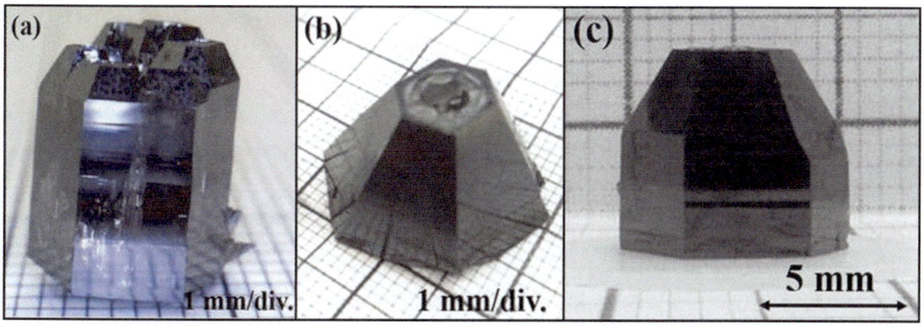

Fig. 5.2 Seed crystals are made from spontaneous nucleated GaN single crystal, Na Flux method grows centimeter-grade GaN single crystal: **a** a-direction 0.85 cm, c-direction 1 cm [3]; **b** a-direction 12 mm, c-direction > 20 mm [2]; **c** a-direction 9 mm, c-direction 7.5 mm [2]

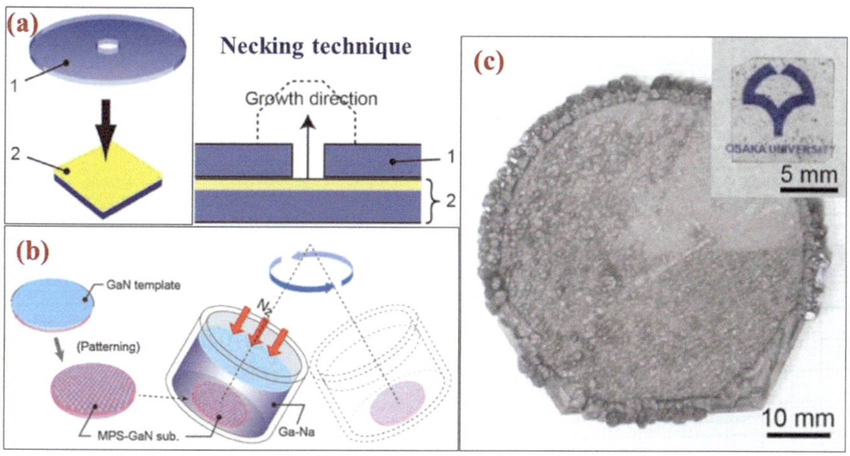

Fig. 5.3 **a** The diagram of the Necking Technique [3]; **b** using melt mixing technology [4]; **c** epitaxial-combined growth of 2 inch GaN [5]

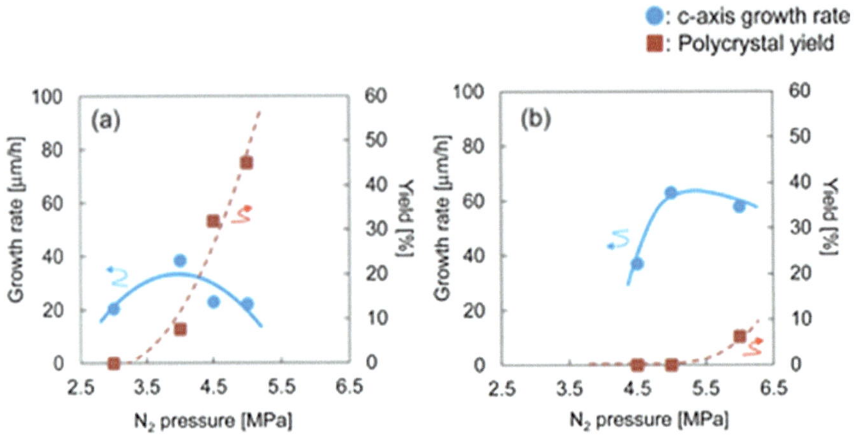

Fig. 5.4 Effects of c-axis growth rate and polycrystalline yield on N$_2$ pressure in the growth with **a** graphite and **b** methane growth [7]

yield of using methane as an additive is greatly improved compared to graphitic carbon.

In 2017, Mori et al. introduced a Tiling technique [8] that is, by using an initial substrate (composed of multiple HVPE seed crystals adhered to the substrate) as a bulk seed crystal to obtain 175 mm Large diameter single crystal of GaN.

5.2 Basic Principle and Control Method

5.2.1 Basic Principle

The schematic diagram of Na-flux growing GaN single crystal is shown in Fig. 5.5. Under a specified growth temperature and pressure conditions, Na in Ga–Na molten liquid ionizes nitrogen at the gas–liquid interface, forming N^{3-}. Although the solubility of nitrogen in Ga metal and Na metal is very low, the solubility of nitrogen in the molten is increased by nearly a thousand times by the ionized N^{3-}. Moreover, the ionized N^{3-} can exist stably in the Ga-Na metal ion system. Driven by temperature gradient or concentration gradient, N^{3-} ions are constantly transported downward. When the solubility of nitrogen in Ga-Na melt exceeds the critical value of nitrogen required for GaN crystal growth, spontaneous nucleation of GaN is formed, or N^{3-} ions are transported downward to the seed crystal for liquid phase epitaxy (LPE) growth on GaN seed crystal. By precise control of growth conditions, continuous

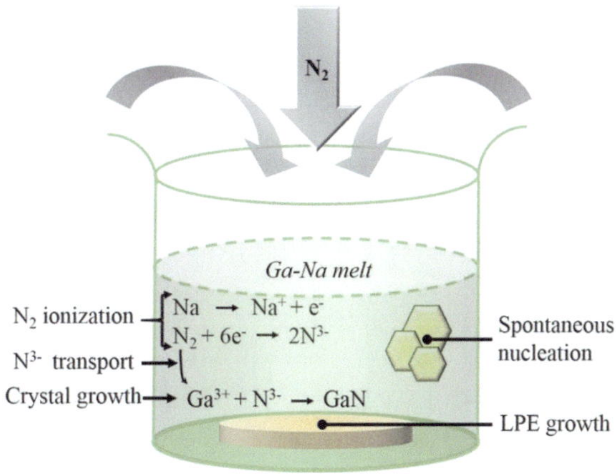

Fig. 5.5 Schematic diagram of the reaction process of growing GaN single crystal by Na Flux method

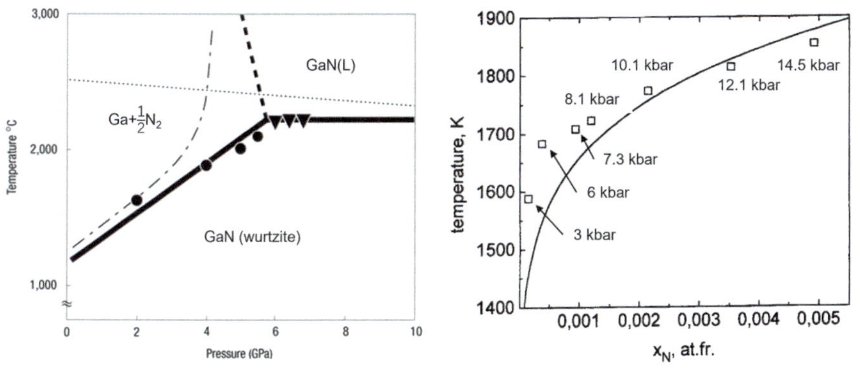

Fig. 5.6 **a** GaN phase diagram at high temperature and high pressure [9]; **b** The solubility curve in Ga melt, solid line is the ideal melt approximate calculation value of the value of the GaN [10]

and effective crystal growth can be obtained, and then large size, high quality GaN single crystals can be obtained.

5.2.2 Supersaturation Regulation

Another key technology for the growth of GaN by Na-flux method is how to control the supersaturation of the GaN solution to realize the crystallization of GaN single crystals, so as to obtain high-quality GaN bulk single crystals. We all know that the Na-flux method is a growth method in a near thermodynamic equilibrium state, so it is easier to obtain a higher crystal quality. Because GaN has a very high melting point [9], the melting point is 2200 °C at 6 GPa (as shown in Fig. 5.6a, the traditional Czochralski method, Bridgeman method and other single crystal growth methods are

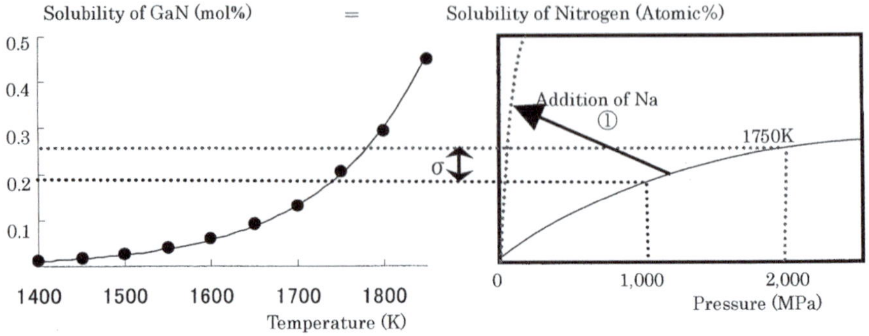

Fig. 5.7 GaN solubility curve in Ga (or Ga-Na) melt (left) and nitrogen solubility and pressure relationship curve in Ga melt (right) diagram [12]

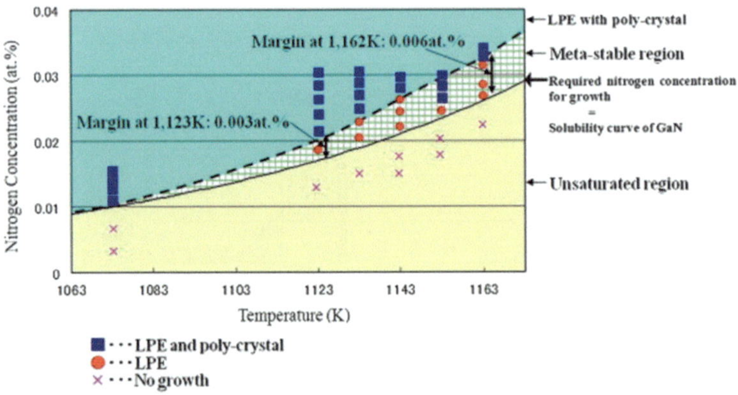

Fig. 5.8 The metastable zone (dotted line) and solubility curve (solid line) of GaN show that the metastable region of GaN expands with the increase of growth temperature [13]

not suitable for growing GaN single crystals. However, the high pressure solution method is extremely demanding on equipment and growth conditions, and the solubility of nitrogen in Ga melt is very low [10], (1500 °C, 1 GPa is only 0.2 mol%), the growth rate is only a few microns per hour. The introduction of Na source greatly improves the solubility of nitrogen in Ga melt (as shown in Fig. 5.6b), although the solubility of nitrogen in Na is also very small (only 7.1×10^{-9} mol% at 600 °C). But Na can cause N_2 to undergo ionization process to form N^{3-}, increase the concentration of N source in the system, so that GaN single crystal can be obtained under lower pressure (<10 MPa) and temperature (<900 °C) [11] (as shown in Fig. 5.7).

Through the Na flux method to study the metastable region in the growth of GaN, we can see that, due to its growth of the small controllable temperature zone, the large degree of supersaturation of the local spontaneous nucleation [13], easy to produce polycrystalline, reducing the growth rate of GaN single crystals, as well as the single-crystal yield. Therefore, the control of supersaturation is particularly important to avoid the generation of polycrystals and improve the single crystal yield. Usually, the addition of carbon as an additive to reduce the supersaturation of the gas–liquid interface, effectively inhibit the spontaneous nucleation process at the gas–liquid interface [1], so that GaN melt seed crystal surface crystallisation precipitation for liquid-phase epitaxial growth.

In 2015, Mori et al. introduced the Na flux dipping method [6], which controls the surface supersaturation of the GaN seed crystal to maintain a high level during the growth process using this seed crystal dipping technology. The GaN seed crystal was prevented from remelting in the initial stage of growth, and a thick and low dislocation density GaN crystal was successfully prepared, as shown in Fig. 5.9.

In 2016, Mori et al. introduced Na flux dipping technology in the Na-flux growth system in order to grow GaN single crystals under low supersaturation conditions. A 2-inch $\{11\bar{2}2\}$ plane GaN single crystal was successfully fabricated by the system, which avoided the re-dissolution of the seed crystal, and the grown GaN had high transparency and crystallinity, as shown in Fig. 5.10.

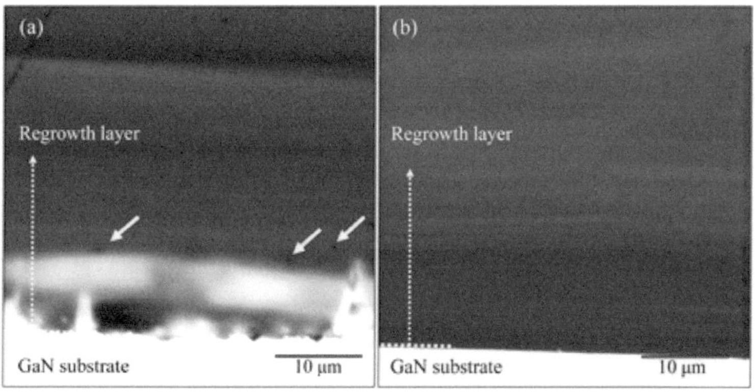

Fig. 5.9 **a** without the dipping technique and **b** with the dipping technique [6]

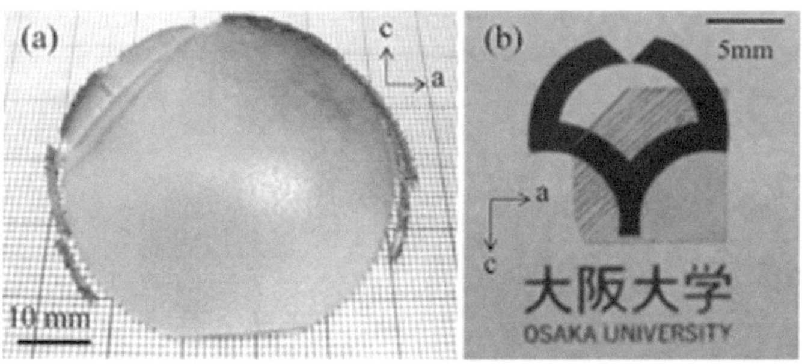

Fig. 5.10 **a** Optical photographs of 2-inch $11\bar{2}2$ GaN crystals with high transparency grown using dipping technique. **b** Photographs of $11\bar{2}2$ GaN substrates polished with CMP on both sides. Use this substrate to measure light transmittance

5.2.3 Defect Control

Crystal defects generally refer to the phenomenon of repetitive arrangement of various deviations from the periodicity of the crystal structure, and the reason for this deviation from the periodicity is the factor that produces the defect. At the same time, it can be divided into three types: point defects, line defects, and planar defects according to the degree of extension. Usually, we focus on the suppression of point defects and line defects during the growth process.

1. Point Defects In the process of growing GaN crystals by the flux method, point defects are introduced due to changes in temperature, pressure, and concentration of medium components. The existence and uneven distribution of these point defects will lead to changes in material properties. Point defects in semiconductors can generally be divided into intrinsic point defects, point defects introduced by unintentional doping, and point defects obtained by deliberate doping. There are six intrinsic defects in GaN crystals, which are: N-vacancy, Ga-vacancy, N-interstitial, Ga-interstitial, Ga substitute N, and N substitute Ga. In the Na-flux system, these six intrinsic defects all exist in the grown GaN crystals, and N vacancies (V_N) and Ga vacancies (V_{Ga}) are two of the more common intrinsic defects. Especially in GaN samples after ion implantation and irradiation treatment, there are generally more vacancy defects. As a shallow donor impurity in p-type materials, N vacancies often appear as compensation in the crystal. In Mg-doped p-type materials, it is generally believed that the concentration of N vacancies in the crystal is positively correlated with the concentration of Mg impurities, but after annealing, the concentration of N vacancies will also be effectively reduced (as shown in Fig. 5.11) [14–16].

Similarly, the corresponding Ga vacancy is a common compensating impurity in n-type materials. In the system, it often increases with the increase of the concentration of doped donor-type impurities, and is often accompanied by the formation of non-radiative recombination centers. It is generally believed that the Ga vacancies in the system are easy to further combine with O impurities, forming a V_{Ga}-O complex, which will also adversely affect the luminescence characteristics of the material (The yellow luminescence and green luminescence in the material's emission spectrum are generally considered to be related to this part of the complex). Unlike N vacancies, Ga vacancies are relatively stable in the crystal (isolated V_{Ga} is difficult to exist alone in an environment above 300 °C, generally in the form of V_{Ga}-O complex), and it is difficult to reduce by annealing treatment. Therefore, it is necessary to minimize these intrinsic defects through certain adjustments in the actual growth process, so as to improve the crystal quality and further control the resistivity of the obtained material.

In the Na-flux method growth system, the control ratio of the growth materials, the pressure and temperature during the growth process are generally adjusted to suppress the formation of this part of the intrinsic defects. It has also been found in existing reports that during the Na-flux growth process, appropriately increasing the corresponding growth temperature can effectively inhibit the generation of vacancy-type defects in the crystal, (as shown in Fig. 5.12, it is generally considered that the

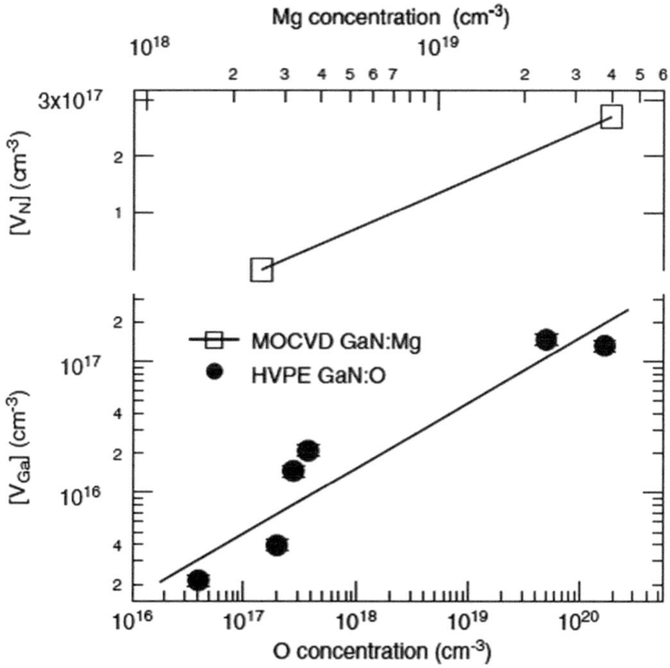

Fig. 5.11 The N vacancy defect in the GaN crystal, the dependence of the concentration of Ga vacancy defect with the concentration of other impurities [14]

sharp peak at 653 cm^{-1} that appears in the Raman scattering spectrum of GaN is related to the vacancy concentration in the material, and after increasing the growth temperature, it can be seen that the sharp peak disappears) [17]. Similarly, during the growth process, increasing the N_2 gas pressure of the system by an appropriate amount can generally effectively inhibit the concentration of N vacancies.

In addition to the intrinsic point defects analyzed above, there are also a considerable number of point defects introduced by unintentional doping in the Na-flux method growth system. Due to the growth characteristics of the Na-flux method itself, the autoclave wall in the system has a certain adsorption effect on impurities; in addition, Na and Ga are relatively active in the high-temperature growth state, and the alkaline volatile substances produced will corrode other parts, thereby making the growth atmosphere contains certain impurities; and the impurities brought by the raw materials in the growth system, etc., resulting in the existence of more point defects introduced by unintentional doping in the grown crystals, mainly including Al, Fe, Cr, Mn, Mo, C, O, etc. (as shown in Fig. 5.13).

Among these impurities, part of the metal impurities may originate from the crucible that loaded with growth material. At present, in the growth of flux, the use of alloy crucibles is more common, including alumina crucibles, nickel alloy crucibles, molybdenum alloy crucibles. The use of such crucibles has resulted in the appearance of metallic impurities such as Al, Fe, Cr, Mn, Mo, etc. in GaN crystals

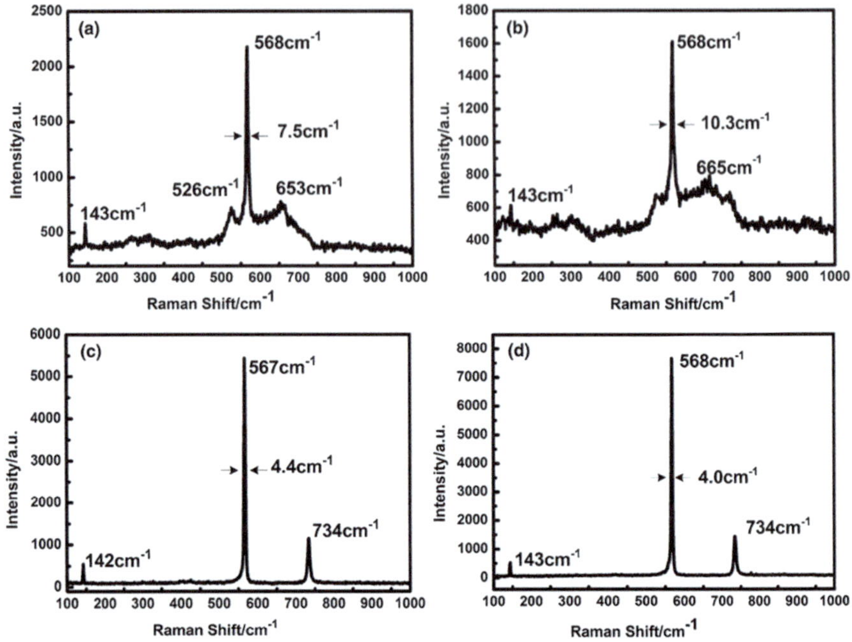

Fig. 5.12 Raman scattering spectrum of the GaN grown at different temperatures: **a** 700 °C, **b** 750 °C, **c** 800 °C, and **d** 850 °C [17]

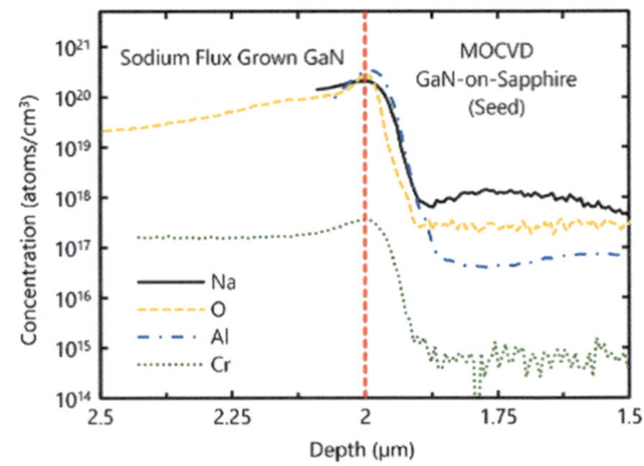

Fig. 5.13 Various impurity concentrations of GaN crystals grown by the Na-flux method [18]

[18]. For GaN crystals, the presence of metal impurities usually has a profound effect on the transparency, optical properties, and resistivity of the material [19]. Based on this thinking, in the current growth of fluxes, crucibles with more stable chemical properties and purer materials are gradually being used to replace the original alloy crucibles, thereby reducing metal impurities introduced by unintentional doping. In the Na-flux method system, the C impurity mainly comes from the C additive in the

growth material. Existing experiments have shown that the presence of C additive in the Na-flux method growth system can effectively inhibit the spontaneous nucleation process (as shown in Fig. 5.14), which is conducive to the liquid phase epitaxial growth on the seed crystal, which can further improve crystal quality [1, 20, 21]. Therefore, in the current Na-flux growth, different types of C additives are widely used, including ordered mesoporous carbon, activated carbon, graphene, graphite powder, etc. On the other hand, C impurity, as the main impurity in the material, has a further influence on the characteristics of the material.

It is generally believed that C impurity has a compensation effect on the carriers in the material, so the presence of C impurity will reduce the electrical properties of the material; at the same time, C impurity is considered to be closely related to the formation of the yellow band in the luminescence of GaN [22–25], there is a bad influence on the light-emitting characteristics of the material, so it is generally hoped that the concentration of this part of the impurity can be reduced. In material growth, the type of C additive, as well as the growth temperature and pressure are generally adjusted to reduce the concentration of C impurity.

O impurity, as the main impurity in the material, has a relatively high content in the GaN crystal grown by Na-flux method, and is very unevenly distributed inside the crystal, which is greatly affected by the growth temperature and growth pressure.

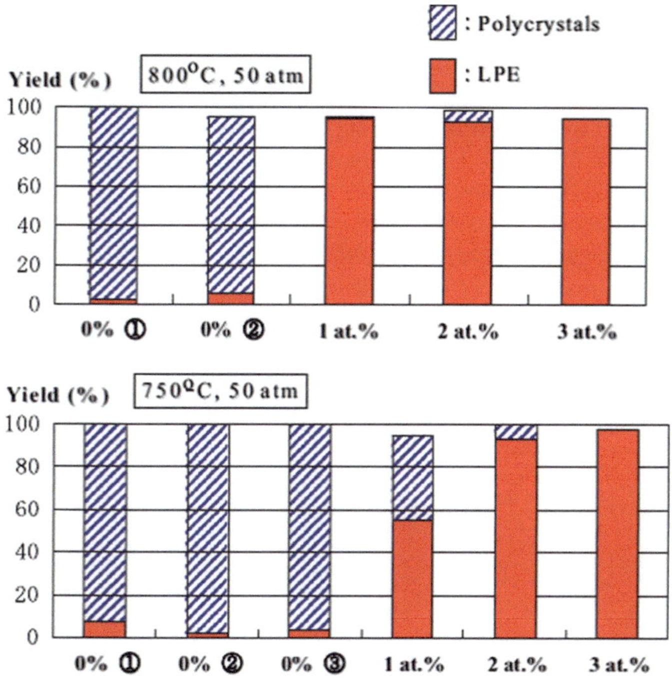

Fig. 5.14 The content of C additives has an effect on the yield of LPE crystals in the Na-flux method [1]

There are many forms of O impurity in the crystal, including substitution, meta position, etc. (O_N, O_{Ga}, O_I), which are related to the growth conditions.

It is generally believed that the introduction of O impurity during the crystal growth process will cause the expansion of the lattice constant, the accumulation of stress and finally the generation of cracks, which will adversely affect the growth of large-sized crystals [26–28]; O impurity has a high affinity with other impurities including Na, Al, Cr, etc., which will promote the enrichment of this part of the impurity in the growing crystal [18, 29]; O impurity has also been found to be associated with Ga in the crystal. Intrinsic defects are further combined to form deep-level defects such as V_{Ga}-O complexes, which will have a great impact on the light-emitting properties of the material itself [30]. At present, in many cases, GaN crystals grown by Na-flux are dark black, showing low transparency, and this black appearance is also widely considered to be related to O impurities (as shown in Fig. 5.15).

On the other hand, O impurity is one of the common donor impurities of GaN materials, which has an important influence on the electrical properties of GaN, including resistivity and mobility. In O-doped GaN crystals, under certain conditions, O will replace the N site to form O_N after being doped in GaN, becoming one of the main sources of shallow donors in n-type GaN (activation energy ~ 29 meV), and its carriers concentration generally shows a positive correlation with the doping concentration of O. In the recently reported GaN crystal grown by oxide vapor phase epitaxy, the carrier concentration is 8.92×10^{19} cm^{-3}, the mobility is 9.03×10 cm^2 V^{-1} s^{-1}, and it exhibits a very low resistivity (7.75×10^{-4} Ω cm) [31], which shows the great potential of O doping and provides a new idea for the preparation of heavily doped n-type substrates. However, O may also exist in the form of deep energy levels that are difficult to ionize in GaN crystals (generally related to gallium vacancies), which ultimately leads to the electrical characteristics of GaN crystals with the same doping concentration have large differences in electrical properties. For unintentionally doped GaN crystals, we should strive to reduce the O impurity concentration as much as possible to improve the quality of the crystal. In the Na-flux system, the main sources of O impurities include: alumina crucibles and the adsorption of the growth environment. In actual growth, the concentration can be further reduced by replacing the crucible material and improving the cleanliness of the growth environment.

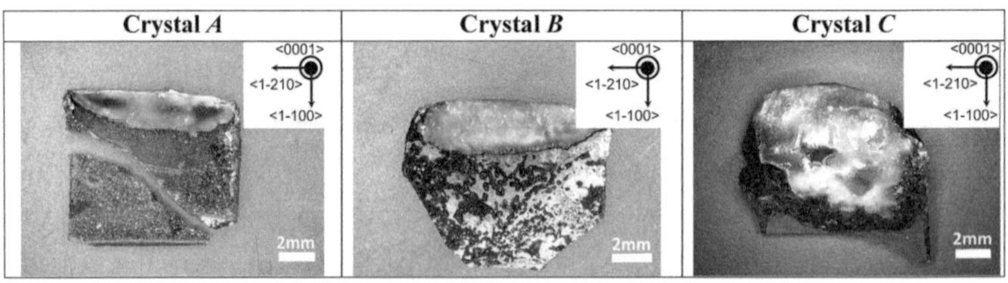

Fig. 5.15 The GaN crystals grown by the Na-flux method are black [27]

In addition to the above-analyzed intrinsic point defects and point defects introduced by unintentional doping, in order to achieve special properties, intentional doping atoms will also be introduced during the growth process to form intentional doping point defects. For example, doping Mg in the growth process to make it p-type conductivity; doping with donor impurities such as Si and O to control the electrical properties of the material, including resistivity, mobility, etc.; doping with impurities such as Fe to make the material appear the characteristics of semi-insulation and so on. At present, in the growth of Na-flux method, there is relatively little research on this part, and it is still in its infancy.

2. Line defects

A line defect is a defect that is small in the two-dimensional scale and large in the third-dimensional scale. It is characterized by a small size in two directions and a longer extension in the other direction. Its main manifestation is a dislocation.

At present, the main methods for growing high-quality and large-size GaN single crystal substrates are: HVPE [32, 33], ammonothermal method [34, 35], Na-flux method [36] and so on.

The HVPE method has a fast growth rate and is easy to obtain large-size crystals. It is currently the main method for commercial provision of GaN single crystal substrates; its disadvantages are high cost, high dislocation density, small radius of curvature, and environmental pollution. The ammonothermal method growth technology is similar to the hydrothermal growth technology. It can grow on multiple seed crystals, is easy to produce on a large scale, and can significantly reduce the cost; its disadvantage is that the growth pressure is high and the growth rate is low. In the Na-flux method, the crystal quality is high, the growth conditions are relatively mild, and the requirements for growth equipment are low. Large-size GaN single crystals can be grown by Na-flux method; its disadvantage is that it is easy to spontaneously nucleate to form polycrystals, and it is difficult to grow thicker GaN. However, in the aspect of line defects, that is, the growth of low dislocation density GaN single crystal has certain advantages, which is also the key point that the Na-flux method as a growth method in a near-thermodynamic equilibrium state is worthy of further research.

Different from other vapor growth methods, in the Na-flux system, due to the characteristics of liquid phase epitaxy, the control of the growth interface is particularly important. Generally, most of the dislocations in the epitaxial single crystal come from the substrate. The dislocation density in the epitaxial single crystal can be effectively reduced through the control of the growth interface. Single crystal epitaxial growth mode, epitaxial crystal quality, epitaxial crystal impurity content, etc. are all affected by the state of the crystal growth interface. Therefore, effective control of the growth interface, especially the state of the initial growth interface, is a prerequisite for obtaining high-quality, large-scale liquid phase epitaxial GaN single crystals. The growth temperature not only affects the viscosity of the system, but also increases the free diffusion length of the atoms on the surface of the seed crystal epitaxial growth. Under certain growth conditions, with the increase of temperature, the growth mode of the growth interface can be changed (from three-dimensional island growth to two-dimensional growth transformation). For the liquid phase epitaxial growth technology, the stable two-dimensional layered growth mode is more conducive to obtaining high-quality epitaxial single crystals. Using the growth mode of the initial growth interface can effectively reduce the dislocations in the early stage of growth and improve the quality of epitaxial crystals.

Similarly, in the Na-flux method growth process, there are many innovative methods to eliminate dislocations.

The introduction of a stirring process and a rocking system. The application of these technologies not only improves the uniformity of the reaction source material system, but also increases the dissolution rate and transport rate of nitrogen, thereby reducing the dislocation density and improving the crystal quality and growth rate.

Single-point seed technique (SPST) and Necking technique are mainly achieved by making small holes on the seed crystal to artificially limit the growth of liquid phase epitaxy and inhibit the increase of dislocations in the early growth stage. Dislocation-free growth in a certain area reduces the overall dislocation density of the crystal [3]. The application of SPST is shown in the Fig. 5.16.

On the basis of SPST, Mori et al. further developed the multipoint seed technology (MPST) (as shown in the Fig. 5.17). They achieved low dislocation density growth in the single-point seed crystal area by making small hole regions of multiple seed crystals. And by further controlling the supersaturation and the flat growth of the crystal surface, the dislocation density of the crystal is reduced, and the large-size (2–6 inch) GaN single crystal is successfully grown, and the single crystal dislocation density can be reduced to the order of ~ 10^2 cm^{-2}.

5.2.4 Stress Control

In the early growth process of GaN single crystals, a heteroepitaxial method is generally used, and the substrate materials used include sapphire, silicon, GaAs, SiC, $LiGaO_2$ and so on [37–41]. Its main advantages are easy availability and low price. However, due to the existence of lattice mismatch and thermal mismatch, a large number of dislocations and stress will be introduced, which will eventually lead to the degradation of the crystal quality of the grown GaN material. Take the Si substrate as an example. On the one hand, because there is a 16.9% lattice constant mismatch between the Si substrate and the epitaxial GaN material, this part of the lattice constant mismatch will introduce a large number of dislocations during the epitaxial growth process. On the other hand, the thermal expansion coefficient mismatch between Si and GaN materials is as high as 56%. In the cooling process after the growth, this part of the thermal expansion coefficient mismatch will cause the accumulation of stress in the epitaxial layer, resulting in the epitaxial wafer warpage occurs, and even cracks on the entire surface of the epitaxial layer [42]. Therefore, if we want to further grow high-quality GaN crystals, heteroepitaxial growth is not desirable.

According to existing studies, compared with heteroepitaxial, homoepitaxial epitaxy based on high-quality GaN single crystal substrates can effectively avoid

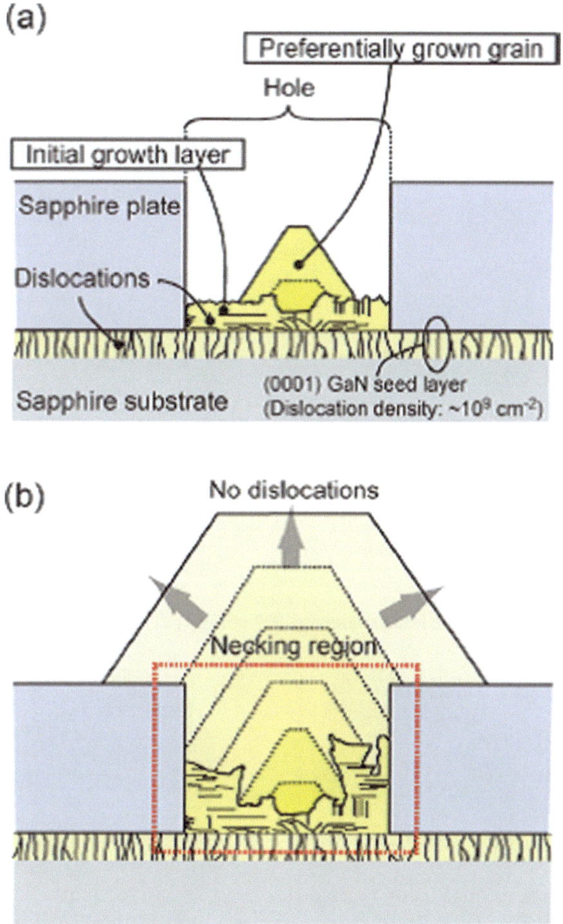

Fig. 5.16 Single-point nucleation of the growth of the Na-flux method [3]

thermal mismatch or lattice mismatch, and further effectively reduce the residual stress and dislocation density in the epitaxial crystal. The Na-flux method for growing GaN crystals is exactly a method based on homogeneous seed crystal liquid phase epitaxial growth, so it has significant advantages in terms of stress control.

Studies have shown that in GaN crystals, both compressive stress and tensile stress will cause the blue shift or red shift of each phonon peak. Among them, E_2(high) is the most sensitive to the stress response, so Raman scattering spectroscopy is generally used to measure the relative movement of the phonon peak to obtain the stress in the crystal, where E_2(high) phonon Raman peak shift ($\Delta\omega$) corresponding biaxial stress σ can be expressed as [43]:$\sigma_{xx=yy} = \Delta\omega/K$. When the crystal is under stress during the growth process, it usually releases the stress through dislocations, holes or cracks. Figure 5.18 shows cracked and crack-free GaN single crystals, and XRC analysis is performed on them. The crack-free GaN single crystals have higher crystal quality.

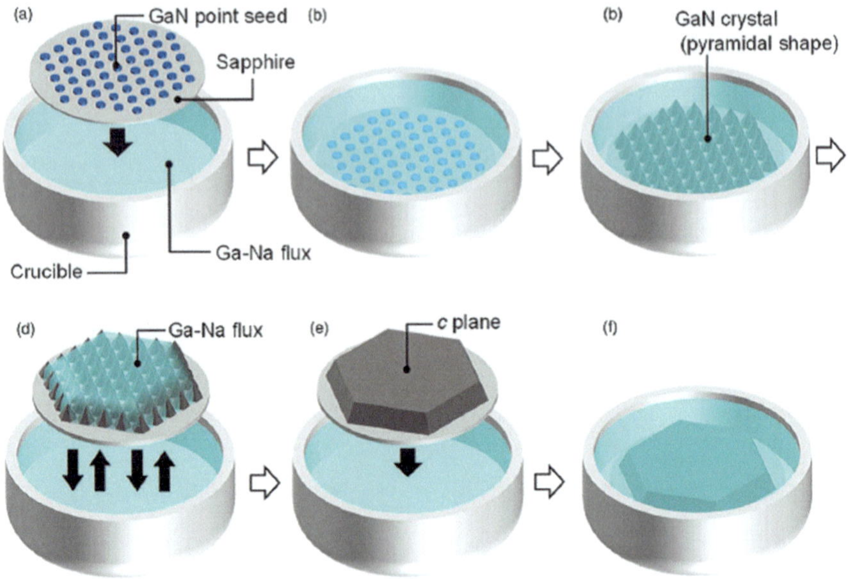

Fig. 5.17 The multipoint seed technology [21]

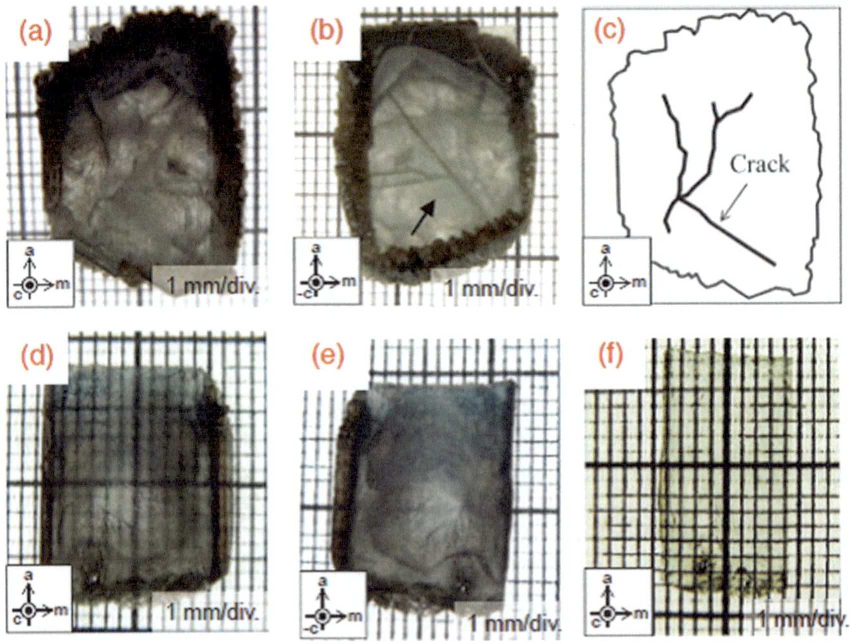

Fig. 5.18 **a–c** Cracks in GaN crystals grown by Na-flux method; **d–e** Non-cracking Na-flux GaN crystals [44]

In homoepitaxial crystals, the stress derived from the mismatch of the substrate is relatively small, and the growth stress is generally released in various ways during the growth process. For the HVPE growth method, lateral epitaxial growth (ELOG) is generally used during its growth process, that is, by bending the dislocation lines and forming a void structure to enhance dislocation annihilation and reduce dislocation density, thereby releasing growth stress and thermal stress, and finally reduce dislocations and stress.

According to existing research findings, there are still certain stresses and cracks in the GaN crystal grown by Na-flux method. Although it is homoepitaxial growth, due to the differences in the lattice constants at the interface, the incorporation of impurities, and the accumulation of thermal stress, when the size of the crystal increases, it often leads to corresponding cracks and eventually causes the crystal to break. Therefore, for the large-scale and industrial development of the Na-flux method, the realization of stress control and crack-free growth is the top priority to be solved urgently.

In the Na-flux method system, the main sources of stresses include: thermal stresses in crystal growth, which need to be further reduced through the regulation of the growth technique; stresses brought by impurity doping in crystals, especially O impurities, which, as the main impurity in Na-flux GaN crystals, are enriched in the crystal growth process, causing the expansion of lattice constants, the accumulation of stresses, and ultimately leading to cracks. O impurities will cause lattice constant expansion, stress accumulation and ultimately crack generation. Therefore, in the actual growth process, it is necessary to improve the cleanliness of the growth environment and raw materials, eliminate the adsorption of impurities, and study the elimination methods of O impurities and stress.

Meanwhile, during the growth process, stresses and cracks mostly arise during the cooling process of crystal growth, which is related to the accumulation of stresses and the difference in thermal expansion coefficients between the seed crystals and single crystals. Therefore, it is necessary to further control the cooling rate during growth. The origin and distribution of these stresses and cracks need to be further investigated systematically.

5.3 Liquid Phase Epitaxial Growth Technology

In 2003, Mori et al. used the liquid phase epitaxy (LPE) method to grow a 12 $\times$ 5 $\times$ 0.8 mm GaN single crystal by Na-flux method. A metal organic chemical vapor deposition (MOCVD) method was used to synthesize a 3 μm thick GaN film on sapphire, and a GaN single crystal was grown on the film by Na-flux method is shown in Fig. 5.19. The results show that the MOCVD-GaN film on the sapphire substrate has a seed crystal effect in the Na-flux method. Using mixed nitrogen containing 40% ammonia instead of pure N_2 gas, it is also possible to grow 10 μm thick GaN homoepitaxial films on MOCVD-GaN films at 5 atm, which is the lowest pressure for growing GaN in Na flux. PL measurement results show that the peak intensity of GaN single crystal grown by LPE is 40 times that of MOCVD-GaN thin film. In addition, LPE growth technology can significantly reduce the dislocation density of bulk GaN.

In 2008, Mori et al. used the liquid phase epitaxy (LPE) growth method to grow a 2 inch, 3 mm thick GaN single crystal for the first time (as shown in Fig. 5.20); In 2011, Mori cooperated with Toyota central R&D Lab/Sanken Electric co. and successfully developed the first 4 inch bulk single crystal; In 2015, Toyota Gosei obtained a 6 inch GaN single crystal; The dislocation densities of the bulk single crystal GaN obtained by the LPE technology above are all 10^4–10^6 cm^{-2} (as shown in Fig. 5.21).

The addition of carbon can inhibit the generation of polycrystals, and at the same time can increase the liquid phase epitaxial growth rate [46]. Since the C-N bond energy is greater than the Ga-N bond energy, the formation of C-N bonds inhibits the combination of gallium atoms and nitrogen atoms, thereby inhibiting the generation of polycrystalline at the gas–liquid interface (as shown in Fig. 5.22). After the carbon and nitrogen atoms are combined, they reach the surface of LPE.

The Ga coordination number around the C-N bond increases and the dissociation energy decreases. Nitrogen atoms are incorporated into the kinks on the surface of LPE and combine with Ga^{3+} to form GaN (as shown in Fig. 5.23) [47].

Xu Ke's research group used self-developed high-temperature and high-pressure growth equipment to explore the Na-flux method liquid phase epitaxial growth technology. In the process of GaN single crystal growth by Na-flux method liquid phase epitaxy, the control of the growth interface is particularly important. Generally, most of the dislocations in the epitaxial single crystal come from the substrate. Through the control of the growth interface, the dislocation density in the epitaxial single crystal can be effectively reduced. Single crystal epitaxial growth mode, epitaxial crystal quality, epitaxial crystal impurity content, etc. are all affected by the state of the crystal growth interface. Therefore, effective control of the growth interface, especially the state of the initial growth interface, is a prerequisite for obtaining high-quality, large-scale liquid phase epitaxial GaN single crystals. The growth temperature not only

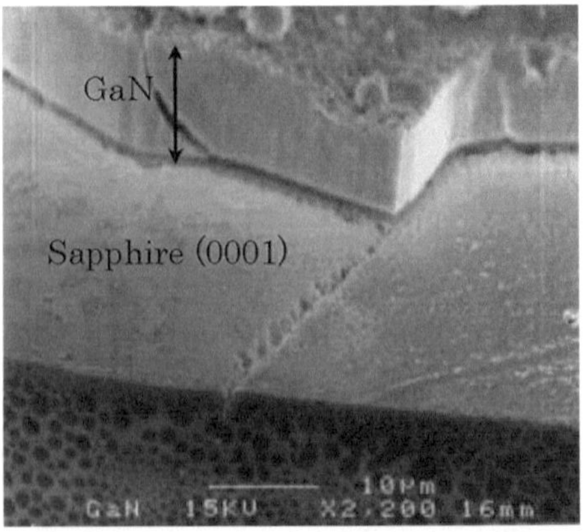

Fig. 5.19 3 μm thick MOCVD-GaN film is seed crystal, and the LPE method is used to grow 10 μm thick GaN film in 40% ammonia mixed nitrogen [45]

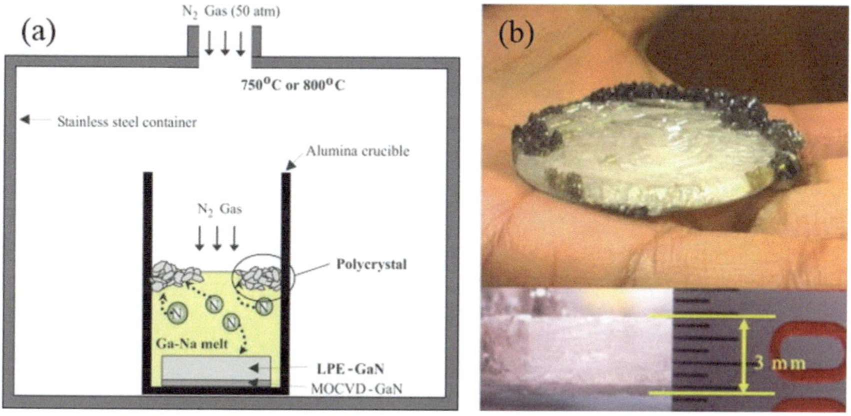

Fig. 5.20 **a** A diagram of the growth of LPE single crystal and polycrystalline in a stainless steel growth furnace and **b** an optical photograph of LPE growth of GaN single crystal [1]

Fig. 5.21 Using HVPE-GaN single crystal as seed crystal, Na-Flux LPE growth **a** 2 inch, **b** 4 inch and **c** 6 inch GaN single crystal [2]

affects the viscosity of the system, but also increases the free diffusion length of the surface atoms of the seed crystal epitaxial growth. Under certain growth conditions, as the temperature rises, the growth mode of the growth interface can be changed from three-dimensional island growth to two-dimensional growth. For the liquid phase epitaxial growth technology, the stable two-dimensional layered growth mode is more conducive to obtaining high-quality epitaxial single crystals. The surface of the grown GaN is attached with GaN polycrystals produced during the cooling process. After grinding, the crystals are colorless and transparent. After polishing, the cathodoluminescence (CL) characterization shows that the dislocation density is approximately 3×10^5 cm^{-2}. Further through the precise control of the Na-flux method liquid phase epitaxial growth process of GaN, a liquid phase epitaxial GaN single crystal with a diameter of 2 inch and a thickness of more than 2 mm was obtained. The growth rate exceeded 25 μm/h. After cutting the bulk single crystal, multiple GaN single crystal substrates can be obtained (as shown in Fig. 5.24).

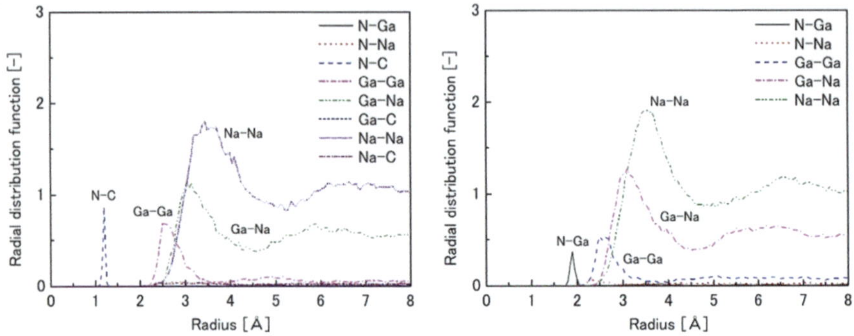

Fig. 5.22 The diameter distribution functions of the Na-Ga melt are **a** model I (no C model) and **b** model II (with C model) [46]

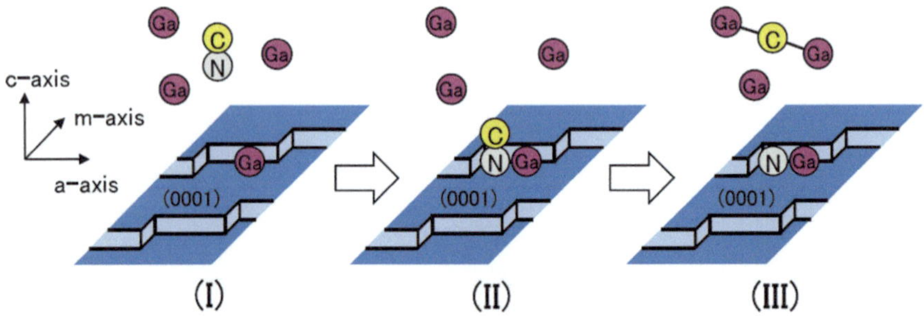

Fig. 5.23 Nitrogen transport process of CN ion media [47]

5.4 Large-Size Single Crystal Growth

5.4.1 Multi-point Seed Crystal Combination Method

In 2014, Mori et al. used an autoclave with stirring function combined with multi-point seed crystal growth technology to obtain a 2 inch high-quality single crystal with a dislocation density of 10^2 cm^{-2} in 75% area as shown in Fig. 5.25. In the future, we believe that 6–8 inch GaN single crystals can be grown by arranging more seed crystals. Therefore, the combined growth of multi-point seed crystals may become a key technology for preparing large-size dislocation-free GaN single crystals.

5.4.2 Tiling Technique

In 2017, Mori et al. used HVPE GaN to fabricate φ111-mm tiling crystals as shown in Fig. 5.26a, and then the Na-flux method was used to grow GaN single crystals to

Fig. 5.24 **a** HVPE GaN seed crystal; **b** Na-flux GaN after growth (GaN polycrystalline attached during cooling of the surface); **c** grinding GaN crystal image; **d** diagram of multiple GaN substrates cut from Na-flux GaN crystals; **e** 2 mm GaN bulk single crystal, **f** line cut GaN substrates [48]

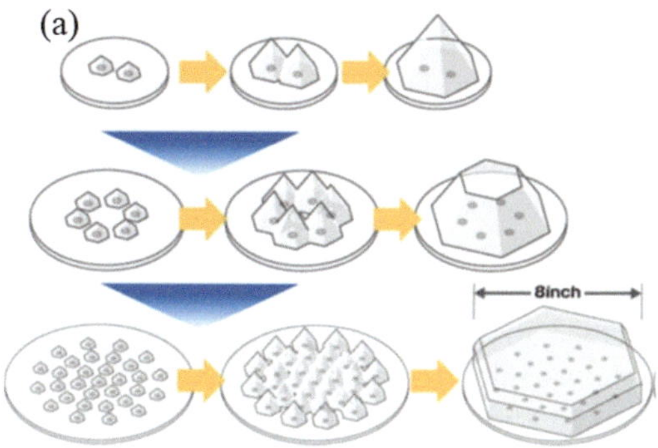

Fig. 5.25 **a** Multi-point seed crystal diagram **b** Multi-point seed crystal growth GaN single crystal [4]

prepare Na-flux GaN single crystals with an approximately 4-mm thickness on the φ111-mm tiling crystal as shown in Fig. 5.26b.

5.4.3 Research Progress of This Group

In the process of growing GaN bulk single crystals by the Na-flux method, heteroepitaxial GaN or GaN single crystals on sapphire or silicon carbide are usually used

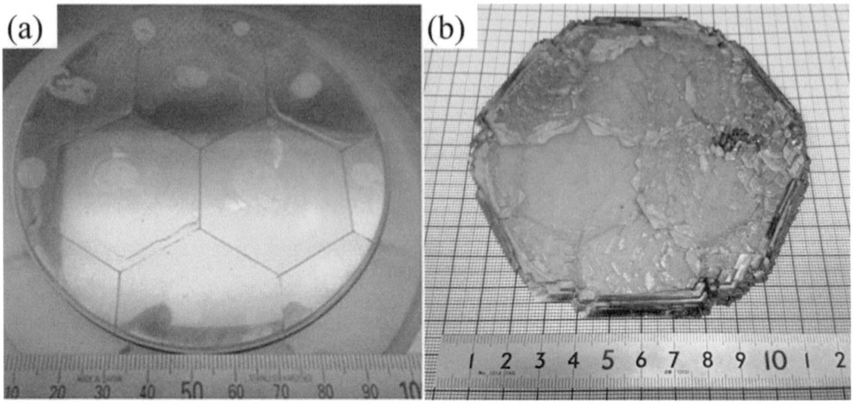

Fig. 5.26 **a** φ111-mm tiling GaN crystal. **b** Image of the Na-flux grown GaN crystal with an approximately 4-mm thickness on the φ111-mm tiling crystal [8]

as the seed crystal. Therefore, during the liquid phase growth process, the growth control of the interface between the seed crystal and the epitaxially grown GaN is particularly important. Generally, due to the remelting of the growth material on the surface of the GaN seed crystal, unevenness or holes in the epitaxial growth interface will appear, which will lead to the disorder of the subsequent steady-state growth and easily cause dislocations. By optimizing the growth process, a good growth interface and high-quality GaN single crystal are realized. The Xu Ke' research group further researched and found that optimizing the growth conditions at the epitaxial growth interface can effectively reduce the growth stress and the dislocations in the seed crystal are effectively suppressed at the interface. This lays an important foundation for the subsequent acquisition of large-size GaN single crystals.

We used 2-inch GaN single crystals and 3-inch heteroepitaxial GaN films as seed crystals, combined with liquid phase epitaxial growth technology, and optimized growth conditions to obtain 2-inch GaN bulk single crystal with a thickness of 2 mm and 3-inch GaN bulk single crystal with a thickness of 1 mm (Fig. 5.27). This laid an important experimental foundation for further obtaining high-quality and large-size GaN bulk single crystals.

5.5 Development Trend

The Na-flux method provides a method that can obtain high quality (dislocation density $10^{2-4}/cm^2$), large size (2 inch ~ 6 inch) GaN single crystal under relatively mild growth conditions ($T < 950$ °C, $P < 10$ MPa). The current international report has a growth rate of ~ 60 μm/h, which provides a feasibility guarantee for the industrialized mass production of Na-flux method. Although the Na-flux method has achieved rapid development in recent years, it has also encountered greater challenges: (1)

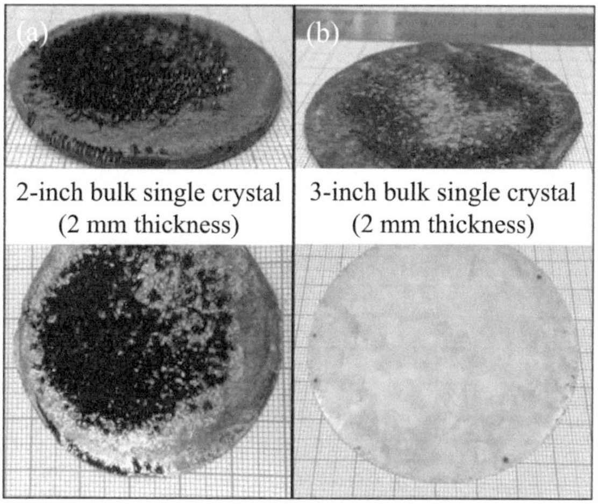

Fig. 5.27 **a** 2 inch GaN single crystal and **b** 3 inch GaN single crystal grown by Na-flux [49]

Na and Ga are very active under high temperature growth, which has high requirements on the corrosion resistance and stability of crucible materials, and affects the reduction of growth costs. (2) Polycrystalline is easy to form during the growth process, and it is a major challenge to improve the effective utilization of raw materials. (3) Due to various factors such as nitrogen source mass transfer, polycrystalline nucleation, and raw material supply, it is difficult to grow thick GaN bulk single crystals, it is difficult to grow thicker GaN bulk single crystals. In addition, the current Na-flux method has a low degree of automation in the growth process, which has also become a factor affecting its industrialization process. In addition, the current Na-flux growth equipment is not sufficiently automated, which is also a factor affecting the Na-flux GaN industrialization process. In short, the Na-flux method is a promising method to realize the industrialization of GaN substrates, but it needs to overcome many problems currently faced. In addition, to prepare high-quality GaN single crystal substrates, the Na-flux method can be used to combine multi-point seed crystals to grow large-size and high-quality GaN seed crystals, and then use HVPE technology for homoepitaxial thickening. Comprehensive utilization of the advantages of each growth method to solve their own problems may be an effective solution for low-cost production of high-quality GaN bulk single crystals. Currently, Professor Mori of Osaka University and Professor Amano of Nagoya University have carried out in-depth cooperation. Professor Mori uses the Flux-Film-Coated technique combined with multi-point seed technology to grow high-quality GaN, and then uses the obtained GaN as a seed crystal to grow high-quality GaN bulk single crystal by the HVPE method. A 2 inch HVPE GaN bulk single crystal with no cracks, no V-pit, dislocation density of $10^{4-5}/cm^2$ and a radius of curvature of more than 50 m was grown [50]. Therefore, with the continuous development and maturity of large-size and multi-chip technologies, the era of mass production and application of lower-cost, higher-quality GaN single crystal substrates is coming.

References

1. Kawamura F, Morishita M, Tanpo M, Imade M, Yoshimura M, Kitaoka Y, Mori Y, Sasaki T (2008) Effect of carbon additive on increases in the growth rate of 2 in GaN single crystals in the Na flux method. J Cryst Growth 310:3946–3949
2. Mori Y, Imade M, Maruyama M, Yoshimura M (2013) Growth of GaN crystals by Na flux method. Ecs J Solid State Sc 2:N3068–N3071
3. Imade M, Murakami K, Matsuo D, Imabayashi H, Takazawa H, Todoroki Y, Kitamoto A, Maruyama M, Yoshimura M, Mori Y (2012) Centimeter-sized bulk GaN single crystals grown by the Na-Flux method with a necking technique. Cryst Growth Des 12:3799–3805
4. Imade M, Imanishi M, Todoroki Y, Imabayashi H, Matsuo D, Murakami K, Takazawa H, Kitamoto A, Maruyama M, Yoshimura M, Mori Y (2014) Fabrication of low-curvature 2 in. GaN wafers by Na-flux coalescence growth technique. Appl Phys Exp 7:035503
5. Maruyama IMM, Yoshimura M, Mori Y (2014) Growth of bulk GaN crystals by the Na-flux point seed technique. Jpn J Appl Phys 53:05FA06
6. Sato T, Nakamura K, Imanishi M, Murakami K, Imabayashi H, Takazawa H, Todoroki Y, Matsuo D, Imade M, Maruyama M, Yoshimura M, Mori Y (2015) Homoepitaxial growth of GaN crystals by Na-flux dipping method. Jpn J Appl Phys 54:105501
7. Murakami K, Ogawa S, Imanishi M, Imade M, Maruyama M, Yoshimura M, Mori Y (2017) Increase in the growth rate of GaN crystals by using gaseous methane in the Na flux method. Jpn J Appl Phys 56:055502
8. Yoshida T, Imanishi M, Kitamura T, Otaka K, Imade M, Shibata M, Mori Y (2017) Development of GaN substrate with a large diameter and small orientation deviation. Phys Status Solidi B 254:1600671
9. Utsumi W, Saitoh H, Kaneko H, Watanuki T, Aoki K, Shimomura O (2003) Congruent melting of gallium nitride at 6 GPa and its application to single-crystal growth. Nat Mater 2:735–738
10. Porowski S, Grzegory I (1997) Thermodynamical properties of III-V nitrides and crystal growth of GaN at high N-2 pressure. J Cryst Growth 178:174–188
11. Yamane H, Kinno D, Shimada M, Sekiguchi T, Disalvo FJ (2000) GaN single crystal growth from a Na-Ga melt. J Mater Sci 35:801–808
12. Kawamura F, Morishita M, Omae K, Yoshimura M, Mori Y, Sasaki T (2005) The effects of Na and some additives on nitrogen dissolution in the Ga-Na system: A growth mechanism of GaN in the Na flux method. J Mater Sci Mater Electron 16:29–34
13. Kawamura F, Morishita M, Miyoshi N, Imade M, Yoshimura M, Kitaoka Y, Mori Y, Sasaki T (2009) Study of the metastable region in the growth of GaN using the Na flux method. J Cryst Growth 311:4647–4651
14. Hautakangas S, Ranki V, Makkonen I, Puska MJ, Saarinen K, Liszkay L, Seghier D, Gislason HP, Freitas JJA, Henry RL, Xu X, Look DC (2006) Gallium and nitrogen vacancies in GaN: Impurity decoration effects. Physica B 376–377:424–427
15. Uedono A, Imanishi M, Imade M, Yoshimura M, Ishibashi S, Sumiya M, Mori Y (2017) Vacancy-type defects in bulk GaN grown by the Na-flux method probed using positron annihilation. J Cryst Growth 475:261–265
16. Uedono A, Takashima S, Edo M, Ueno K, Matsuyama H, Egger W, Koschine T, Hugenschmidt C, Dickmann M, Kojima K, Chichibu SF, Ishibashi S (2018) Carrier Trapping by vacancy-type defects in mg-implanted gan studied using monoenergetic positron beams. Phys Status Sol (b), 255:1700521
17. Wu X, Hao H, Li Z, Fan S, Xu Z (2017) Growth temperature dependence of morphology of GaN single crystals in the Na-Li-Ca flux method. J Electron Mater 47:1569–1574
18. Von Dollen P, Pimputkar S, Alreesh MA, Albrithen H, Suihkonen S, Nakamura S, Speck JS (2016) A new system for sodium flux growth of bulk GaN. Part I: system development. J Cryst Growth 456:58–66 (2016)
19. Reshchikov MA, Morkoç H (2005) Luminescence properties of defects in GaN. J Appl Phys 97:061301

20. Liu Z, Ren G, Shi L, Su X, Wang J, Xu K (2015) Effect of carbon types on the generation and morphology of GaN polycrystals grown using the Na flux method. CrystEngComm 17:1030–1036
21. Mori Y, Imanishi M, Murakami K, Yoshimura M (2019) Recent progress of Na-flux method for GaN crystal growth. Jap J Appl Phys 58:SC0803
22. Christenson SG, Xie W, Sun YY, Zhang SB (2015) Carbon as a source for yellow luminescence in GaN: Isolated CN defect or its complexes. J Appl Phys 118:135708
23. Reshchikov MA, Demchenko DO, Usikov A, Helava H, Makarov Y (2014) Carbon defects as sources of the green and yellow luminescence bands in undoped GaN. Phys Rev B 90:235203
24. Lyons JL, Janotti A, Van de Walle CG (2010) Carbon impurities and the yellow luminescence in GaN. Appl Phys Lett 97:152108
25. Demchenko DO, Diallo IC, Reshchikov MA (2013) Yellow luminescence of gallium nitride generated by carbon defect complexes. Phys Rev Lett 110:087404
26. Imanishi M, Yoshida T, Kitamura T, Murakami K, Imade M, Yoshimura M, Shibata M, Tsusaka Y, Matsui J, Mori Y (2017) Homoepitaxial hydride Vapor phase epitaxy growth on Gan wafers manufactured by the Na-Flux method. Cryst Growth Des 17:3806–3811
27. Abo Alreesh M, Von Dollen P, Malkowski TF, Mates T, Albrithen H, DenBaars S, Nakamura S, Speck JS (2019) Investigation of oxygen and other impurities and their effect on the transparency of a Na flux grown GaN crystal. J Cryst Growth 508:50–57
28. Imanishi M, Murakami K, Yamada T, Kakinouchi K, Nakamura K, Kitamura T, Okumura K, Yoshimura M, Mori Y (2019) Promotion of lateral growth of GaN crystals on point seeds by extraction of substrates from melt in the Na-flux method. Appl Phys Express 12:045508
29. Tingberg T, Ive T, Larsson A (2017) Investigation of Si and O donor impurities in unintentionally doped MBE-Grown GaN on SiC(0001) substrate. J Electron Mater 46:4898–4902
30. You W, Zhang XD, Zhang LM, Yang Z, Bian H, Ge Q, Guo WX, Wang WX, Liu ZM (2008) Effects of different ions implantation on yellow luminescence from GaN. Physica B 403:2666–2670
31. Takino J, Sumi T, Okayama Y, Nobuoka M, Kitamoto A, Imanishi M, Yoshimura M, Mori Y (2019) Development of a 2-inch GaN wafer by using the oxide vapor phase epitaxy method. Jap J Appl Phys 58:SC1043
32. Bockowski M, Grzegory I, Borysiuk J, Kamler G, Lucznik B, Wroblewski M, Kwiatkowski P, Jasik K, Krukowski S, Porowski S (2005) Growth of GaN on patterned GaN/sapphire substrates by high pressure solution method. J Cryst Growth 281:11–16
33. Hennig C, Richter E, Weyers M, Trankle G (2008) Freestanding 2-in GaN layers using lateral overgrowth with HVPE. J Cryst Growth 310:911–915
34. Bockowski M (2007) Bulk growth of gallium nitride: challenges and difficulties. Cryst Res Technol 42:1162–1175
35. Kudrawiec R, Misiewicz J, Rudzinski M, Zajac M (2008) Contactless electroreflectance of GaN bulk crystals grown by ammonothermal method and GaN epilayers grown on these crystals. Appl Phys Lett 93
36. Yamane H, Shimada M, Clarke SJ, DiSalvo FJ (1997) Preparation of GaN single crystals using a Na flux. Chem Mater 9:413–420
37. Cho SI, Chang K, Kwon MS (2007) Strain analysis of a GaN epilayer grown on a c-plane sapphire substrate with different growth times. J Mater Sci 42:3569–3572
38. Tian Y, Shao Y, Wu Y, Hao X, Zhang L, Dai Y, Huo Q (2015) Direct growth of freestanding GaN on C-face SiC by HVPE. Sci Rep 5:10748
39. Wang FC, Cheng CL, Chen YF, Huang CF, Yang CC (2007) Residual thermal strain in thick GaN epifilms revealed by cross-sectional Raman scattering and cathodoluminescence spectra. Semicond Sci Technol 22:896–899
40. Zhou C, Jiang Q, Huang S, Chen KJ (2012) Vertical leakage/breakdown mechanisms in AlGaN/GaN-on-Si devices. IEEE Electron Device Lett 33:1132–1134
41. Yang W, Wang W, Lin Y, Liu Z, Zhou S, Qian H, Gao F, Zhang S, Li G (2014) Deposition of nonpolar m-plane InGaN/GaN multiple quantum wells on LiGaO2(100) substrates. J Mater Chem C 2:801–805

42. Dirk Ehrentraut EM (2010) A brief review on the Na-Flux method toward growth of large-size GaN crystal springer series in materials science 235
43. Qi L, Xu Y, Li Z, Zhao E, Yang S, Cao B, Zhang J, Wang J, Xu K (2016) Stress analysis of transferable crack-free gallium nitride microrods grown on graphene/SiC substrate. Mater Lett 185:315–318
44. Yamada T, Imanishi M, Nakamura K, Murakami K, Imabayashi H, Matsuo D, Honjo M, Maruyama M, Imade M, Yoshimura M, Mori Y (2016) Crack-free GaN substrates grown by the Na-flux method with a sapphire dissolution technique. Appl Phys Express 9:071002
45. Kawamura F, Iwahashi T, Omae K, Morishita M, Yoshimura M, Mori Y, Sasaki T (2003) Growth of a large gan single crystal using the liquid phase epitaxy (LPE) technique. Jpn J Appl Phys 42:L4–L6
46. Kawamura T, Imabayashi H, Yamada Y, Maruyama M, Imade M, Yoshimura M, Mori Y, Morikawa Y (2013) Structural analysis of carbon-added Na–Ga melts in Na Flux GaN growth by first-principles calculation. Jpn J Appl Phys 52:08JA04
47. Kawamura T, Imabayashi H, Maruyama M, Imade M, Yoshimura M, Mori Y, Morikawa Y (2015) First-principles investigation of the GaN growth process in carbon-added Na-flux method. Phys Status Solidi B 252:1084–1088
48. LZ-lCD-mSX-jXK Guo-qiang R, Jian-feng W (2019) Research progress on Gan single crystal growth. J Synthetic Crystals 48:1588–1598
49. Ren Guoqiang LZ, Tengkun L, Ke X (2020) Liquid phase growth of GaN single crystal. J Synthet Crystal 49:2024–2037
50. Imanishi M, Usami S, Maruyama M, Yoshimura M, Mori Y (2021) Growth of a high quality GaN wafer from point seeds by the na-flux method. In: 2021 28th international workshop on active-matrix flatpanel displays and devices (AM-FPD). IEEE, (2021) 70–72

Chapter 6
Homoepitaxy Technology of Gallium Nitride Single Crystals

6.1 Characteristics of Homoepitaxy

6.1.1 Advantages of Homoepitaxy

The pioneering work in nitride semiconductors was developed in the 1980s based on heteroepitaxial technology. The GaN-based quantum structures of heteroepitaxy realized the manufacture of efficient blue light-emitting diodes. This work was completed by three Japanese professors: Isamu Akasaki, Hiroshi Amano, and Shuji Nakamura, and was awarded the Nobel Prize in Physics in 2014. Today, GaN and its alloys with indium and aluminum have become the basic materials for manufacturing optoelectronic devices such as light-emitting diodes and laser diodes. In addition, GaN is also suitable for high-frequency and high-power electronic devices due to its wide bandgap, high carrier mobility, high breakdown strength, and thermal conductivity. Currently, most GaN-based devices are prepared by epitaxial growth on hetero-substrates (such as: sapphire, silicon, or silicon carbide).

However, due to the differences in lattice constants and thermal expansion coefficients between the substrate and the epitaxial material used in heteroepitaxy, there is often a high dislocation density and residual stress caused by lattice mismatch and thermal mismatch after epitaxial growth. This can severely affect the crystal quality of the epitaxial material and even cause the bending or crack of the epitaxial wafer. This can prevent Group III nitride-based devices from exhibiting their excellent performance due to the degradation of crystal quality, severely limiting the further improvement of device performance. For heteroepitaxy, many growth processes have been applied to reduce these crystal defects, such as low-temperature buffer layer technology [1], epitaxial lateral overgrowth (ELOG) technology [2], and patterned substrate [3], etc. Although these processes have improved the crystal quality of the epitaxial material, they also cause the long epitaxial growth time of the device and the complex process. In addition, most hetero-substrates have low electrical and thermal conductivity, which affects the excellent performance of GaN and other epitaxial

K. Xu et al., *GaN Single Crystal Growth and Application*, Wide Bandgap Semiconductors, https://doi.org/10.1007/978-981-96-7572-2_6

materials, and various lift-off methods not only increase the preparation cost but also cause waste of epitaxial materials.

The most fundamental solution to solve the above problems is to use GaN substrates for homoepitaxy. The advantage of homoepitaxial growth of GaN is that the lattice and thermal expansion coefficients match with substrate, so it can eliminate crystal defects and residual stress caused by lattice mismatch and thermal mismatch to the greatest extent, and GaN substrates can improve the vertical electrical conductivity and thermal conductivity, greatly enhancing the performance of GaN-based devices. In addition, since a low-temperature buffer layer is not required during epitaxy, the GaN epitaxial growth time can be greatly shortened, simplifying the epitaxial growth steps. Therefore, using GaN substrates and developing GaN homoepitaxial growth processe are the inevitable trend of GaN epitaxial growth development in the future [4, 5]. In recent years, with the development of GaN substrate fabrication technology, several units have obtained free-standing GaN substrates, and many large companies and research institutions around the world have invested a lot of manpower and resources in the technology of homoepitaxial GaN devices. Therefore, it can be predicted that GaN substrates will be more widely used in the fields of semiconductor optoelectronics and microelectronics.

Currently, GaN substrates can be obtained by three methods: hydride vapor phase epitaxy, flux method, and ammonothermal method. The biggest advantage of homoepitaxial growth of GaN devices on GaN substrates is the low dislocation density. This is mainly due to the relatively low dislocation density of GaN substrates. The dislocation density of GaN substrates prepared by hydride vapor phase epitaxy is generally $10^5 - 10^6$ cm^{-2}, and the dislocation density of GaN substrates prepared by the ammonothermal method is generally 10^4 cm^{-2}. Screw dislocations or mixed dislocations are the main sources of device leakage, therefore, for high-power density GaN-based devices, the dislocation density will affect the device current density, mobility, and lifetime, etc. The impact of dislocation density on device performance is shown in Fig. 6.1.

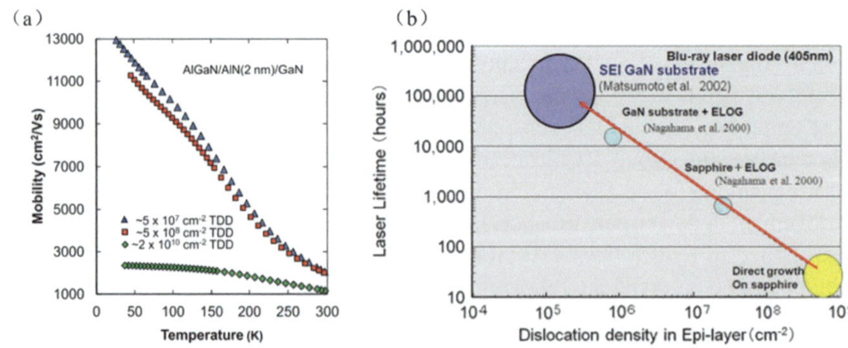

Fig. 6.1 The impact of dislocation density on device performance. **a** The impact of dislocation density on the mobility of HEMT devices [6]. **b** The impact of dislocation density on the lifetime of blue laser diodes [7]

Another advantage of homogeneous epitaxial growth GaN devices is the small stress. During heterogeneous epitaxy, large stress exists at the interface due to lattice and thermal mismatch. The lattice mismatches of sapphire, silicon, and silicon carbide substrates relative to GaN are −17%, −17%, and + 4%, respectively, and the thermal mismatches are + 25%, − 56%, and − 33.1%, respectively, where "–" indicates that the mismatch will produce tensile stress, and " + " indicates that the mismatch will produce compressive stress. As can be seen, heterogeneous epitaxial growth of GaN on silicon and silicon carbide substrates will produce large tensile stress, and heterogeneous epitaxial growth of GaN on sapphire substrates will produce large compressive stress. However, during homogeneous epitaxy, the lattice mismatch is very small, and there is basically no thermal mismatch. Of course, when there are InGaN or AlGaN alloys in GaN devices, there is also stress in homogeneous epitaxy, but it is relatively small compared to heterogeneous epitaxy.

6.1.2 The Specialty of Homoepitaxy

The specialty of homogeneous epitaxial growth is due to the specialty of GaN substrates. Currently, the 2-inch or 4-inch GaN substrates available on the market are basically prepared by the hydride vapor phase epitaxy growth method. This growth method is generally epitaxially grown a thick film on a heterogeneous epitaxy substrate (such as sapphire or gallium arsenide substrate), and then a free-standing GaN substrate is obtained after removing the substrate. This obtained GaN substrate has a large bending of crystal plane, because the main type of dislocation in GaN are mixed dislocation. Mixed dislocation is a dislocation with a certain tilt angle, and the curvature of the crystal face is related to the dislocation density and the tilt angle of the dislocation. Compared with other substrates (silicon, gallium arsenide, silicon carbide), the dislocation density of GaN substrates is about a thousand times that of gallium arsenide or silicon nitride, and about a hundred thousand times that of silicon, and its curvature of crystal plane is generally between a few meters and tens of meters. The curvature of the GaN crystal plane leads to nonuniformity of surface orientation, that is, nonuniformity of the miscut angle of GaN substrate (as shown in Fig. 6.2), which leads to nonuniformity of doping concentration of ternary alloys during the homogeneous epitaxy device process and unintentional impurity doping. Usually, the deviation of the miscut angle of the 2-inch GaN substrate should not exceed 0.15°, that is, the corresponding curvature radius of GaN substrate cannot be less than 15 m.

From a microscopic perspective, Nonuniformity of the miscut angles lead to the nonuniformity of atomic step density on the GaN substrate surface. For epitaxial growth, the atomic steps on the substrate surface are the nucleation points for epitaxy the atomic steps on the (0001) surface of GaN have a clear alternation of double hanging bonds and single hanging bonds, which are key to the nucleation of new atoms during the epitaxy process. As shown in Fig. 6.3, the larger the miscut angle, the higher the surface atomic step density; the smaller the miscut angle,

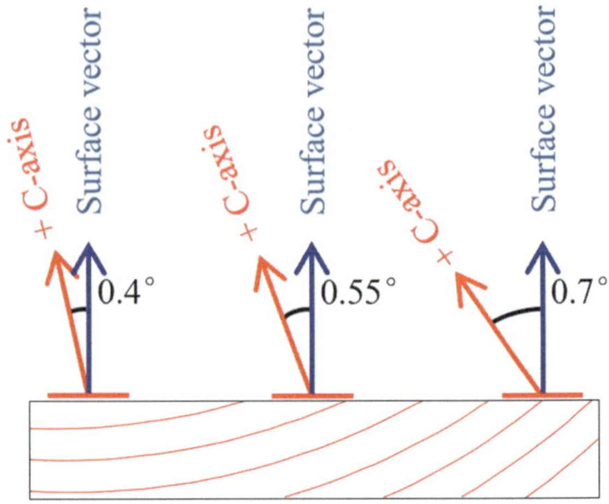

Fig. 6.2 Schematic diagram of GaN substrate curvature of crystal plane and miscut angle deviation (with 0.55 degrees as the center value of C off surface normals)

the lower the surface atomic step density. Under different miscut angle conditions, the surface atomic step density is different, the density of hanging bonds at the step edge is different, which leads to different doping concentrations of ternary alloys or unintentional doping.

The preparation of GaN substrates requires an appropriate wafering process: grinding, lapping, mechanical, and chemical–mechanical polishing. The general wafering process involves both chemical and physical effects, especially during the chemical–mechanical polishing stage. For GaN, the process is mainly physical, which leads to scratches or damage on the GaN substrate after polishing. Secondly, the surface of GaN is easily contaminated, whether it is lapping, polishing or cleaning, it can easily lead to the presence of an oxide layer or silicon adsorption on the sample surface. The presence of a damaged layer or contamination layer has a significant impact on GaN-based devices, especially for GaN-based power electronic devices, which require the substrate to be semi-insulating, with a resistivity > 10^6 Ω*cm, and also require the device interface to be semi-insulating, in order to obtain high-quality devices with no leakage layer at the growth interface.

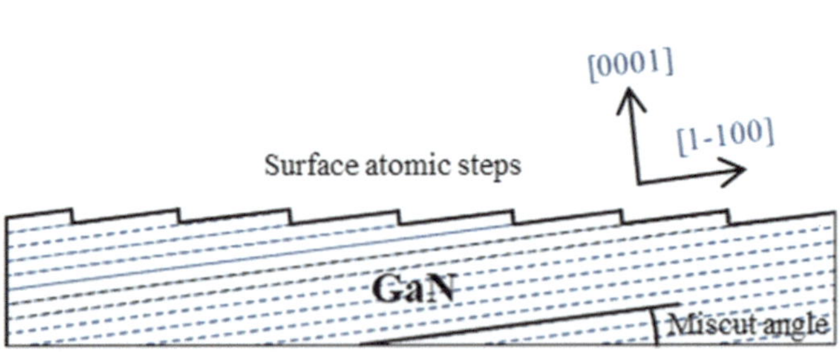

Fig. 6.3 Schematic diagram of the relationship between GaN substrate surface atomic steps and miscut angle

Table 6.1 Differences between homoepitaxy and heteroepitaxy

Epitaxy type	Heteroepitaxy			Homoepitaxy
Substrate	GaN/Sapphire	GaN/SiC	GaN/Si	GaN/GaN
Dislocation density (cm^{-2})	1–5E8	1–5E8	1–5E9	1E5–3E6
Lattice mismatch	− 17%	+ 4%	− 17%	0%
Thermal mismatch	+ 25%	− 33%	− 56%	0%
Thickness (μm)	< 300	< 10	< 5	
Device type	Horizontal	Horizontal	Horizontal	Horizontal or vertical

In summary, the advantages of homoepitaxy are very obvious, with a dislocation density 2–3 orders of magnitude lower, lattice matching during epitaxial growth, which can eliminate crystal defects caused by lattice mismatch to the greatest extent; at the same time, there is no thermal mismatch during epitaxial growth, which can effectively reduce the stress caused by lattice mismatch and thermal mismatch. GaN substrates can also improve the electrical conductivity and thermal conductivity in the vertical direction, realize vertical structure GaN-based devices, and have no restrictions on the thickness of the GaN epitaxial layer (as shown in Table 6.1), thus greatly improving the performance of GaN-based devices. In addition, because there is no need for a low-temperature buffer layer during homoepitaxy, it can greatly shorten the time for GaN epitaxial growth and simplify the epitaxial growth steps. With the continuous improvement of GaN substrate development technology, homoepitaxy of GaN is the inevitable trend for the development of GaN-based device epitaxial growth in the future.

6.2 Epitaxial Substrate Surface Treatment

6.2.1 Chemical Mechanical Polishing (CMP)

Chemical Mechanical Polishing (CMP) can be used to prepare the epitaxial growth surface and is an important part of the process of preparing high-quality GaN substrates. Generally, the combination of chemistry and mechanics is used to achieve surface planarization and remove surface defects, leaving a flat, undamaged surface. CMP is a complex process that includes chemical and mechanical effects, and its material removal rate (MRR) is controlled by the slower process of CMP. The results of CMP depend on the PH conditions of the solution, the abrasive material concentration of the polishing pad, the polishing pressure, and the flow rate of the polishing fluid.

Since GaN can react with alkali at room temperature, NaOH or KOH solution can be used to enhance the CMP effect on the GaN surface. Weyher et al. [8] reported the use of KOH solution for chemical polishing during the diamond slurry grinding process to flatten the surface. Although chemical polishing in KOH solution with a soft polishing pad can achieve a damage-free surface, it requires a higher pressure (2–6 kg/cm^2).

Among the several candidate materials for CMP slurry, colloidal silica is the most popular and is used in actual industry for CMP of silicon, glass, and sapphire substrates. Therefore, colloidal silica is a good choice as a slurry for CMP of GaN substrates. Tavernier et al. [9] first reported the study of CMP processing of GaN using colloidal silica. Because GaN is polar, the colloidal slurry based on silica can quickly and effectively polish the N face of GaN, but the polishing rate for the Ga face is extremely low. Aida et al. [10] proved that when colloidal silica is used for the polishing of Ga face GaN, the MRR of GaN is 17 nm/h, and an atomic-level flat surface with Ra = 0.1 nm is obtained after CMP.

So far, in order to improve the CMP efficiency of GaN, great efforts have been made. The main direction of research is to find strong oxidants that can chemically oxidize GaN faster. By using hydroxyl radicals (OH·) or sulfate radicals (SO_4·) as oxidants, the highest MRR of about 122 nm/h can be achieved on the Ga face using SiO_2 abrasives [11]. However, this MRR is still far below the level of μm/h in the etching process. Therefore, it is necessary to further explore the direction that can effectively promote the increase of MRR along the path related to chemical oxidation, and combine mechanical polishing to obtain a more efficient CMP scheme.

6.2.2 Dry and Wet Etching

In traditional processes, GaN is usually etched by dry methods. Dry etching technology has advantages such as anisotropy, high selectivity for different materials, good uniformity and repeatability, and continuous production. The main techniques include inductively coupled plasma etching (ICP), reactive ion beam etching (RIE), electron cyclotron resonance (ECR) plasma etching, chemically assisted ion beam etching (CAIBE), and magnetically controlled reactive ion etching (MIE), etc. The equipment used in these dry etching processes is generally large-scale, and the overall process cost is very high. Moreover, the operation of these large-scale equipment is also complex, and some processes require the use of toxic and hazardous gases (e.g., chlorine). In addition, during dry etching, high-energy plasma bombardment of the GaN surface may cause significant damage to the GaN surface, resulting in a rough surface morphology, and high-energy plasma will enter the GaN surface, causing a non-stoichiometric surface, which affects the device performance.

The GaN epitaxial process also involves wet etching. Defects on the GaN epitaxial layer have lower surface energy, so they are easily etched. The equipment used in the wet etching process is relatively simple, the operation is relatively convenient, and the raw materials used are less harmful to the human body. The damage of GaN during the process is very low. It is a very convenient process method, so wet etching is an important device manufacturing process for supplement of dry etching technology. So far, wet etching is a very effective and low-cost method to remove surface damage caused by dry etching.

6.3 Homogeneous Epitaxial Interface Control

6.3.1 Main Impurity Types of Homogeneous Epitaxial Interface

Storm et al. [12] used SIMS to assess impurity concentrations of Si, C, O, and H at the homoepitaxial interface, finding that the concentrations of Si, C, and O were significantly higher at the interface than in the substrate by 2–3 orders of magnitude. Further comparison of various surface treatment methods showed minimal variation in interface impurity levels, indicating that the CMP process is not a primary source of impurity atoms on the GaN substrate surface, as shown in Table 6.2. A specialized cleaning process effectively reduced interface impurity concentrations, lowering C and O levels by approximately 2 orders of magnitude, reaching 8E16 cm^{-3} and 3E18 cm^{-3}, respectively. The Si impurity concentration was also reduced by about 25% of its initial value, reaching 8E17 cm^{-3}. However, Si and O impurity levels at the interface remained about an order of magnitude higher than those in the substrate. Multiple studies have reported that Si impurities at the interface are an inherent issue in homoepitaxial growth [12–14].

6.3.2 Source of Impurities at the Homoepitaxial Interface

The GaN substrate surface tendency to adsorb impurity atoms is a primary source of impurities at the homoepitaxial interface. In 1999, Zywietz et al. [15] used density functional theory (DFT) calculations to study oxygen atoms adsorption on Ga-polar and N-polar surfaces of GaN. Their findings showed that as the coverage area increased, the adsorption energy per atom on the Ga-polar surface rose significantly, changing from − 0.325 to + 0.5 eV as oxygen atom coverage increased from 0.25 layer to a full monolayer. In contrast, the adsorption energy per atom on the N-polar surface remained negative, ranging from − 0.3 to − 0.35 eV. This suggests that oxygen atoms saturate quickly on the Ga-polar surface but can accumulate extensively on the N-polar surface, resulting in oxygen atom concentrations on the N-polar surface that are 10–100 times higher than on the Ga-polar surface.

Selvanathan et al. [16] used XPS technology to study the relationship between the oxidation of the Ga face of GaN and the time exposed to air after cleaning. They found that 24 h after the cleaned GaN substrate surface was exposed to air, the surface oxygen signal was equivalent to that without cleaning, and the surface oxygen signal

Table 6.2 Main impurity concentrations at the homoepitaxial interface

The processing method of substrate	Epitaxy methods	Impurity concentration of the interface (cm^{-3})				References
		Si	C	O	H	
CMP + Organic solvent degrease + Acid etch	MBE	~ 1E18	~ 2E18	~ 4E19	Close to the detection limit	[12]
CMP + RIE + Organic solvent degrease + Acid etch	MBE	~ 1E18	~ 5E18	~ 4E19	Close to the detection limit	[12]
1100°C Ultra-high vacuum heat treatment	MBE	3E18	7E18	1E20	Close to the detection limit	[12]
Cleaning + In situ Ga deposition	MBE	8E17	8E16	3E18	Close to the detection limit	[12]
Reactive ion etching or CMP	OMVPE	2E19	3E17	2E19	5E17	[13]
Reactive ion etching or CMP	MBE	4E18	7E18	3E19	3E19	[13]
None	MOCVD	2E19	The same as the substrate	The same as the substrate	Not measured	[14]

one hour after exposure to air was close to half of that after 24 h. This shows that the GaN surface easily reacts with oxygen.

To distinguish impurities adsorbed on the GaN wafer surface from residual impurities introduced by surface polishing, a comparative SIMS analysis of impurity distribution was conducted on homoepitaxial GaN layers grown on surfaces exposed to CMP and RIE treatments [12]. Selective chlorine-based etching removed 700 nm from one half of the Ga-polar GaN wafer surface, prepared by HVPE and subjected to CMP, while the other half was shielded from etching with photoresist, which was later dissolved using a solvent. The wafer was then thoroughly cleaned using an organic solvent degreaser, followed by acid and alkali etching, and subsequently introduced into an ultra-high vacuum MBE system. Following a 30-min degassing process at 600 °C in the preparation chamber, the wafer was transferred to the deposition chamber, where standard growth conditions were employed to deposit a 1-μm-thick GaN layer. SIMS analysis was performed on the epitaxial layer and regrowth interface across both etched and unetched areas of the wafer. As illustrated in Fig. 6.4, aside from chlorine, the impurity distribution at the interface of the epitaxial layer

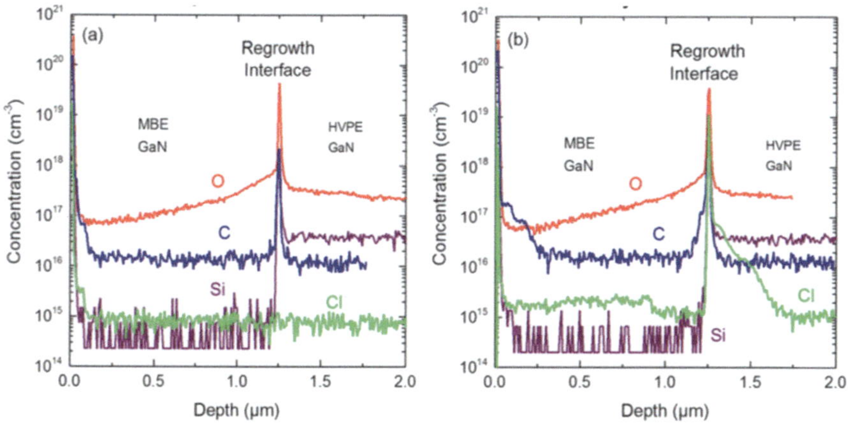

Fig. 6.4 SIMS distribution of impurities at the GaN regrowth interface under different process conditions

was similar for both treated and untreated regions. Notably, no substantial differences in oxygen or silicon distribution were observed between the two regions, while the carbon concentration near the regrowth interface on the etched surface showed only a marginal increase. These findings indicate that impurities near the regrowth interface are independent of the CMP process but strongly suggest that GaN surfaces adsorb impurities during brief environmental exposure between cleaning and chamber loading. Furthermore, a comparative analysis of chlorine profiles revealed a higher Cl content beneath the etched substrate area, consistent with high-energy ion implantation from inductively coupled plasma. Conversely, the chlorine concentration measured from the unetched area of the sample was below the detection threshold, suggesting that the dilute hydrochloric acid used in wafer cleaning does not leave detectable Cl residues at the regrowth interface.

To further investigate whether interface impurities originate from the growth chamber or the cleaning process, Pickrell et al. [17] used MOCVD to conduct a controlled study on the sources of regrowth interface impurities (Fig. 6.5). In one sample, a 0.8-μm thick, unintentionally doped GaN epitaxial layer was initially grown on a HVPE-grown GaN single-crystal substrate, followed by two regrowth processes. For the first regrowth, the sample was cooled to room temperature, transferred to the loading chamber, and subsequently moved into the growth chamber, where a 0.8-μm undoped GaN layer was regrown. For the second regrowth, after cooling to room temperature, the sample remained in place, was reheated, and an additional 0.8-μm undoped GaN layer was grown. In a second sample, a 1-μm thick, silicon-doped GaN layer was grown on a HVPE GaN substrate, then cooled to room temperature, removed from the chamber, chemically cleaned, and subsequently regrown with a 1.2-μm lightly doped GaN layer. SIMS analysis of the interface layers revealed no significant O impurities at either sample's interfaces (O concentration below the detection limit). The first sample exhibited trace Si impurities (~ 5E16 cm^{-3})

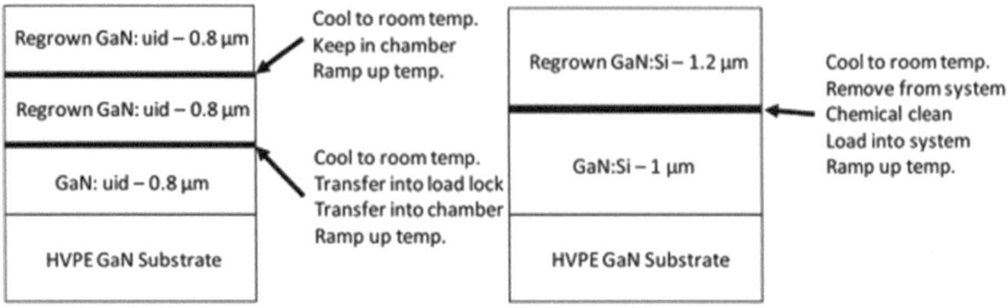

Fig. 6.5 Schematic diagram of GaN structure under different regrowth conditions

at the interface, whereas the second sample displayed markedly higher Si levels (~ 2E17 cm^{-3}), an order of magnitude above normal. These findings indicate that the growth chamber and the thermal cycling process do not introduce significant contamination, highlighting that contamination predominantly occurs outside the growth chamber, particularly during the cleaning process.

In summary, the impurities present at the GaN homoepitaxial interface mainly come from the substrate cleaning process. At present, choosing an appropriate cleaning process can effectively improve the interface impurities, but there is no fundamental solution to eliminate impurities, and there is an urgent need to develop surface cleaning technology.

6.3.3 *Effect of Homoepitaxial Interface Impurities on Devices*

The interface impurity problem is a unique interface issue in homoepitaxy, where the main impurities, Si and O, are donor impurities [18], thus forming a conductive layer at the interface. Although this interface issue does not cause significant increase of defects such as interfacial dislocations [12], and does not affect the performance of devices prepared on conductive substrates by homoepitaxy. When Schottky diodes are prepared on conductive substrates by homoepitaxy, the interface layer becomes a leakage channel, greatly affecting the performance of the device, especially reverse leakage current, thereby reducing the breakdown voltage [17]. Additionally, in HEMT devices on semi-insulating substrates grown by homoepitaxy, interface impurities affect the two-dimensional electron gas and decrease its mobility due to the impurity scattering [19–21].

6.3.4 Solution to Interface Impurity in Homoepitaxy

To solve the interface impurity problem, the primary method is to remove surface impurities from the substrate through improved process or cleaning conditions. Bermudez et al. [22] used metal Ga deposition and thermal desorption to clean C and O impurities adsorbed on the surface of GaN substrates. Khan et al. [23] used a similar method, exposing MOCVD-grown GaN thin films to Ga flax at a wafer temperature of 900°C to remove surface contaminants. Auger peak intensity ratios of C/N and O/N were found to decrease from ~ 25% to less than 2% and below the detection limit, respectively, through Auger electron peak characterization tests, it indicats that this method can effectively remove surface C and O impurities.

Surface contamination can be reduced by adopting wet chemical cleaning or thermal cleaning under high vacuum. However, there is an increasing consensus that, in addition to acid or alkali etching, a combination of ex-situ wet chemical cleaning with organic solvent degreasing and in-situ thermal cleaning is most effective for removing contaminants from the GaN surface. But the GaN surface becomes rough at high temperatures. For example, Gangopadhyay et al. [24] reported that oxygen impurity contamination on the GaN surface can be effectively reduced by high temperatures and exposuring to an active nitrogen plasma atmosphere. Similarly, Hao et al. [25] cleaned the samples with a degreaser in an organic solvent, followed by boiling KOH cleaning, then cleaning with a mixture of sulfuric acid and hydrogen peroxide, and finally thermal cleaning for MOCVD growth. They studied the effect of the order of acid etching and alkali etching on the concentration of C and O on the surface, and they found that the concentration of both impurities had a small but significant reduction when the surface was etched in an alkali solution before acid washing.

Since interface impurities are donor impurities, deep-level impurity compensation can be used to solve the interface problem. Lee et al. [26] found that by comparing different growth thicknesses and concentrations of Fe-doped GaN compensation layers, the interface can be effectively compensated by using a 280 nm thick Fe-doped layer with a doping concentration of 5E17 cm^{-3}, achieving a high-power field-effect transistor mobility increase from 1380 to 1750 cm^2/Vs. Cordier et al. [27] found that effective interface compensation can also be achieved by doping a high concentration of Fe (1E20 cm^{-3}) above the substrate surface, achieving a device mobility of 2000 cm^2/Vs or above, but the surface morphology of the device was deteriorated. By modulating the iron doping concentration and mode (continuous or pulsed doping), the surface morphology can be improved. Further analysis of the impact of different growth methods (MOVPE and MBE) on device performance [28] found that devices grown by MOVPE have higher mobility and lower device leakage current density, and the device leakage current density is also related to the iron doping mode. Devices grown on substrates with subsurface pulsed doping have the optimal leakage density (1E5 mA/mm). In addition, through the ultraviolet photochemical etching method, the interface Si impurity concentration can be effectively reduced from 6.3E18 cm^{-3} to 1.2E18 cm^{-3}, and then the device interface charge [29] can be resolved through the iron-doped compensation layer. High-temperature annealing can also effectively eliminate interface C and O impurities, and then the compensation layer can be achieved through C doping [20], solving the device interface charge.

So far, for the problem of GaN interface impurities, especially the presence of interface Si impurities, it is still difficult to handle completely. The scheme of iron doping or carbon doping compensation is a relatively effective method to solve the device performance problem, but this method is not an essential solution.

6.4 Homoepitaxy Defect Control

The advantage of homoepitaxial growth of GaN lies in the matching of lattice and thermal expansion coefficients, which can eliminate crystal defects and residual stress caused by lattice and thermal mismatch to the greatest extent. However, when homoepitaxy is carried out on a homogeneous substrate, the dislocations in the substrate often enter the epitaxial material along with the growth interface, which greatly reduces the crystal quality of the GaN material. In order to obtain GaN materials with higher crystal quality, new technologies and processes need to be developed in GaN homoepitaxy technology to reduce the crystal defects of epitaxial GaN, obtain GaN films with higher crystal quality, and prepare GaN-based optoelectronic devices with better performance. During homoepitaxy, some processes are often needed to control the defects produced by epitaxy. Common methods are as follows:

1. Homogeneous substrate surface treatment

During homoepitaxy, the substrate surface treatment process has a significant impact on subsequent homogeneous epitaxial interface dislocations. The GaN substrate must be grinded and polished to make the surface smooth. However, grinding will cause damage (scratches, etc.) of the surface and subsurface of the GaN substrate [30]. The polishing process can effectively reduce damage, but it is difficult to completely remove damage. Chemical assisted ion beam etching (CAIBE) [31], reactive ion etching (RIE) [32], and inductively coupled plasma (ICP) etching [33] can effectively solve the surface damage of GaN substrate caused by grinding and polishing, thereby reducing the epitaxial defects caused by GaN substrate surface damage.

2. Lateral epitaxy

Nagahama and others [34] carried out GaN homoepitaxy on free-standing GaN substrates through the method of epitaxial lateral overgrowth (ELOG), and found that this could effectively reduce the dislocations widely distributed on the entire surface. The dislocation densities of the GaN layer above the window area of the SiO_2 mask were reduced to 5E7cm^{-2}, and the dislocation densities of the GaN layer above the SiO_2 mask region were reduced to 7E5 cm^{-2}.

The GaN thin film was grown using the MOCVD method. First, a 2.5 μm thick GaN layer was grown, then a 0.1 μm thick SiO_2 mask was etched to form an 8 μm wide stripe window with a 20 μm periodicity on the [1–100] crystal direction of GaN, and a 15 μm thick GaN was grown on this SiO_2 mask pattern. The substrate obtained at this stage was called the ELOG substrate. Next, a 200 μm thick GaN layer was grown on the ELOG substrate using a hydride vapor phase epitaxy (HVPE) system with a horizontal quartz reactor. Then, the sapphire substrate and the ELOG part of HVPE-GaN were removed by mechanical peeling to obtain an independent GaN substrate with a thickness of about 150 μm. The substrate obtained at this stage is called the free-standing GaN substrate. Then, the ELOG method was used again to grow on the GaN substrate, as previously described. The substrate obtained at this stage is called the GaN substrate with ELOG. The growth schematic is shown in Fig. 6.6.

After the three substrates mentioned above were homoepitaxially grown on GaN, the dislocation density was tested. Figure 6.6a showed that through the ELOG method, the dislocation densities of GaN above the SiO_2 mask region was reduced to 1E6 cm^{-2}. However, in the GaN above the window area, there are a large number of high-density dislocations, with the densities about 5E7 cm^{-2}. As shown in Fig. 6.6c, on the GaN substrate with ELOG, the dislocation density of GaN above the SiO_2 mask region was 7E5 cm^{-2}. This showed that homogeneous epitaxial growth on the GaN substrate using ELOG could effectively reduce the thread dislocation density on the entire surface, improving the crystal quality of the GaN layer.

However, ELOG technology also has limitations, such as only being able to locally reduce the thread dislocation density. Moreover, the dielectric layer acting as a growth mask can also generate stress and be incorporated as impurities into the GaN epitaxial film [35].

Yang et al. [36] reported that, on a GaN template patterned with nanopores prepared by nanoimprint lithography, a GaN epitaxial layer is grown by PAMBE. By optimizing the depth and periodicity of the nanopores etched on the GaN template, the thread dislocation density of the regrown GaN could be effectively reduced. The mechanism of thread dislocation reduction was shown in Fig. 6.6. Figure 6.7 It clearly confirmed that thread dislocations could terminated and disappeared at the voids, while other thread dislocations could extend to the surface. Since regrowth occured inside the nanopores, shallower pores (100 nm) may lead to incomplete void formation, where thread dislocations continued to extend upward. On the other hand, if the depth of the nanopores was further increased to 300 mm or more, the nanopores were affected by dry etching during the nanoimprint process, leading to an increase of defects. By optimizing the depth of the nanopores (200 nm) and the nanopore area

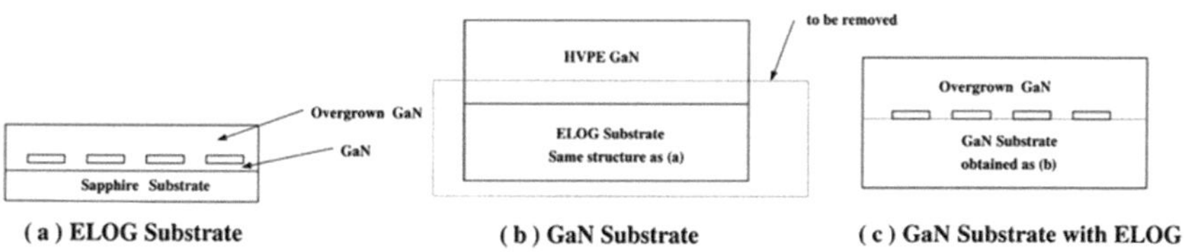

Fig. 6.6 Schematic of ELOG growth structure **a** ELOG substrate, **b** GaN substrate, and **c** Schematic structure of GaN substrate with ELOG

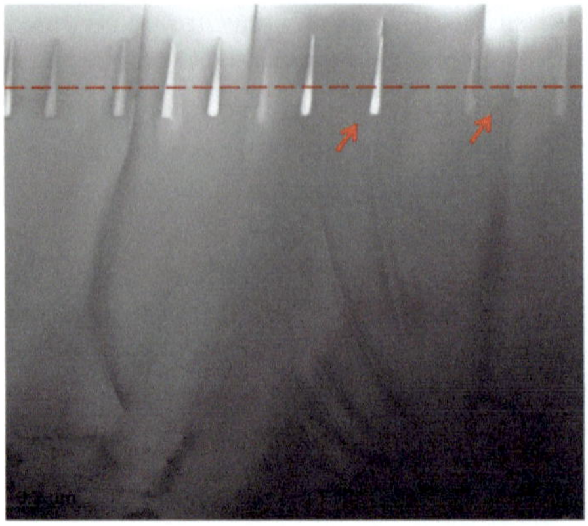

Fig. 6.7 Cross-sectional TEM bright field images show two dislocations terminating at blank spots, as indicated by the arrows. The dashed line represents the regrowth interface between the template surface and the homoepitaxial GaN layer

coverage rate (25%), high-quality GaN was successfully obtained with a nanomask, resulting in a low dislocation densities of 1.2E7 cm^{-2}. HRXRD also found that the reduction of screw or mixed dislocations were more significant compared to edge dislocations.

By treating the surface of the homoepitaxial substrate and using traditional ELOG techniques, the crystal defects of homoepitaxial GaN can be effectively controlled, obtaining higher quality GaN thin films, which can be used for GaN-based optoelectronic devices, such as GaN lasers. The dislocation density directly affects the device life, power density, etc.

References

1. Vennegues P, Beaumont B, Bousquet V, Vaille M, Gibart P (2000) Reduction mechanisms for defect densities in GaN using one- or two-step epitaxial lateral overgrowth methods. J Appl Phys 87:4175–4181
2. Chen CQ, Yang JW, Wang HM, Zhang JP, Adivarahan V, Gaevski M, Kuokstis E, Gong Z, Su M, Khan MA (2003) Lateral epitaxial overgrowth of fully coalesced a-plane GaN on r-plane sapphire. Jap J Appl Phys Part 2-Letters 42:L640-L642
3. Wang L, Jin J, Hao ZB, Luo Y, Sun CZ, Han YJ, Xiong B, Wang J, Li HT (2017) V-shaped semipolar InGaN/GaN multi-quantum-well light-emitting diodes directly grown on c-plane patterned sapphire substrates. Phys Stat Solidi a-Appl Mater Sci 214:6
4. Cao XA, Lu H, Kaminsky EB, Arthur SD, Grandusky JR, Shahedipour-Sandvik F (2007) Homoepitaxial growth and electrical characterization of GaN-based Schottky and light-emitting diodes. J Cryst Growth 300:382–386
5. Chen KM, Yeh YH, Wu YH, Chiang CH, Yang DR, Gao ZS, Chao CL, Chi TW, Fang YH, Tsay JD, Lee WI (2010) Stress and defect distribution of thick Gan film homoepitaxially regrown on free-standing GaN by hydride Vapor phase epitaxy. Jpn J Appl Phys 49:6
6. Stephen WK, Peter GB, Man HW, Erin CHK, Umesh KM, James SS (2012) Effect of dislocations on electron mobility in AlGaN/GaN and AlGaN/AlN/GaN heterostructures. Appl Phys Lett 101:262102
7. Nakamura T, Motoki K (2013) GaN substrate technologies for optical devices. Proc IEEE 101:2221

8. Weyher JL, Muller S, Grzegory I, Porowski S (1997) Chemical polishing of bulk and epitaxial GaN. J Cryst Growth 182:17–22
9. Tavernier PR, Margalith T, Coldren LA, DenBaars SP, Clarke DR (2002) Chemical mechanical polishing of gallium nitride. Electrochem Solid State Lett 5:G61–G64
10. Aida H, Takeda H, Koyama K, Katakura H, Sunakawa K, Doi T (2011) Chemical mechanical polishing of gallium nitride with colloidal silica. J Electrochem Soc 158:H1206–H1212
11. Wang J, Wang T, Pan G, Lu X (2016) Effect of photocatalytic oxidation technology on GaN CMP. Appl Surf Sci 361:18–24
12. Storm DF, Hardy MT, Katzer DS, Nepal N, Downey BP, Meyer DJ, McConkie TO, Zhou L, Smith DJ (2016) Critical issues for homoepitaxial GaN growth by molecular beam epitaxy on hydride vapor-phase epitaxy-grown GaN substrates. J Cryst Growth 456:121–132
13. Murthy M, Freitas JA Jr, Kim J, Glaser ER, Storm D (2007) Residual impurities in GaN substrates and epitaxial layers grown by various techniques. J Cryst Growth 305:393–398
14. Hite JK, Anderson TJ, Luna LE, Gallagher JC, Mastro MA, Freitas JA, Eddy CR Jr (2018) Influence of HVPE substrates on homoepitaxy of GaN grown by MOCVD. J Cryst Growth 498:352–356
15. Zywietz TK, Neugebauer J, Scheffler M (1999) The adsorption of oxygen at GaN surfaces. Appl Phys Lett 74:1695–1697
16. Selvanathan D, Mohammed FM, Bae JO, Adesida I, Bogart KHA (2005) Investigation of surface treatment schemes on n-type GaN and Al0.20Ga0.80N. J Vacuum Sci Technol B 23:2538–2544
17. Pickrell GW, Armstrong AM, Allerman AA, Crawford MH, Cross KC, Glaser CE, Abate VM (2019) Regrown vertical GaN p-n diodes with low reverse leakage current. J Electron Mater 48:3311–3316
18. Freitas JA, Moore WJ, Shanabrook BV, Braga GCB, Lee SK, Park SS, Han JY, Koleske DD (2002) Donors in hydride-vapor-phase epitaxial GaN. J Cryst Growth 246:307–314
19. Fu K, Fu H, Huang X, Yang T-H, Cheng C-Y, Peri PR, Chen H, Montes J, Yang C, Zhou J, Deng X, Qi X, Smith DJ, Goodnick SM, Zhao Y (2020) Reverse leakage analysis for as-grown and regrown vertical GaN-on-GaN schottky barrier diodes. IEEE J Electron Dev Soc 8
20. Koblmueller G, Chu RM, Raman A, Mishra UK, Speck JS (2010) High-temperature molecular beam epitaxial growth of AlGaN/GaN on GaN templates with reduced interface impurity levels. J Appl Phys 107
21. Yang F, He L, Zheng Y, Li L, Chen Z, Zhou D, He Z, Yao Y, Ni Y, Shen Z, Zhang X, He L, Wu Z, Zhang B, Liu Y (2016) Influence of interface contamination on transport properties of two-dimensional electron gas in selective area growth AlGaN/GaN heterostructure. J Mater Sci-Mater Electron 27:9061–9066
22. Bermudez VM (1996) Study of oxygen chemisorption on the GaN(0001)-(1x1) surface. J Appl Phys 80:1190–1200
23. Khan MA, Kuznia JN, Olson DT, Kaplan R (1993) Deposition and surface characterization of high-quality single-crystal Gan layers. J Appl Phys 73:3108–3110
24. Gangopadhyay S, Schmidt T, Kruse C, Figge S, Hommel D, Falta J (2014) Surface oxidation of GaN(0001): nitrogen plasma-assisted cleaning for ultrahigh vacuum applications. J Vacuum Sci Technol A 32
25. Hao G, Zhang Y, Jin M, Feng C, Chen X, Chang B (2015) The effect of surface cleaning on quantum efficiency in AlGaN photocathode. Appl Surf Sci 324:590–593
26. Lee W, Ryou JH, Yoo D, Limb J, Dupuis RD, Hanser D, Preble NM, Williams KE (2007) Optimization of Fe doping at the regrowth interface of GaN for applications to III-nitride-based heterostructure field-effect transistors. Appl Phys Lett 90
27. Cordier Y, Azize M, Baron N, Chenot S, Tottereau O, Massies J (2007) AlGaN/GaN HEMTs regrown by MBE on epi-ready semi-insulating GaN-on-sapphire with inhibited interface contamination. J Cryst Growth 309:1–7
28. Cordier Y, Azize M, Baron N, Bougrioua Z, Chenot S, Tottereau O, Massies J, Gibart P (2008) Subsurface Fe-doped semi-insulating GaN templates for inhibition of regrowth interface pollution in AlGaN/GaN HEMT structures. J Cryst Growth 310:948–954

29. Liu JP, Ryou JH, Yoo D, Zhang Y, Limb J, Horne CA, Shen SC, Dupuis RD, Hanser AD, Preble EA, Evans KR (2008) III-nitride heterostructure field-effect transistors grown on semi-insulating GaN substrate without regrowth interface charge. Appl Phys Lett 92
30. Xu XP, Vaudo RP, Brandes GR (2003) Fabrication of GaN wafers for electronic and optoelectronic devices. Opt Mater 23:1–5
31. Schauler M, Eberhard F, Kirchner C, Schwegler V, Pelzmann A, Kamp M, Ebeling KJ, Bertram F, Riemann T, Christen J, Prystawko P, Leszczynski M, Grzegory I, Porowski S (1999) Dry etching of GaN substrates for high-quality homoepitaxy. Appl Phys Lett 74:1123–1125
32. Weyher JL, Kamler G, Nowak G, Borysiuk J, Lucznik B, Krysko M, Grzegory I, Porowski S (2005) Defects in GaN single crystals and homoepitaxial structures. J Cryst Growth 281:135–142
33. Chen KM, Wu YH, Yeh YH, Chiang CH, Chen KY, Lee WI (2011) Homoepitaxy on GaN substrate with various treatments by metalorganic vapor phase epitaxy. J Cryst Growth 318:454–459
34. Nagahama S, Iwasa N, Senoh M, Matsushita T, Sugimoto Y, Kiyoku H, Kozaki T, Sano M, Matsumura H, Umemoto H, Chocho K, Mukai T (2000) High-power and long-lifetime InGaN multi-quantum-well laser diodes grown on low-dislocation-density GaN substrates. Jap J Appl Phys Part 2-Lett 39:L647-L650.
35. Li X, Bohn PW, Coleman JJ (1999) Impurity states are the origin of yellow-band emission in GaN structures produced by epitaxial lateral overgrowth. Appl Phys Lett 75:4049–4051
36. Yang WC, Chen KY, Cheng KY, Wang YL, Hsieh KC, Cheng KY (2015) Dislocation reduction in GaN grown on nano-patterned templates. J Cryst Growth 425:141–144

Chapter 7
Progress in Optoelectronic Device Applications

7.1 Market Analysis

Gallium Nitride (GaN), as a new type of semiconductor material for the development of microelectronic and optoelectronic devices, along with SiC, diamond and other semiconductor materials, is hailed as the third-generation semiconductor material following the first-generation Ge, Si semiconductor materials, and the second-generation GaAs, InP compound semiconductor materials. As a representative of the third-generation semiconductor materials, GaN-based semiconductor materials are the core materials and basic devices of the emerging semiconductor optoelectronic industry. They not only bring about a revolution in digital storage technology in the IT industry, but also completely change the history of traditional human lighting, and will promote the development of communication technology. GaN-based semiconductor materials can be made into high-efficiency blue, green, purple, and white light-emitting diodes (LEDs) and lasers (LDs) due to their high internal and external quantum efficiency, high luminous efficiency, high thermal conductivity, high temperature resistance, radiation resistance, acid and alkali resistance, high strength and high hardness. They have a wide range of application prospects and research value in the field of optoelectronics.

7.1.1 Light Emitting Diode (LED) Application Field

The emergence of GaN-based blue and green LED products fundamentally solves the problem of color deficiency in light-emitting diodes, and the LED product have become an indispensable key device for full-color display. Blue and green LEDs have the characteristics of small size, cold light source, short response time, high luminous efficiency, explosion-proof, energy-saving, and long service life (Service life can reach more than 100,000 h). Therefore, GaN-based light-emitting diodes

K. Xu et al., *GaN Single Crystal Growth and Application*, Wide Bandgap Semiconductors, https://doi.org/10.1007/978-981-96-7572-2_7

have a lot to offer in fields such as large-screen color display, vehicles and traffic, multimedia imaging, LCD backlight source, optical fiber communication, satellite communication, and marine optical communication.

In terms of white light illumination, compared with traditional fluorescent lamps, LED lamps are more energy-saving and environmentally friendly, with greater brightness and longer service life. At present, LEDs have already entered thousands of households and are widely used in daily equipment such as mobile phones, cameras, cars, refrigerators, and street lights, including large-scale full-color display screens on the exterior of buildings, which are also made of LED pixels. According to statistics, LEDs have helped more than 1.5 billion people around the world say goodbye to the era without lights [1]. Since 1/4 of the world's electricity is used for lighting purposes, LEDs have largely replaced traditional light sources, greatly saving electricity and reducing environmental pollution. In the current situation where thermal power generation is still the main source of electricity, it is equivalent to reducing the emission of several hundred million tons of CO_2, SO_2 and other gases each year.

The 2014 Nobel Prize in Physics was awarded to Japanese scientists Isamu Akasaki, Hiroshi Amano, and Japanese-American scientist Shuji Nakamura) Fig. 7.1 to commend them for inventing the blue light-emitting diode (LED). The pioneering work of the three Nobel laureates in blue LEDs has sparked a revolution in lighting technology, profoundly affecting people's lives, and LEDs illuminate the twenty-first century. The Nobel Prize in Physics aims to reward breakthroughs in human understanding of natural laws and significant improvements in human life. The award of this Nobel Prize in Physics is well-deserved, and it fully reflects the affirmation of scientific and technological achievements with a clear chain from basic research to industrial application [2].

7.1.2 Applications of Laser Diodes (LD)

GaN-based lasers can cover a wide spectrum range, capable of producing lasers with blue, green, and ultraviolet wavelength radiation. Gallium nitride-based lasers have broad application prospects and huge market demand in fields such as increasing the optical storage density of information, laser printing, deep-sea communication, and atmospheric environment detection. Violet lasers can be used to manufacture high-capacity optical discs, with data storage space 20 times greater than that of Blu-ray discs. In addition, violet lasers can also be used for medical disinfection, fluorescence excitation light sources, and other applications. Blue lasers can be used with existing red lasers and frequency-doubled solid-state green lasers to achieve true color display, enabling widespread use of laser televisions. The deep sea has a window in the blue light range, and gallium nitride-based lasers can be used for deep-sea exploration and communication, which is of profound significance in the field of defense.

Fig. 7.1 Physics Prize winners (from left) Lifetime Professor Isamu Akasaki of Meijo University, Professor Hiroshi Amano of Nagoya University, Professor Shuji Nakamura of the University of California, Santa Barbara

This chapter will mainly introduce the basic principles of GaN-based LEDs and GaN-based LDs, chip structure, quantum efficiency improvement, problems encountered in the preparation process, and solutions, etc.

7.2 Light Emitting Diode

7.2.1 Introduction to GaN-Based LED

LED is a type of semiconductor diode, as shown in Fig. 7.2, its principle of light emission is: When current flows through the LED, electrons and holes recombine internally to release the excess energy in the form of photons, also known as electroluminescence.

1. GaN p–n junction

Using different doping processes, through diffusion, GaN p-type semiconductors and GaN n-type semiconductors are made on the same semiconductor substrate, and the space charge region formed at their interface is called a p–n junction. The p–n junction has unidirectional conductivity, which is a characteristic utilized by many devices in electronic technology, such as semiconductor diodes and bipolar transistors.

For GaN p–n junction diodes, the main dopant for n-type doping is Si, with an electron concentration reaching $10^{15} \sim 10^{20}$ cm^{-3}, and a mobility exceeding 300 cm^2/

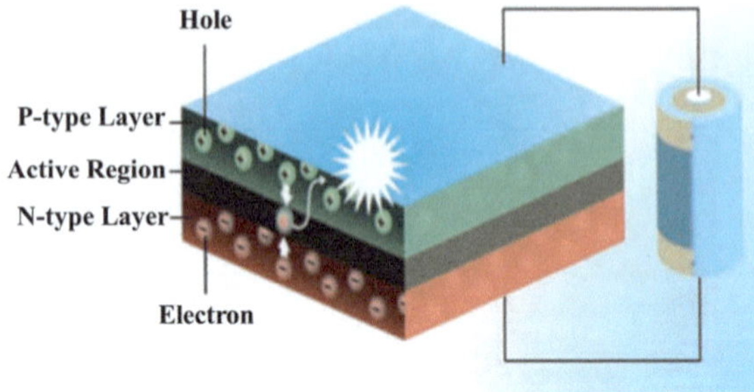

Fig. 7.2 Schematic diagram of LED principle [3]

(V·s) at room temperature [4]. Typical p-type doping is achieved using Mg as the acceptor dopant in the H_2 atmosphere. When Mg-doped GaN epitaxy is grown by MOCVD process, because Mg has a larger covalent radius compared to Ga, and Mg is a deep acceptor, the carrier gas H_2, the reactants NH_3 and the H decomposed from the organic source combine with Mg to form Mg-H complexes, passivating the Mg acceptors in the GaN epitaxial layer. Therefore, the resistivity of GaN directly doped with Mg is as high as 10^8 Ω·cm, showing as n-type. The immature development of p-type GaN severely affects the performance of GaN-based optoelectronic devices such as light-emitting diodes (LEDs) and laser diodes (LDs), restricting the research and development of such devices. It has been found through research that low-resistance p-type GaN can be obtained by treating GaN doped with Mg with low-energy electron beam irradiation (LEEBI). Low-energy electron beam irradiation generates a large number of electron–hole pairs in GaN, these electrons, holes and their excitons tend to gather in the $H^+ - Mg^-$ donor–acceptor pair [5]. The gap ion H^+ is neutralized by electrons to form neutral H atoms, some of which diffuse to the surface of the GaN film and combine with other neutral H atoms to form H_2 molecules and escape, leaving a large number of activated acceptors. This can reduce the resistivity by several orders of magnitude. The low-resistance Mg-doped p-GaN obtained through LEEBI depends on the penetration depth of the LEEBI, and only a very thin area on the surface of the GaN film shows strong p-type characteristics. Therefore, the p-type area of the p–n junction in commercial GaN wafers is very thin, usually on the order of tens to hundreds of nanometers.

2. GaN Quantum Well Structure

A quantum well (QW) is a structure grown in sandwich style from two different semiconductor materials, using a wider bandgap layer to sandwich a narrow and extremely thin layer. The potential of the narrower bandgap layer is lower than that of the surrounding (wider bandgap layer), thus forming a potential well. A potential well for electrons or holes with a significant quantum confinement effect is called a QW. The most basic feature of a QW is that the carrier wave function is localized in one dimension due to the limitation of the QW width (a QW can only be formed when the well width is small enough). In the QW, the thickness of the source layer is only within the average free path of the electron, the well wall has a strong confinement effect, causing the carrier to have a two-dimensional degree of freedom only in the plane parallel to the well wall, and the conduction band and valence band are split into sub-bands in the vertical direction. In a multilayer structure formed by alternating growth of thin layers of two different semiconductor materials, if the barrier layer is thick enough so that the coupling between the carrier wave functions in adjacent potential wells is very small, the multilayer structure will form many separated QWs, called multiple quantum wells (MQWs). If the barrier layer is very thin and the coupling between adjacent wells is very strong, the discrete energy levels in each QW will extend into energy bands (microbands), the width and position of the energy bands are related to the depth, width of the potential well and the thickness of the barrier, and such a multilayer structure is called a superlattice. Structures with superlattice characteristics are sometimes referred to as coupled MQWs.

3. Principle of GaN-based LED Emission

In the GaN-pn junction, the carrier (electron) migration rate of the n-GaN layer is much higher than that of the p-GaN layer (holes). According to the Fermi energy level, the electron energy level in the n-region is higher than the hole energy level in the p-region. Under normal conditions, the two cannot naturally recombine because of the barrier layer of the pn junction (the built-in electric field formed by the space charge region). The space charge region is formed when the carriers in the n-layer and p-layer drift towards each other, and they will first gather at the junction of the p-region and n-region. At the junction of the region, an internal electric field E that hinders the drift of carriers is formed in the junction area. When a positive voltage is applied to the pn junction, the space charge region narrows, making the diffusion movement of carriers stronger than the drift movement, causing the valence band (low energy) holes in the p region to be injected into the n region, and the conduction band (high energy) electrons in the n region to be injected into the p region. Therefore, near the p–n junction and biased towards the p region, valence band electrons and conduction band holes recombine, releasing excess energy and manifesting in the form of light. The reason why the luminescent region is close to the p region is that the carrier migration rate in the n region is much higher than that in the p region, and the diffusion of carriers in the n region is stronger than that in the p region.

The role of the quantum well is to confine the carriers within the active layer, reducing the range of activity for minority carriers and thereby increasing carrier concentration and enhancing carrier recombination, achieving the effect of increasing the light intensity. Meanwhile, the band width of the quantum well determines the final wavelength of light emitted by the LED device.

7.2.2 History of GaN-Based LED Development

Figure 7.3 shows the development history of LEDs [6]. In 1963, Jr.N. Holonyak and S.F. Bevacqua made a p–n junction using Ga(As1-xPx), achieving red LED lighting [7]. In 1965, D.G. Thomas, J.J. Hopfield, and C.J. Frosch found that doping N, Zn-0 impurities in GaP can effectively improve the luminescent characteristics of the

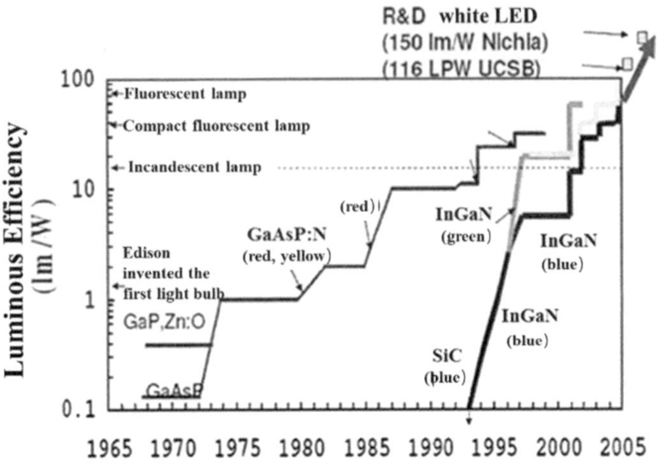

Fig. 7.3 History of LED development [6]

crystal [8]. In 1968, F.A. Ponce and D.P. Bour reported that doping N elements in GaP can generate green light. In 1969, H.M. Manasevit and W.I. Simpson prepared GaN compounds using MOCVD growth [9]. In 1971, J.I. Pankove, E.A. Miller, D. Richman, J.E. Berkeyheiser et al. observed a luminescent peak at 2.6 eV in the electroluminescent spectrum of Zn-doped GaN [10]. The same year, they made the first GaN-based blue diode [11]. In 1978, R.D. Dupuis and P.D. Dapkus prepared GaAlAs/GaAs heterojunction lasers using MOCVD [12]. In 1983, J. Nishizawa and M. Koike made a red LED using AlGaAs. In 1985, J.S. Yuan, C.C. Hsu, R.M. Cohen, and G.B. Stringfellow et al.made a four-color visible light LED using AlGalnP [13]. In the early development of GaN, GaN-based blue LEDs were mainly achieved by doping Zn [14]. The emission wavelength of such LEDs is generally between 485 and 520 nm. In 1992, GaN-based LEDs made a breakthrough, low-resistance p-n GaN LEDs were successfully made, with an efficiency of 1.5% [15]. In 1994, Shuji Nakamura, Takashi Mukai, and Masayuki Senoh made GaInN/AlGaN double heterojunction LEDs. In the following years, p-i-n type GaN LEDs, AlGaN/InGaN/GaN LEDs were successively made, and the external quantum efficiency had reached 3.9% at 20 mA. In the 1990s, GaN-based quantum wells, due to their insensitivity to temperature and injection current, were widely used in the field of short-wave optoelectronic devices [16–18]. Shuji Nakamura et al. also successfully made GaN-based single quantum wells that can emit blue, green, and yellow light. In 1997, K.H. Huang and T.P. Chen reported on multiple quantum wells [19]. By this time, the external quantum efficiencies of blue and green MQW LEDs had reached 10% and 7% respectively. By 2000, the efficiency of LEDs using flip-chip technology had reached as high as 20% [20]. Entering the twenty-first century, the development of blue LEDs reached its peak. White light illumination was achieved through red, green, and blue primary colors, and the field of human lighting entered a new stage.

7.2.3 LED Substrate Materials and Epitaxial Growth Techniques

GaN-based materials refer to GaN, AlN, InN, and their multicomponent alloy materials, such as $In_x Ga_{1-x} N$ and $Al_x Ga_{1-x} N$ materials. The growth techniques of GaN-based materials mainly include: Molecular Beam Epitaxy (MBE), Metal Organic Chemical Vapor Deposition (MOCVD), and Hydride Vapor Phase Epitaxy (HVPE). Additionally, the development of material growth technology has led to the emergence of Epitaxial Lateral Overgrowth (ELOG) technology [21].

MBE technology is a process in which gallium (Ga), nitrogen (N), or other atoms are sprayed onto the heated substrate surface in the form of atomic or molecular beams under high vacuum conditions. After complex physical and chemical processes on the substrate surface, GaN materials are formed. The advantages of this technology are low growth temperature, high purity single crystal can be obtained, and the growth thickness of ultra-fine structures can be controlled. The disadvantage is that the growth rate is slow, which cannot meet the needs of large-scale production.

The main process of MOCVD epitaxial growth of thin films is to transport gaseous substances to the substrate surface, heat the substrate to enable Group III metal organics to react with NH_3 to form a thin film with a certain crystal structure. This method makes it easy to control the thickness of the crystal thin film and ensures good uniformity in large-area growth, making it an ideal device for producing LEDs, LDs, and HEMTs. In addition, it can also grow superlattices, quantum wells, and other microstructure materials. Most of the key technologies and patents for MOCVD equipment manufacturing are held by foreign companies Aixtron, Veeco, and Nichia.

HVPE technology involves passing HCl into a Ga boat to react and generate GaCl, which is then transported by carrier gas H_2 or N_2 to the vicinity of the heated substrate to react with NH_3 to generate GaN and deposit on the substrate. The advantages of this technology are its simplicity, fast growth rate, up to 300 μm/h, making it the most commonly used and practical method for growing thick GaN films or even GaN ingots, and favored by many GaN researchers [22]. However, there are some disadvantages such as difficulty in precisely controlling film thickness, corrosiveness of the reaction gas affecting the purity of the GaN material, and poor repeatability of epitaxial growth. Currently, companies that can produce GaN substrates internationally basically use HVPE technology. The method is to first grow a thick GaN single crystal on a sapphire substrate using HVPE equipment, and then obtain a GaN self-supporting substrate using a laser lift-off method.

The substrate material has a significant impact on the quality of the epitaxially grown crystal material and the performance of the LED chip. Therefore, the choice of substrate is crucial in the epitaxial growth process. In LED In the chip epitaxial growth process, the selection of the substrate usually requires the crystal structure of the substrate material to be consistent with the epitaxial material, the lattice constants to match, the thermal expansion coefficients to match, and the substrate material to have a high thermal conductivity and electrical conductivity, be easy to make large sizes, and have affordable prices. The substrate materials currently used for

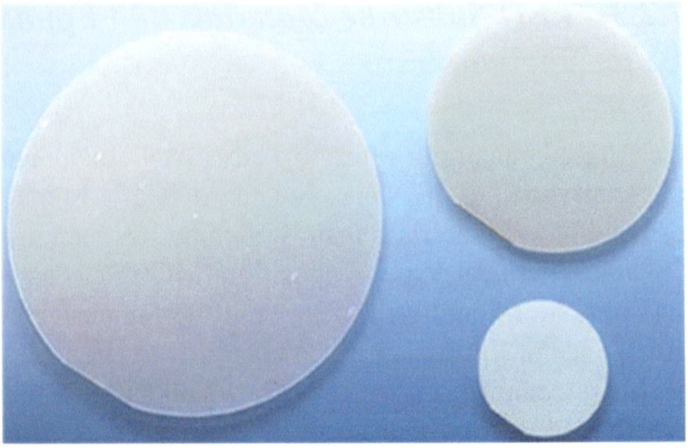

Fig. 7.4 LED epitaxial wafers grown on sapphire substrates of different sizes [21]

epitaxial growth of LED chips mainly include sapphire substrates (Sapphire), silicon carbide (SiC) substrates, silicon (Silicon) substrates, and gallium nitride (GaN) homogeneous substrates. The existing commercialized high-light-efficiency high-power GaN-based blue LED chips are usually grown on sapphire or SiC substrates. Because the sapphire substrate is very stable at the high temperature required for the epitaxial growth of GaN material, and the preparation process is mature and the cost is relatively low, sapphire material is widely used in LED chip substrate materials. Japan's Nichia Corporation has monopolized the patent technology for epitaxial growth of GaN-based LEDs on sapphire substrates. Figure 7.4 shows LED epitaxial wafers grown on 2-inch, 4-inch, and 6-inch sapphire substrates.

From the perspective of lattice matching and conductivity, and thermal conductivity, sapphire material is not the most ideal substrate for heteroepitaxial growth of GaN material. Compared with sapphire substrates, silicon carbide (SiC) has very good physical properties, good high-temperature stability, a crystal structure has multiple configurations such as 4H, 6H, 3C, and thermal conductivity reaches 3.7 $W\cdot(cm\cdot K)^{-1}$, which is 10 times that of sapphire, and is more suitable for high-power LED substrates. The lattice constant of 4H-SiC commonly used for LED epitaxial substrates is only about 3% different from GaN. At the same time, silicon carbide is a semiconductor material, and n-type material can be achieved by doping. This material has a large bandgap width and is transparent to visible light. Therefore, SiC is very suitable for vertical structure LEDs without having to peel off the substrate [3]. However, high-quality SiC substrate materials are difficult to obtain. At present, the suppliers of SiC substrate materials on the market are mainly American Cree Company and a few Japanese companies. The growth of SiC substrates mainly uses the PVT method, which requires a high growth temperature, a small growth rate, and it is not easy to control the microtube defects in the crystal, so the price of high-quality SiC substrates is high. The State Key Laboratory of Crystal Materials at Shandong University has achieved a series of results in the research and development of SiC single crystal growth technology [23]. At present, the mainstream LED manufacturers only have the American Cree Company using SiC as the epitaxial substrate of LEDs. The company is the best in the industry in terms of the reported electro-optical conversion efficiency, and the target reached in the laboratory in 2014 exceeded 300 $lm \cdot W^{-1}$ [24].

Compared with sapphire substrates and SiC substrates, silicon substrates have the advantages of low cost, large size, and mature silicon processing technology. They can use three-dimensional integration technology to integrate LED driving circuits, chips, heat dissipation structures, electrostatic protection, etc., on large-size silicon, improving the integration of silicon substrate LED devices and application systems. In recent years, the research on epitaxial growth of GaN thin films on silicon substrates has received widespread attention. The research group of Professor Jiang Fengyi from Nanchang University in China has made progress in silicon substrate LED. The technology has made a breakthrough, successfully developing silicon substrate LED chips [25]. Currently, the luminous efficiency of silicon substrate LED chips is lower than that of sapphire substrate LED chips and SiC substrate LED chips. Once a breakthrough is made in the luminous efficiency of GaN materials grown epitaxially on silicon substrates, it will greatly expand the application field of GaN-based LED devices.

Generally speaking, homoepitaxy is the best choice for device epitaxial growth [3]. Traditional homoepitaxial devices with substrates made of materials such as GaAs and InP have reliable performance. The defect density of homogeneous substrate materials is low, there is no lattice mismatch and thermal mismatch, and the epitaxial materials have good quality, thereby achieving better device performance. The epitaxial growth of GaN-based thin film materials on GaN homogeneous substrates can greatly improve the crystal quality of the epitaxial layer, reduce the dislocation density, and provide a basis for further improving the luminous efficiency and reliability of LED devices. The key to GaN homoepitaxy is the growth of GaN bulk materials. Due to the very high vapor pressure of equilibrium GaN and N_2, as shown in Fig. 7.5, it is very difficult to prepare GaN crystals using the usual Bridgman method or Czochralski method for growing GaAs. The methods for growing bulk material GaN mainly include high-pressure nitrogen solution method (HPNS), sodium flux method (Na flux), hydride vapor phase epitaxy (HVPE), ammonothermal method (Ammonothermal), etc. (as shown in Table 7.1), the dislocation density of GaN bulk materials grown by the currently commonly used HVPE method is usually 10^6 cm^{-2}.

Currently, the market reports that American company Soraa and Korean company Samsung are developing GaN bulk materials for LEDs. The MR16 type LED prepared by Soraa using a GaN substrate has a light output power that is 5 to 10 times higher than traditional LEDs based on sapphire substrates [27], due to the substrate's dislocation density being three orders of magnitude lower, resulting in less device heating and simpler packaging technology. In recent years, non-polar/semi-polar GaN epitaxy has also received a lot of attention, mainly used to eliminate or weaken the built-in polarization field in polar GaN semiconductors. The next section will focus on this issue. Although using bulk material GaN as a substrate for epitaxial growth of LEDs can achieve good device performance, the high price still hinders the industrialization of GaN substrates as LED epitaxial substrates in practical applications.

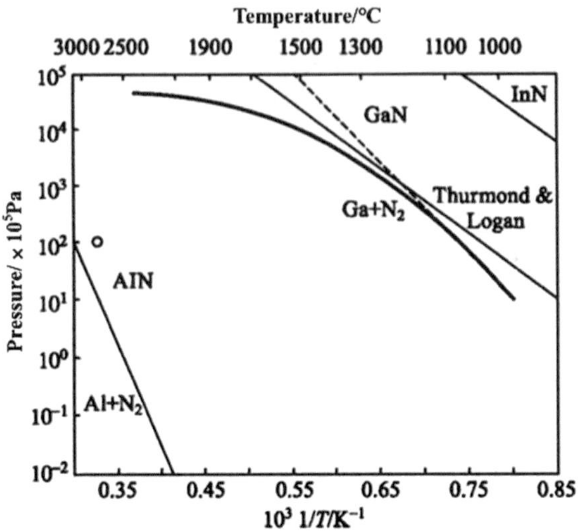

Fig. 7.5 Equilibrium vapor pressure of AlN, GaN, InN [20]

Table 7.1 Comparison of GaN bulk material preparation methods [26]

Growth method	Growth pressure/ GPε	Growth temperature/°C	Growth rate / $(\mu m \cdot h^{-1})$	Dislocation density/cm^{-2}
High-pressure nitrogen solution method	1–2	–1700	1–3	10^2
Na-flux method	3–5	800	10–40	10^2–10^4
Hydride Vapor Phase Epitaxy	10^{-4}	1020–1050	100–200	10^4–10^6
Ammonothermal method	100–600	500–750	1–30	10^3

In addition, many scholars have studied the epitaxial growth of GaN materials on substrates such as AlN, ZnO, MgO, and $MgAl_2O_4$. Table 7.2 lists the characteristics of the mainstream substrate materials currently used to manufacture blue light LED chips. In addition to the above common substrate materials, nitride epitaxial substrate materials that can be seen in the literature also include $LiGaO_2$, $MgAlO_3$, and metal substrates, but these substrate materials are rarely used in actual production. In recent years, researchers have explored the possibility of epitaxial growth of GaN materials on glass or metal substrates, hoping to further reduce the cost of LEDs, but the performance is not yet ideal [21].

Table 7.2 Properties of Blu-ray LED chip substrate materials [21]

Substrate material	Lattice constant/nm	Coefficient of thermal expansion/K^{-1}	Thermal conductivity/ [w/(cm·K)]
Sapphire (Hexagonal)	$a = 0.4785$ $c = 1.2991$	5.0×10^{-6} 6.6×10^{-6}	0.35
SiC (Hexagonal)	$a = 0.3073$ $c = 1.5118$	4.3×10^{-6} 4.7×10^{-6}	4.9
Si (Cubic)	$a = 0.54301$	3.59×10^{-6}	1.5
GaN (Hexagonal)	$a = 0.3189$ $c = 0.5185$	5.59×10^{-6} 7.75×10^{-6}	2.2

7.2.4 LED Chip Structure

1. Horizontal Structure

The horizontal structure LED chip refers to the positive and negative electrodes of the GaN-based LED chip being on the same side of the epitaxy. Figure 7.6 is a schematic diagram of a horizontal structure LED chip. In the chip manufacturing process, the p-type electrode is directly evaporated on the surface of the p-GaN epitaxial layer to complete the ohmic contact metal, while the n-type electrode uses ICP to etch away the epitaxial wafer until the n-GaN material is exposed, and then the ohmic contact metal is evaporated on the n-GaN [21].

The current of this structure flows laterally, and a current aggregation effect will occur at the chip step; a part of the front light is absorbed by the P-type ohmic contact electrode, N-type ohmic contact electrode, and electrode pad. As the thermal conductivity of sapphire material is low, which affects the thermal reliability of the device. For high-power LED chips, it will limit the further increase of the injection current.

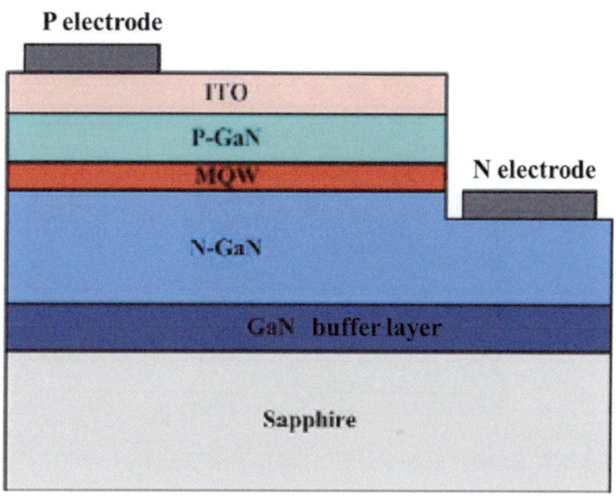

Fig. 7.6 Horizontal structure of the LED chip [21]

2. Inverted Structure

The inverted structure LED chip is to invert the horizontal structure LED chip, the p-type electrode uses a metal film with a high reflectivity, so that the light is emitted from the sapphire substrate, avoiding the absorption of light by the p-type electrode metal [21]. The inverted technology can reduce the thermal resistance and improve the heat dissipation performance of the chip by directly contacting the electrode (or micro bumps) with the Si substrate of the package. Wierer et al. first proposed the AlGaInN inverted structure LED chip in 2001. This technology first bonds the LED upside down to a metal substrate with good thermal conductivity, so that the light is emitted from the sapphire substrate surface. The results show that the light extraction efficiency of the inverted structure LED chip is 1.6 times that of the ordinary horizontal structure LED chip. The refractive index of sapphire is 1.78, which is equivalent to adding a refractive index buffer layer between the high refractive index GaN and air, increasing the chance of photons emitting into the air, and at the same time, the inverted structure LED chip avoids the absorption of light by the electrodes, transparent conductive layer, etc. of the horizontal structure LED chip [28]. In 2006, Shchekin et al. used laser lift-off technology to remove the sapphire substrate and thin the intrinsic GaN material under n-GaN to make a thin-film inverted structure LED chip (Thin-film Flip-chip), making the light power of the LED chip twice that of the ordinary inverted structure LED chip. Under a current drive of 350 mA, the external quantum efficiency of the thin-film inverted structure LED chip reached 36% [29]. Figure 7.7 is a schematic diagram of the inverted structure LED chip manufactured by Lumileds. In China, Crystal Science and Technology Co., Ltd. uses gold bump technology to bond high-power blue LED chips upside down to a silicon substrate. Under a current drive of 350 mA, the light power output by the inverted structure LED chip reaches 370mW.

3. Vertical Structure

The vertical structure LED chip [21], as shown in Fig. 7.8. The American company Cree uses SiC as a substrate to manufacture GaN-based LED chips with a vertical electrode structure, which emit light from the n-face, effectively solving the problems of heat dissipation and light blocking. Moreover, vertical conductivity is beneficial

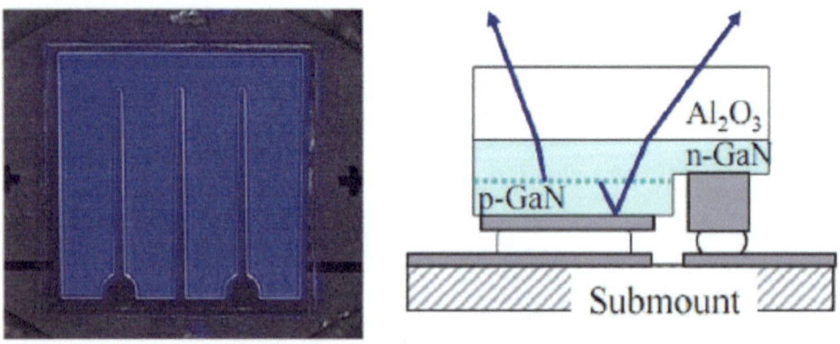

Fig. 7.7 Inverted structure LED chip produced by Lumileds [21]

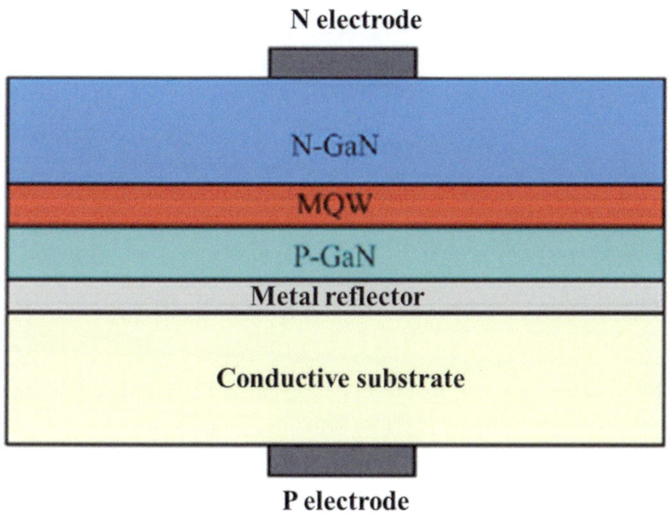

Fig. 7.8 Schematic diagram of the vertical structure LED chip [21]

for carrier injection, improving the recombination efficiency of carriers. However, because SiC substrates are relatively expensive and more difficult to process, the price of LED chips manufactured on SiC substrates is higher. Currently, the academic and industrial communities mainly use wafer bonding technology and laser lift-off technology to transfer the GaN epitaxial wafer from the sapphire substrate to a substrate material with good electrical and thermal conductivity (such as Si, ZnO substrates), with the device electrodes vertically distributed above and below, thereby completely solving a series of problems such as heat dissipation, uneven current distribution, reliability, etc., caused by the planar distribution of electrodes and lateral injection of current in the normal and inverted structure GaN-based LED chips [30].

Wafer bonding technology combines the LED epitaxial wafer grown on the sapphire substrate with other materials with good conductivity and heat dissipation performance. The key parameters of the bonding process are the temperature and pressure during bonding. Vertical structure LED chips are often prepared using metal eutectic bonding. Common eutectic bonding includes Au–Sn, Au–Si, Au-In, Pb–Sn, Au–Ge, and Pd-In metal alloys. When using metal eutectic bonding, the surface roughness and wafer warpage must be controlled. The bonding temperature required for metal eutectic alloys is usually distributed between 250 and 390 °C [21].

Laser lift-off technology uses high-power lasers to peel off the sapphire substrate from the epitaxial layer grown on it. In the experimental production process, the LED epitaxial wafer is first bonded to a substrate material with good conductivity and thermal conductivity, and then the bottom surface of the sapphire substrate is irradiated with a KrF pulsed excimer laser. Because the sapphire substrate is transparent to the KrF pulsed excimer laser, and the GaN material absorbs the KrF pulsed excimer laser, the GaN material at the interface between the sapphire and GaN are thermally

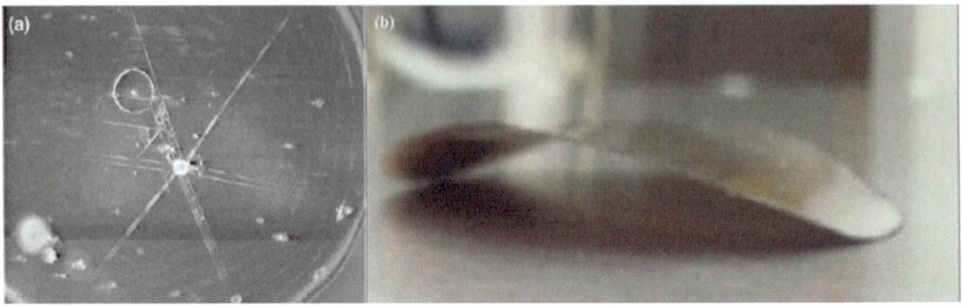

Fig. 7.9 Defects and warping caused by laser lift-off process [21, 34, 33]

decompose due to the absorption of laser energy. Finally, heating causes the sapphire substrate and the GaN epitaxial layer to separate, thereby achieving the peeling off of the sapphire substrate [31, 32]. In the laser lift-off process, whether the GaN epitaxial layer can be completely transferred to other substrate materials depends on the peel-off process parameters and the physical and geometric parameters of the GaN epitaxial layer. If the peel-off process parameters are used improperly, there will be a large thermal stress at the interface between the GaN epitaxial layer and the sapphire substrate, which will cause serious warping, cracking or even breaking of the LED wafer after the peeling process is completed [33], as shown in Fig. 7.9.

4. AC LED chips

Conventional DC-driven LED chips cannot be used directly under high-voltage AC, and must be stepped down by a transformer and switching power supply, converting alternating current (AC) into direct current (DC), DC), then transformed into a constant current source to drive the LED light source. The lifespan of a switching power supply is generally only 20,000 h, but the lifespan of an LED chip can reach 50,000–100,000 h. Therefore, the LED light source driven by direct current usually fails due to power supply failure, which becomes a bottleneck restricting the lifespan of LED light source products and cannot reflect the long lifespan of LEDs [21].

Compared to DC-LED chips, AC-LED chips do not require AC/DC conversion and can directly utilize 220 V (or 110 V) alternating current for operation. Currently, Seoul Semiconductor in South Korea, the Industrial Technology Research Institute in Taiwan, China, and Epistar Corporation, as well as 3N (III-N Technology) in the United States, are leading in the research of AC-LED chip technology. In 2008, the Industrial Technology Research Institute in Taiwan and its partner, Epistar Corporation, proposed a practical application system plan and developed AC LED products that can work directly under 110 V AC. Seoul Semiconductor began mass production of AC LED products with a luminous efficacy of 100 lm/W in the first quarter of 2010. Jung et al. integrated inverted structure LED chips on a silicon substrate with a bridge rectifier circuit using packaging technology to form an AC-LED module. When the input power is 8.3W, the light output power of the AC-LED module reaches 830 mW [35]. Onushkin et al. used microfabrication technology to manufacture an

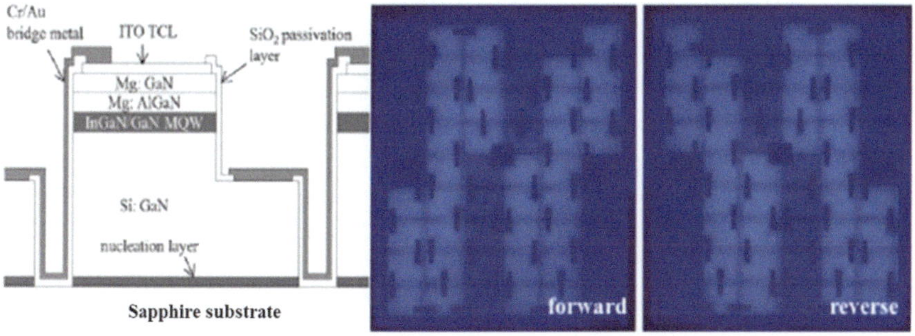

Fig. 7.10 AC-LED chip with integrated microchip array on a single chip [37]

AC-LED chip with dimensions of 4.2 mm × 4.2 mm and coated it with yellow phosphor to emit white light. Under 220 V AC drive, the consumed power is 4W, and the light output is 320 lm [36]. Yen et al. integrated 34 microchips on a single chip to make an AC-LED chip. The light output power under 220 V AC power drive reaches 388 mW [37], as shown in Fig. 7.10.

The production of AC-LEDs is achieved by integrating multiple microchips on a single chip through microfabrication technology, which is high-power single-chip LED technology, and using a staggered matrix arrangement process to form a bridge circuit. The key technologies in the manufacturing process of AC-LEDs include low-damage ICP deep trench etching technology for isolating LED microchips from each other, the production of high mechanical strength interconnect bridges for connecting LED microchips, and the electrode structure design of integrated array microchips. Although AC-LED has its unique advantages, its manufacturing process is more complex than that of DC-driven LED chips. Therefore, the large-scale production of AC-LEDs needs to break through the above key technologies, reduce the manufacturing cost of AC-LEDs, and improve the reliability of AC-LED chips [21].

7.2.5 LED Chip Light Extraction Efficiency

The factors that affect the luminous efficiency of LED chips are mainly two: the efficiency of electrons converting into photons, known as internal quantum efficiency; and the photons generated from the active layer. The efficiency of light emission within an LED is known as light extraction efficiency, and together they determine the luminous efficiency of the LED chip. The internal quantum efficiency is usually closely related to the form of the carrier, transport (scattering), and the recombination mechanism at a certain injection current level, so it strongly depends on the band structure of the material and working conditions. The light extraction efficiency mainly depends on the effects received during the photon transmission process, especially interface scattering and refraction. Due to the significant difference in refractive index between GaN material and air and sapphire substrate materials (the refractive index of GaN is 2.5, air is 1, and sapphire is 1.78), when the light generated by the active layer of the LED chip is emitted from the front of the chip to the air, total reflection occurs at the interface of GaN material and air, with a total reflection angle of 23.6 degrees. Photons with an incident angle less than the total reflection angle can be emitted to the air through the top of the chip, while photons with an incident angle greater than the total reflection angle are reflected back into the LED chip, oscillating inside the LED chip, and finally absorbed by the active layer and electrode or emitted to free space through the side and bottom of the chip. When the light generated by the active layer of the LED chip is emitted from the back of the chip, it may undergo total reflection at the interface of GaN material and sapphire substrate material and cannot be extracted, resulting in light loss. If the light emission from the side and bottom of the chip is ignored, only about 8% of the light generated by the active layer of the chip can be extracted from the front of the chip. Using various technologies to break the total reflection interface inside the LED chip is an effective way to improve the light extraction efficiency of the LED chip. The main methods currently used are as follows: patterned sapphire substrate (micro-pattern and nano-pattern) technology, surface roughening technology, patterned ITO technology, photonic crystal technology, total internal reflection mirror technology, shaped chip technology, and surface microstructure technology [21].

1. Patterned Sapphire Substrate Technology

Sapphire material is the most common substrate material for heteroepitaxial growth of GaN-based LED chips. Due to the large lattice constant mismatch and thermal expansion coefficient mismatch between the sapphire substrate material and GaN epitaxial layer, the GaN epitaxial layer grown on the sapphire substrate using MOCVD epitaxial growth technology has large stress and crystal defects, which reduces the performance of the LED chip [21]. Patterned sapphire substrate technology is used to create micro or nano periodic microstructure patterns on the sapphire substrate, which can improve the crystal quality of the GaN epitaxial layer, reduce defects, and improve the internal quantum efficiency of the LED chip. Moreover, the concave-convex patterns on the sapphire substrate will scatter or refract the incident light, thereby increasing the light extraction efficiency of the LED chip [38, 39]. Gao et al. used a wet etching method to form microstructure and nanostructure patterns on the sapphire substrate, and analyzed the effect of pattern size on the LED. The impact of chip light extraction efficiency shows that forming microstructure patterns on a sapphire substrate can increase the light power output of the LED chip by 29%, and nanostructure patterns can increase the light output power of the LED chip by 48% [40]. Chang et al. performed a thermal reflow process on the photoresist mask to form a hemispherical shape on the photoresist, then used dry etching technology to transfer the pattern to the sapphire substrate, thereby forming a periodic hemispherical pattern on the sapphire substrate. After growing the LED epitaxial layer on it and making it into a chip using microfabrication technology, it was found that the light power of the LED chip increased by 44% [41]. Su et al. used nanoimprint technology to form periodic nanostructure patterns on the sapphire substrate and studied the impact of the aspect ratio of the pattern on chip performance. The experimental results show that when the aspect ratio of the pattern is 2.0 and 2.5, the light power of the chip increased by 11% and 27% respectively, as shown in Fig. 7.11 [42].

To form patterns on a sapphire substrate, there are usually two methods: wet etching and dry etching. When using wet etching technology to pattern the sapphire substrate, silicon dioxide (SiO_2) or silicon nitride (SiN_x) materials are generally used as masks, and a mixture of H_2SO_4 and H_3PO_4 (H_2SO_4: H_3PO_4 = 3:1) is used to etch the sapphire substrate at high temperatures. Because wet etching technology is anisotropic, the pattern formed on the sapphire substrate using wet etching technology is related to the crystal structure of the sapphire material, and its pattern is a triangular pyramid structure. When using dry etching technology to form patterns on sapphire, inductively coupled plasma based on BCl_3 is usually used to etch the sapphire. The

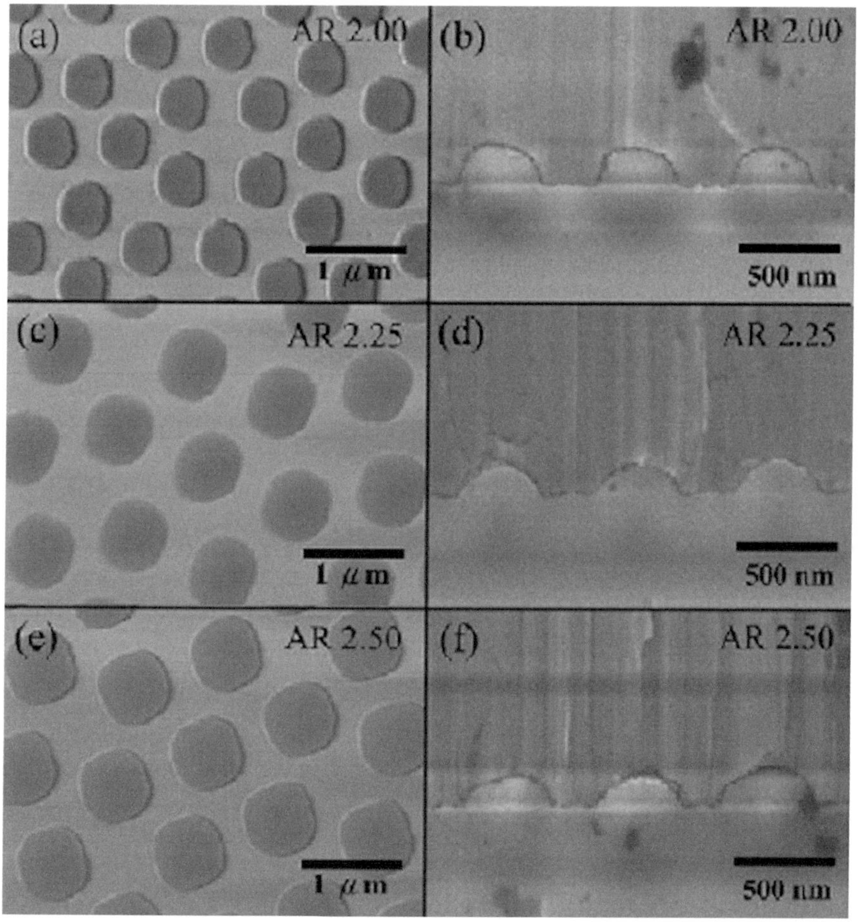

Fig. 7.11 Sapphire substrate with different aspect ratios made using nanoimprint technology [42]

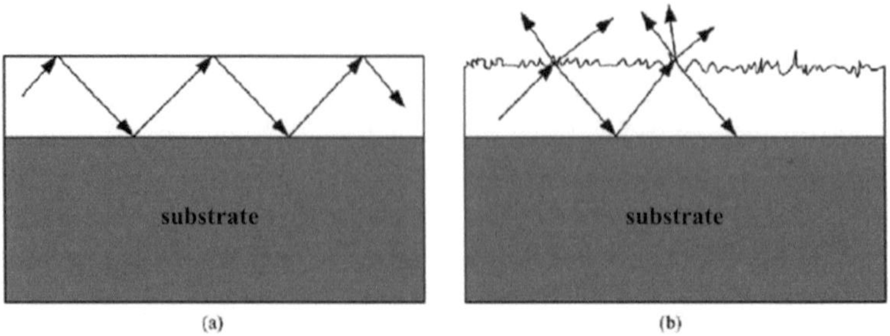

Fig. 7.12 Principal diagram of extracting photons using surface roughening technology [21]

B atoms in the etching can combine with the O atoms in the sapphire (Al_2O_3) to form volatile BOClx etching products [21].

2. Surface roughening technology

The technology of surface roughening mainly involves roughening the p-GaN layer, ITO transparent conductive layer, and n-GaN surface of the horizontal structure LED chip and the vertical structure chip. The roughened surface can effectively reduce the total reflection effect that occurs during the transmission of photons in the chip, causing the photons to scatter at the roughened interface and be extracted [21]. Figure 7.12 shows the principle diagram of extracting photons using surface roughening technology.

For horizontal structure LED chips, the p-GaN surface or ITO transparent conductive layer can be roughened to enhance the light extraction efficiency of the LED chip [43, 44]. Yang et al. have formed a porous nanostructure on the p-GaN surface using photoelectrochemical etching (PEC), increasing the light power of the LED chip by 26% [45]. Pan et al. have utilized photolithography and dry etching processes to roughen the ITO transparent conductive layer of horizontal structure LED chips, leading to a 16% improvement in light power compared to regular chips without roughening [46]. For vertical structure LED chips, since the top layer is n-GaN with a thickness of about 2-3 μm, the n-GaN layer is usually roughened [47, 48]. Fujii et al. have employed photoelectrochemical etching technology to create a roughened n-GaN layer on a vertical structure LED chip, resulting in an increase in light output power by 2–3 times[49]. Lee et al. proposed in their research that after forming a roughened diffuse reflection surface on the lower surface of the p-GaN layer of the vertical chip and then bonding it with a silicon substrate, a roughened diffuse transmission surface is formed on the upper surface of the chip, making the external quantum efficiency of the LED chip reach 40%, a 136% increase compared to the LED chip without roughening [50]. Horng et al. have used a two-sided roughening technique to improve the luminous efficiency of the LED chip, that is, roughening both the p-GaN surface and the n-GaN surface, making the light efficiency of the LED chip using the two-sided roughening technique 77% higher than that of the LED chip with single-sided roughening [51]. Figure 7.13 displays an SEM image

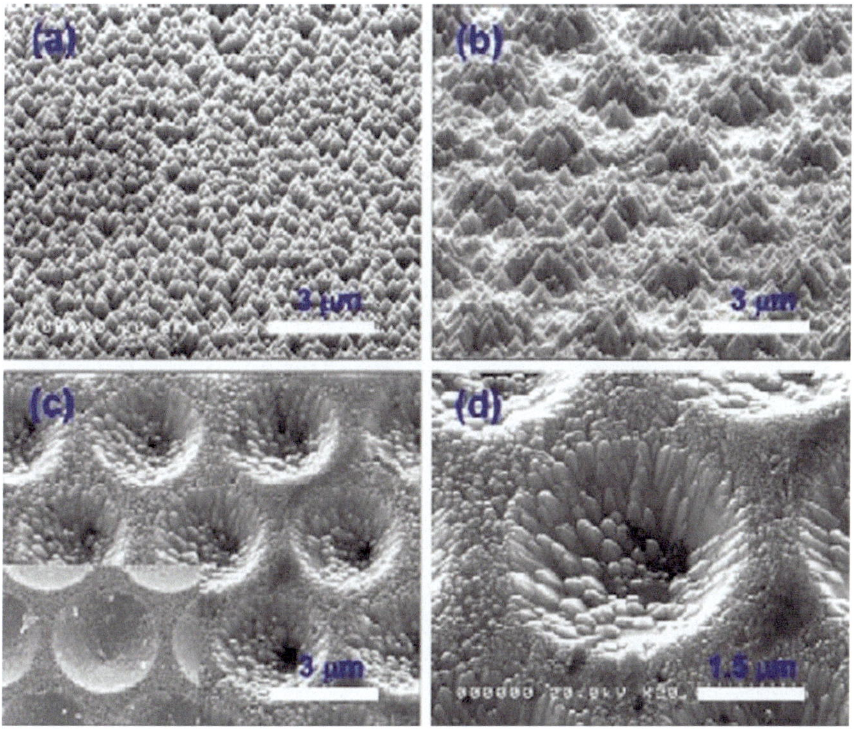

Fig. 7.13 SEM image after roughening the n-GaN surface of the vertical structure chip [51]

illustrating how n-GaN surfaces appear after undergoing roughening treatment for vertical structure LED chips [52].

In China, Dong Zhijiang et al. from Wuhan Deyuan Optoelectronics utilized the ICP dry etching process and natural photolithography technology to roughen the ITO layer on the surface of the LED chip. The results show that the light intensity of the GaN-based LED chip with roughened ITO surface has increased by 70% compared to the traditional chip with a smooth surface under a driving current of 20 mA [53]. He Miao et al. from South China Normal University used Ni nanoparticle masks with ICP dry etching technology to create a roughened surface structure on the ITO layer, resulting in a 30% higher light intensity than non-roughened LED chips under working currents of 20 mA [54]. Fan Jingmei et al. from the Semiconductor Lighting R&D Center of the Chinese Academy of Sciences have used heated KOH solution to corrode the n-GaN surface of the vertical structure LED chip, increasing the light extraction efficiency of the GaN-based LED chip by nearly 100% [55].

Although surface roughening technology can effectively improve the light extraction efficiency of LED chips and the process is relatively simple, there are also shortcomings to the surface roughening technology [21]. First, the existing surface roughening technology cannot effectively control the surface morphology after roughening, and therefore cannot control the emission direction of the light extracted from the LED chip, resulting in the spatial distribution of the emitted light from the roughened LED chip not conforming to Lambert's distribution. Secondly, when packaging the roughened LED chip, if the silicone used for packaging cannot effectively fill the roughened surface, it will cause gaps between the silicone and the chip in the LED packaging module, thereby reducing the light efficiency and reliability of the LED packaging module. The research conducted by Professor Jiang Fengyi's team at Nanchang University reveals that surface roughening leads to a 50% increase in the average output light power of the LED chip, while packaging only results in a 15% increase [56]. This highlights the importance of considering the subsequent packaging process during LED chip manufacturing.. Moreover, for larger high-power LEDs chips, in order to make the current distribution in the active area of the LED chip more uniform, a thin transparent conductive film is usually deposited on the p-GaN surface to enhance the current spreading capability. At this time, if the p-Ga surface is roughened, it will affect the adhesion between the transparent conductive film and the p-GaN surface, thereby affecting the long-term reliability of the device. In horizontal structure chips where the p-GaN layer is thin (usually ranging from 0.1 μm to 0.3 μm), controlling etching depth becomes challenging when using wet or dry etching technologies for roughening purposes. Consequently, horizontal structure chips are usually subjected to roughening on top of their ITO transparent conductive layer.

3. Graphical ITO technology

When manufacturing horizontal structure LED chips, in order to enhance the current spreading capability, a layer of ITO transparent conductive film is usually deposited on the p-GaN material using electron beam evaporation technology. Due to the refractive index of GaN material being approximately 2.48 and that of the ITO transparent conductive layer being 2.08, light experiences total reflection at the interface between GaN material and ITO, resulting in insufficient extraction of light generated by the active layer [21]. By graphically processing the ITO transparent conductive film, the total reflection interface between the GaN material and the ITO material can be broken, thereby improving the light extraction efficiency of the LED chip. Liu et al. used photolithography and ICP dry etching technology to form periodic hexagonal graphics on the ITO transparent conductive layer, which increased the light output power of the LED by 50% [57]. Chang et al. used imprint lithography to form periodic graphics on the ITO transparent conductive layer, which increased the light power of the LED chip by 13.6% [58]. Byeon et al. used nano-imprint lithography to create patterns with different depths of conical structures on the ITO film, which increased the light power of the LED chip by 23% [59]. Figure 7.14 depicts an SEM image after graphical processing is applied to ITO.

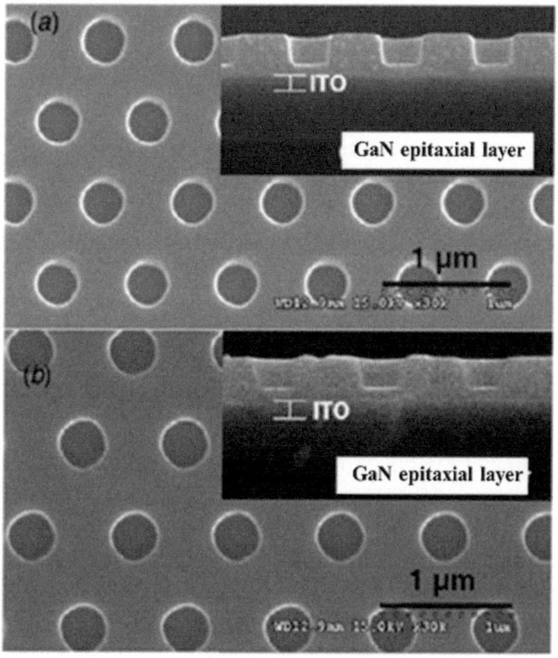

Fig. 7.14 SEM image after graphical processing of the ITO transparent conductive film [21]

Since the graphical ITO technology can effectively improve the light extraction efficiency of the LED chip, and the process is relatively simple, this technology is widely used in the actual production of LED chips.

4. Photonic crystal technology

Since the LED epitaxial layer is a thin film material with a high refractive index, it can serve as a very efficient waveguide, trapping the excited light beam within it. In order to extract light from the waveguide, researchers use photonic crystals to capture photons emitted from the waveguide. The principle is that if the waveguide has a sub-wavelength hole array of suitable size, then no light can pass through the waveguide, and all light can only be emitted in a direction perpendicular to the waveguide. Combining photonic crystals with LEDs can effectively extract photons confined within the LED epitaxial layer through the diffraction effect of photonic crystals [21]. In addition, photonic crystals can also adjust the spatial distribution of the light emitted by the LED chip, change the shape of the light spot, and make its distribution meet different applications. Due to the unique advantages of photonic crystal LEDs, in recent years, the academic and industrial circles have begun to focus on photonic crystal LEDs. Extensive research has been conducted on the technology. Currently, the main methods for creating photonic crystal structures on GaN epitaxial layers are electron beam lithography, laser holography, and imprint lithography. Oder et al. used electron beam lithography and ICP dry etching technology to form photonic crystal structures on GaN epitaxial layers with a diameter of 300 nm and a period of

700 nm. Test results showed that under a driving current of 20 mA, the output light power of the photonic crystal blue LED chip increased by 63% compared to ordinary LED chips [60].

Wierer et al. from Philips Lumileds Lighting Company used dry etching method on vertical structure LED chips to form photonic crystal structures with a depth of 100–300 nm on the n-GaN layer, increasing the light extraction efficiency of the LED chip to 73% [61]. Although electron beam lithography can create patterns of tens of nanometers, its low yield, high cost, and the limitation to only create patterns within a small area make it unsuitable for large-scale applications in the manufacture of photonic crystal LEDs. Compared to electron beam lithography, laser holographic interference technology can create photonic crystal structures over a large area with a higher yield. Kim et al. used laser holographic interference technology to create two-dimensional photonic crystal structures on the surface of GaN epitaxial layers. When the pattern period was 500 nm, the light power of the chip was twice that of the original [62]. Nanoimprint lithography, used to create nanostructure patterns, has the advantages of high resolution, high yield, and low cost. Therefore, imprint lithography can be widely used in the manufacture of LEDs with two-dimensional photonic crystal structures. Truong et al. used nanoimprint technology to create photonic crystals on the surface of GaN, successfully increasing the light power output of the LED chip by 80%, and confirmed their experimental results in simulations [63]. In 2009, Wierer et al. reported in the journal Nature Photonics that they had achieved a light extraction efficiency of 73% for LED chips using photonic crystal technology [64]. Figure 7.15 is a schematic diagram of the photonic crystal LED and comparison photos of the chip before and after lighting.

In China, scholars such as Kang Xiangning from Peking University have been studying photonic crystal LEDs in recent years, using different technologies to create photonic crystal LED chips, increasing the light extraction efficiency of photonic crystal LED chips by 40–100% compared to ordinary LED chips [65]. Hu Haiyang et al. from the Institute of Semiconductors, Chinese Academy of Sciences, used dry

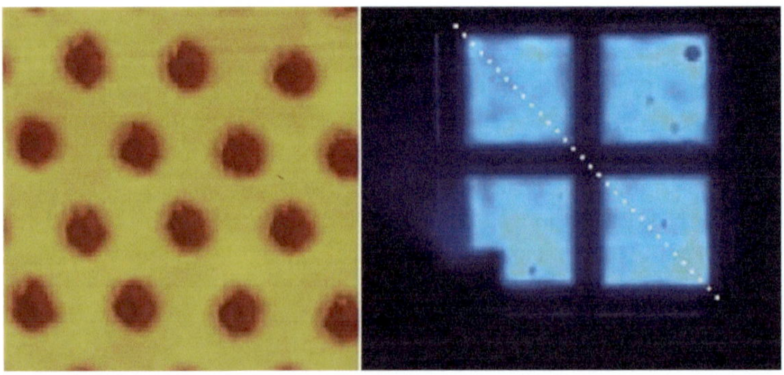

Fig. 7.15 Schematic diagram of the photonic crystal LED structure and comparison photos before and after the chip is lit [64]

etching process to prepare GaN-based two-dimensional flat panel structure photonic crystal blue LED chips. Experimental results showed that the light extraction efficiency of the photonic crystal LED chip is 1.5 times that of the ordinary chip [66]. Wen Feng from Huazhong University of Science and Technology conducted theoretical research on photonic crystal LEDs in his doctoral thesis [67]. Li Xiaoli from Shanghai Jiaotong University showed through experiments that the use of photonic crystal technology can increase the light power of LED chips by 36.65% [68]. Currently, the patents for photonic crystal LEDs are mainly held by a few companies, with Luminus Devices Inc. being the first company to sell LEDs with photonic crystals, and its products are widely used in the projection display market. Although photonic crystal LEDs have their unique advantages, there are also many challenges for their large-scale application. The reason is that the light extraction efficiency of photonic crystal LEDs is very sensitive to the geometric parameters of the photonic crystal structure. However, when growing InGaN semiconductor materials on sapphire or silicon carbide substrates, the growth temperature is higher than 900 °C.

Due to the mismatch of lattice constants and thermal expansion coefficients between the substrate material and the epitaxial material, there is a large thermal stress in the epitaxial layer, causing the LED epitaxial wafer to be uneven and the wafer to warp. This leads to disturbances in the parameters of the photonic crystal structure made on it, affecting the light extraction efficiency of the LED chip. The presence of thermal stress also hinders the use of larger LED epitaxial wafers in the manufacturing process of LED chips, thus unable to effectively reduce the cost of photonic crystal LED chips. In addition, errors in the microfabrication process of LED chips can cause deviations in the geometric structure parameters of photonic crystals, which will inevitably affect the optical characteristic parameters of photonic crystals, thereby affecting the light extraction efficiency of LED chips. On the other hand, during the packaging process of photonic crystal LED chips, the coating of phosphor silicone will also affect the structural parameters of the photonic crystal, leading to a decrease in performance after the photonic crystal LED chip is packaged. Therefore, how to reduce manufacturing costs, reduce wafer warping, and explore suitable packaging structures for photonic crystal LED chips is key to whether photonic crystal LED chips can be widely used [21].

5. Full-angle reflector technology

For high-brightness aluminum gallium indium phosphide (AlGaInP) LED chips, the utilization of MOCVD epitaxial technology to grow distributed Bragg reflector (DBR) layers as reflective surfaces can enhance the luminous intensity of LED chips by 30–60%. However, the reflectivity of DBR decreases rapidly with the increase of photon incidence angle, so there is still high light loss [21]. Therefore, researchers have developed an omni-directional reflector (ODR) with high reflectivity that is independent of the incidence angle, which can have high reflectivity for light incident from any direction. The full-angle reflector can be applied to both horizontal structure chips and inverted structure LED chips. Kim et al. deposited a conductive full-angle reflector (GaN/ITO/Ag) on the substrate of the blue LED chip, resulting in a 31.6% increase in the light output power of the chip compared to ordinary chips with only

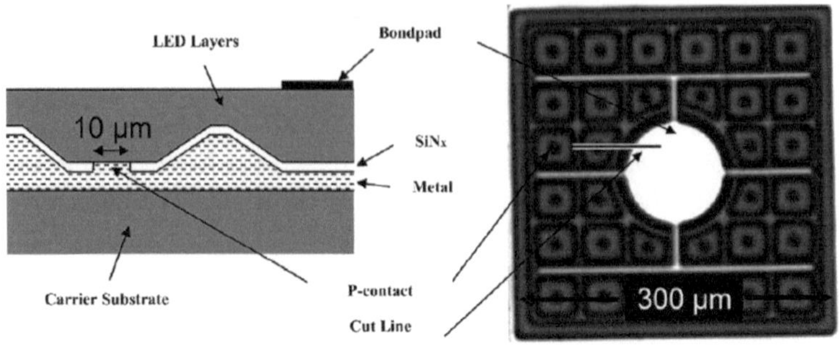

Fig. 7.16 Micro reflector AlGaInP LED chip made by OSRAM [70]

Ag reflective layers, and also reduced the forward voltage of the chip [69]. In 2007, Reiner et al. from OSRAM employed dry etching technology to create a beveled shape on the epitaxial layer of AlGaInP LED chip and subsequently deposited SiNx and metal materials onto it to form a buried reflector. Under a current drive of 20 mA, the external quantum efficiency of the LED chip reached 50% at the 650 nm band [70], as shown in Fig. 7.16.

In China, Zhang Jianming et al. from Beijing University of Technology developed a full-angle reflector for thin-film light-emitting diodes based on AlGaInP material systemwhich nearly doubled the output power of these LEDs [71].

6. Irregular chip technology

The shape of a conventional chip is a cube, with the two sides parallel to each other. Light that undergoes total reflection at the interface of two materials with different refractive indices oscillates between the two ends until it is absorbed by the chip and converted into heat energy, thereby reducing the light output efficiency of the chip [21]. In 1999, Krames et al. processed AlGaInP LEDs into a truncated inverted pyramid (TIP) shape, making the side of the chip at a 35 degree from the vertical direction. Since the sides of the chip are no longer parallel to each other, it allows light hitting the side of the chip to reflect to the top surface and exit at an angle less than the critical angle of total reflection. At the same time, light hitting the top surface at an angle greater than the critical angle can also exit from one side, greatly improving the light output efficiency of the chip and achieving an external quantum efficiency of 55% [72]. Figure 7.17 shows an LED chip with an inverted pyramid shape.

In 2001, Cree Inc. successfully produced GaN-based blue LED chips with the same structure on SiC substrates, and the external quantum efficiency of the chips was increased to 32% by making the SiC substrate into slanted. Figure 7.18 is the SEM micrograph of the LED chip manufactured by Cree Inc.

Due to the stable physical and chemical properties of sapphire substrate materials, neither wet etching nor dry etching processes can efficiently process sapphire materials. However, with the development of laser microprocessing technology, people have started to use laser microprocessing technology to process sapphire substrates,

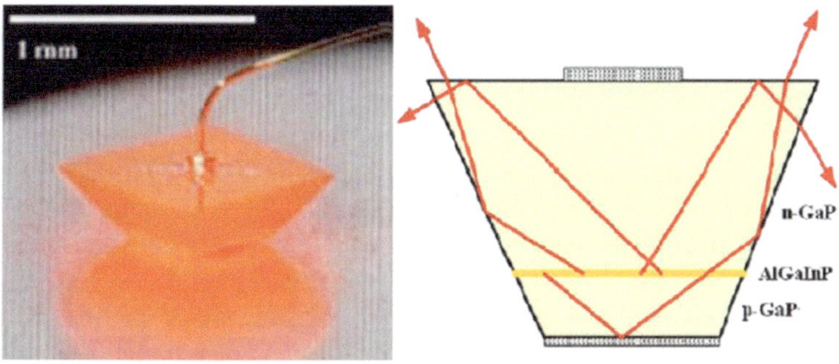

Fig. 7.17 Inverted pyramid structure AlGaInP LED chip [72]

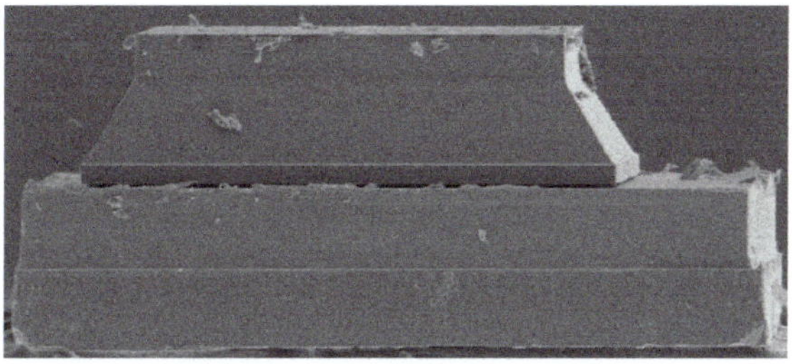

Fig. 7.18 SEM image of the LED chip with a tilted shape manufactured by Cree Inc

making their sidewalls have a certain tilt angle to improve the light extraction efficiency of LED chips[73], as shown in Fig. 7.19.

7. Surface Microstructure Technology

Using ICP etching technology to deeply etch the GaN epitaxial layer and form microstructures with specific shapes on it can effectively improve the light extraction efficiency of LED chips [21]. Lumileds Lighting Company proposed in its patent

Fig. 7.19 Optical microscope image of an LED chip with tilted sidewalls [73]

to create special optical microstructures on the chip surface to improve the light extraction efficiency of LED chips [74]. Hsueh et al. made a microhole array on the upper surface of the horizontal structure chip to improve the light extraction efficiency of the chip. The experimental results showed that when the microhole diameter was 7 μm, the output light power increased by 28% [75]. Lee et al. applied a layer of photoresist about 7 μm thick on the surface of the LED chip, then reflowed the photoresist at different temperatures to make the sidewalls of the photoresist have different tilt angles. The reflowed photoresist was used as the mask material for etching GaN. Using ICP etching technology, the sidewalls of the LED chip were etched into a certain tilt angle, which changed the reflection direction of the light after sidewall reflection and thus improved the light extraction efficiency of the LED chip [76, 77]. Chang et al. made the sidewalls of the GaN epitaxial layer of the LED chip into a wavy curve, which increased the luminous efficiency of the LED chip by 10% [78]. Huang et al. etched the sidewalls of GaN into an inward tilt of 22°, forming an etching shape with a bottom cut, which increased the light extraction efficiency of the LED chip by 70%, as shown in Fig. 7.20 [79].

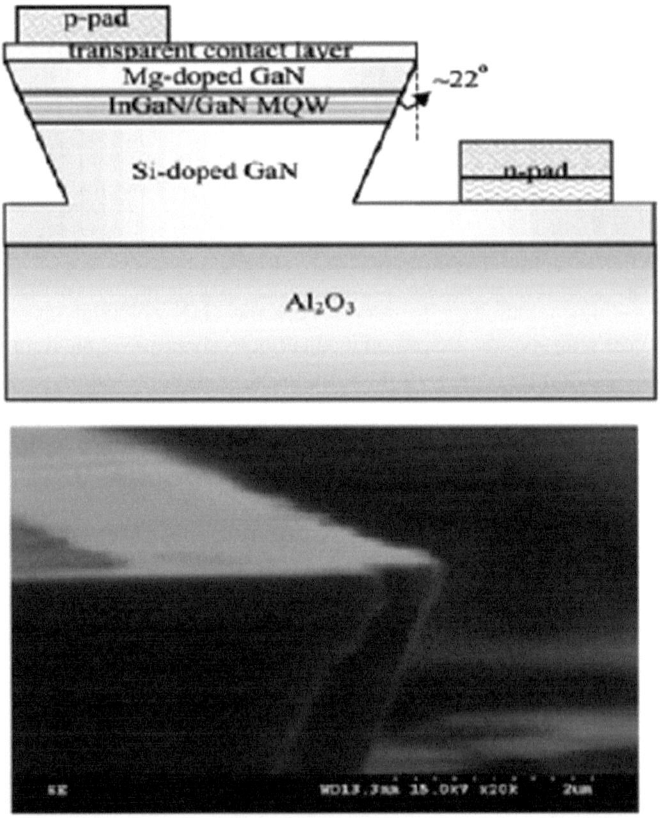

Fig. 7.20 Etching shape with a 22° bottom cut angle [79]

In China, the research group of Liu Sheng at Shanghai Jiao Tong University has improved the light extraction efficiency of high-power LED chips by using ICP dry etching technology to separate the GaN epitaxial layer into a microstructure array. Theoretical analysis showed that processing the microstructure sidewalls into a specific tilt angle can increase the chip's light extraction efficiency by 280%, while significantly reducing the stress of the chip's epitaxial layer [80]. The research groupof Luo Yi at Tsinghua University used ICP etching technology to create a periodic two-dimensional circular hole microstructure on the p-GaN surface, increasing the light power emitted from the front of the LED chip by 38% [81]. The manufacturing process of LED chips based on surface microstructure technology requires deep etching at specific parts of the epitaxial layer, and the etching depth is usually several micrometers. Because the chemical properties of GaN materials are stable, it is difficult to etch them, so the key to mass production of this type of LED chip is to develop high-speed, low-damage GaN etching technology.

7.2.6 Non-polar/Semi-polar LED

The GaN epitaxial film currently used for LEDs usually grows along the polar c-axis. The spontaneous polarization and piezoelectric polarization effects existing in the polar direction generate a strong built-in electric field, which has a significant impact on device performance: as the current density increases, the emission wavelength shifts. At the same time, the probability of radiative recombination decreases, greatly reducing the luminous efficiency of the device [82, 83], which to some extent limits the development of high-power, high-efficiency white LEDs. Using a non-polar GaN epitaxial layer grown in a direction perpendicular to the c-axis can eliminate the adverse effects of polarization on device performance [84–86]: The wavelength shift of the LED with current changes is reduced, and the luminous efficiency of the device is improved. The development of non-polar GaN is regarded by the international industry and academia as an important method to improve the performance of LEDs and has attracted much attention. It has important application prospects in future semiconductor white light illumination engineering. Therefore, the development of large-size, low-cost, and high-performance non-polar GaN epitaxial films has become one of the important trends in the research of high-brightness white LEDs [87].

GaN is a direct bandgap material with a bandgap width of 3.4 eV. The stable GaN at room temperature is a wurtzite structure, without central inversion symmetry. The GaN (0001) plane is a close-packed plane, so the conventionally prepared thin films are grown in the [0001] direction. The growth of GaN bulk single crystal materials is very difficult, and most of the applications in LEDs are GaN heteroepitaxial films. In the case of heteroepitaxy, the lattice mismatch and thermal mismatch between the thin film and the substrate introduce stress, and the spontaneous polarization and piezoelectric effects together change the carrier distribution in the polar GaN thin film and the heterojunction, resulting in a strong built-in electric field (on the order of MV/cm) in the active layer quantum well of the GaN-based device. The energy levels of the quantum well are tilted, and the wave functions of the electrons and holes are separated in space [83], greatly reducing their radiative recombination probability, thereby reducing the quantum efficiency of LEDs based on polar GaN. At the same time, as the excitation power increases, the emission spectrum shifts. Figure 7.21 is a schematic diagram of the polar, non-polar faces of wurtzite GaN. As can be seen from the figure, the non-polar face $(11\bar{2}0)$ and $(1\bar{1}00)$ is perpendicular to the polar face (0001), that is, the non-polar a-face and m-face exist parallel to the [0001] direction on the surface. Looking along the growth direction, the metal Ga atoms and N atoms are evenly distributed on the same crystal plane, so there is no spontaneous polarization electric field at the interface of the potential well and barrier [87]. The band structure of the non-polar quantum well is flat and has a high overlap of electron and hole wave functions, so its carrier emission recombination efficiency is very high, and the time constant of excitation recombination is very small. Figure 7.22a, b are the schematic diagrams of the InGaN/GaN quantum well and its band structure with and without a polarization field.

1. Selection of Substrate

Initially, due to the lack of available non-polar, semi-polar GaN substrates, the growth of non-polar, semi-polar GaN-based nitrides was carried out on hetero-substrates. R-plane sapphire and a-plane SiC were first used as substrates for the growth of a-plane GaN [89, 90], and some research groups obtained good-quality a-plane GaN. Since the lattice mismatch between sapphire and m-plane GaN is large, the growth of m-plane GaN mostly chooses m-plane SiC [91] and (100) oriented γ-$LiAlO_2$ [92]

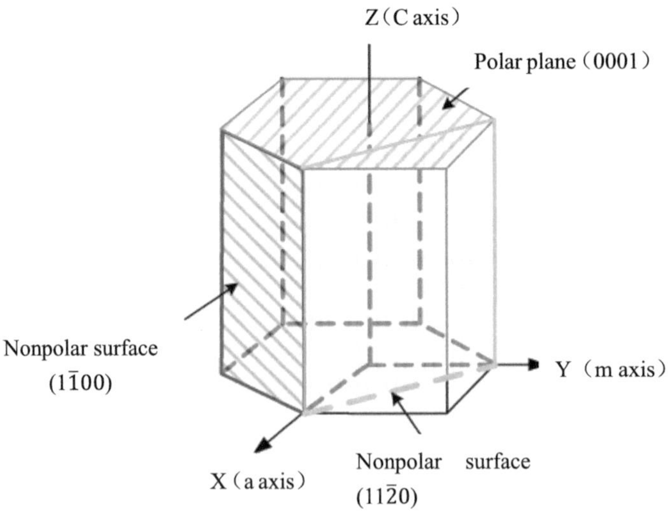

Fig. 7.21 Schematic diagram of the polar (0001) face, non-polar $(11\bar{2}0)$ face and $(1\bar{1}00)$ face in GaN crystal [87]

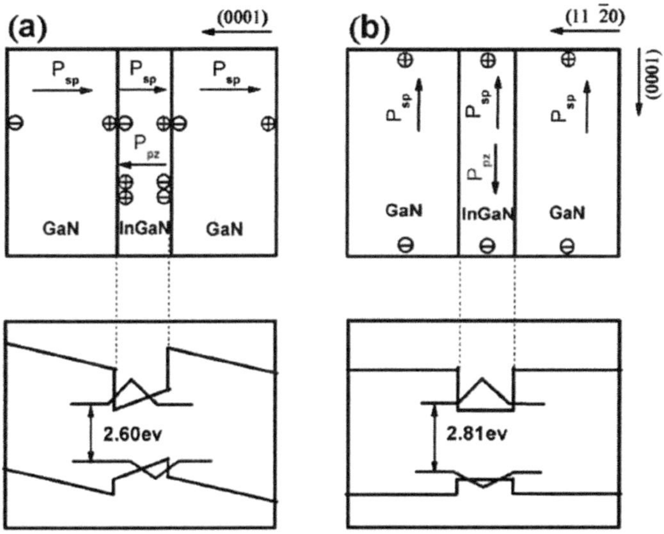

Fig. 7.22 Schematic diagram of the InGaN/GaN quantum well and its band structure with (**a**) and without (**b**) a polarization field [88]

as substrates. The m-plane SiC substrate not only has good electrical and thermal conductivity, but also does not absorb visible light, and has good chemical stability, but also the lattice mismatch between it and the m-plane GaN epitaxial layer is only 3.5%, so it is easy to obtain high-quality m-plane GaN epitaxial layer on SiC substrate using MOVPE technology. However, considering the high price of m-plane SiC, and the (100) oriented γ-$LiAlO_2$ is not very stable under the growth conditions of GaN, the best choice is to grow m-plane GaN on (100) oriented γ-$LiAlO_2$ using HVPE system [93]. In 2004, Chakraborty et al.made the first non-polar LED device. The characteristics of semi-polar materials are more complex than those of non-polar materials, and related research is less [94]. It was not until 2006 that Baker et al. obtained the $(10\overline{1}3)$ and $(11\overline{2}2)$ oriented semi-polar GaN thin films on sapphire substrates for the first time [95].

In recent years, non-polar and semi-polar GaN homogeneous substrates have been widely used in the preparation of non-polar and semi-polar GaN-based nitrides, with good results. Under suitable growth conditions, there is basically no lattice mismatch and thermal mismatch between the GaN homogeneous substrate and the epitaxial layer, and the dislocation and defect density in the epitaxial layer is very small. Therefore, the use of GaN homogeneous substrates can greatly improve the crystal quality of the epitaxial layer and increase the luminous efficiency of LED devices. Moreover, when using GaN homogeneous substrates, there is no need to nitride the substrate, let alone grow a nucleation layer, which greatly simplifies the process. In addition, like SiC substrate, GaN substrate has better thermal and electrical conductivity characteristics, so vertical structure LED devices can be made on GaN substrates, which is beneficial to solve the heat dissipation problem of large-area high-power LED devices [96]. The preparation of GaN substrates is generally first obtained by HVPE technology to grow c-direction GaN, and then cut in different directions according to needs. In 2006, T. Koyama et al. made non-polar LED devices on non-polar m-plane homogeneous substrates [97]. In 2010, Theeradetch Detchprohm et al. grew non-polar blue and green LED structures on non-polar m-plane GaN substrates. Even when the working current increases, the wavelength drift of the LED light is small. Table 7.3 summarizes some relationships between the non-polar and semi-polar LED epitaxial structures and optoelectronic characteristics grown on homogeneous substrates. Although the homogeneous substrates obtained by HVPE technology are suitable for the preparation of non-polar and semi-polar GaN-based nitrides, they are

Table 7.3 Non-polar and semi-polar LED epitaxial structures and optoelectronic characteristics grown on homogeneous substrates

Crystal orientation	Structure	Wavelength (nm)	Radiant flux (mW)	Voltage(V)	Peak shift due to increasing current
a	7-nm 5QW, no EBL	450	.	4.4	There is no [98]
a	3-nm 5QW	522	0.07	…	Redshift caused by thermal effect [99]
m	8-nm 6QW, 2.2–20 nm	407.4	23.7	5.2	Slight blue shift [100, 101]
m	3-nm 5QW	435	1.79	5	Blue shift [102]
m	8-nm 6QW	468	8.9	6.4	Smaller deviation than C-side LEDs [103]
m	2–4 nm 3QW	460–506	Maximum value 6	…	Redshift (mechanism unknown) [104]
$(10\overline{11})$	2.5-nm 6QW	412	20.58	3.93	The displacement is small due to the reduction of the piezoelectric field [105]
$(10\overline{11})$	3-nm 6QW	443.9	13.41	8.29	The displacement is small due to the reduction of the piezoelectric field [106]
$(11\bar{2}2)$, $(11\overline{22})$	3-nm SQW	425,530, 580	1.76,1.91,0.54	3.4,3.8,3.0	Blue shift caused by state filling [107]
$(11\bar{2}2)$	4-nm 6QW	522.4	5.0	6.9	Low energy local state filling [108]
$(11\bar{2}2)$	3.5-nm SQW	562.7	5.9(pulse)	—	Local energy state filling [109]

Unless otherwise specified, all other data are tested under 20 mA conditions, and the structure column gives the quantum well thickness and number of layers [110]

not suitable for commercial production. Therefore, the preparation of substrates that can be used for the growth of non-polar and semi-polar GaN-based nitrides is still a problem to be solved.

2. Epitaxial growth

Chen et al. from the University of South Carolina [111] used MOCVD to grow a-plane GaN/AlGaN quantum well structures on r-plane sapphire substrates, and prepared the first ultraviolet LED. The experimental results show that compared with the polar structure grown on c-plane sapphire, the intensity of the 363 nm ultraviolet emission peak of the non-polar structure quantum well is 30 times higher, and the peak position of the non-polar quantum well emission peak can remain stable with the increase of the driving current, while the polar structure has a 250 meV blue shift.

Chakraborty et al. from the University of California, Santa Barbara [112] reported a blue LED based on GaN/AlGaN quantum well structures. This structure uses HVPE to prepare low defect density a-plane GaN lateral epitaxial overgrowth templates, and uses MOVCD to prepare quantum well structures, as shown in Fig. 7.23. The electroluminescence peak position of the quantum well structure does not depend on the driving current (the peak position does not blue shift), which is attributed to the absence of the built-in electric field caused by the polarization effect in the non-polar a-plane quantum well structure. Figure 7.24 shows the luminescence of the LED grown on the c-plane GaN and the LED grown on the a-plane GaN using sidewall overgrowth at a driving current of 20 mA. The test results show that the intensity of the LED grown on the a-plane is 15 times that of the c-plane LED. J. P. Liu from the Georgia Institute of Technology can prepare InGaN/GaN multi-quantum well LED structures without penetrating dislocations and stacking faults on self-supporting a-plane GaN substrates, and the blue light emission wavelength does not change with the change of driving current [99]. The performance test results of these a-plane GaN-based quantum well LEDs confirm the characteristics of non-polarity, that is, the improvement of external quantum efficiency and the absence of blue shift in the luminescence peak [87].

In the research of LEDs made with m-plane GaN base quantum well structures, including heteroepitaxy on substrates such as patterned sapphire [113] or γ-LiAlO2 [114], and homoepitaxy on self-supporting m-plane GaN substrates [115], the most outstanding results were achieved by the nonpolar GaN materials research team at the University of California, Santa Barbara in May 2007. They used a quantum well with a well thickness of 16 nm and a barrier thickness of 18 nm for 6-layer stacking as the LED active area. The light output power was 28mW at a working current of 20 mA, with an external quantum efficiency as high as 45.4%, and a luminescence wavelength of 402 nm. When the driving current increased to 200 mA, the light output power of the device reached as high as 250mW, and it still maintained a high external quantum efficiency of 41%, as shown in Fig. 7.25. It is worth noting that the significant improvement in LED efficiency is due to the use of high-quality m-plane GaN substrates (these substrate materials were provided by Mitsubishi Chemical of Japan, obtained by slicing c-plane GaN grown by the HVPE method). This LED

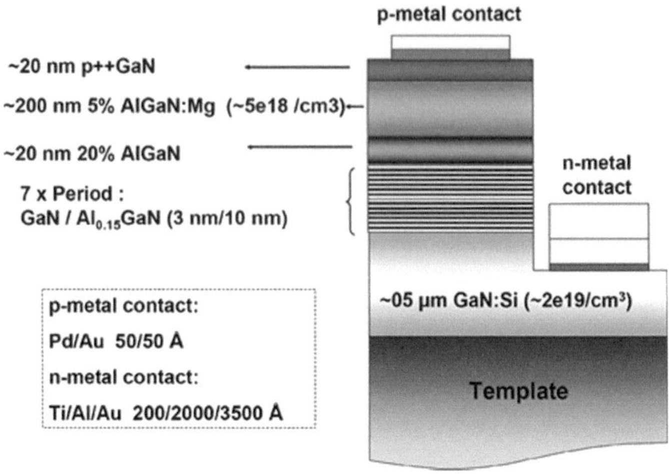

Fig. 7.23 GaN/$Al_{0.15}$ GaN LED device structure [112]

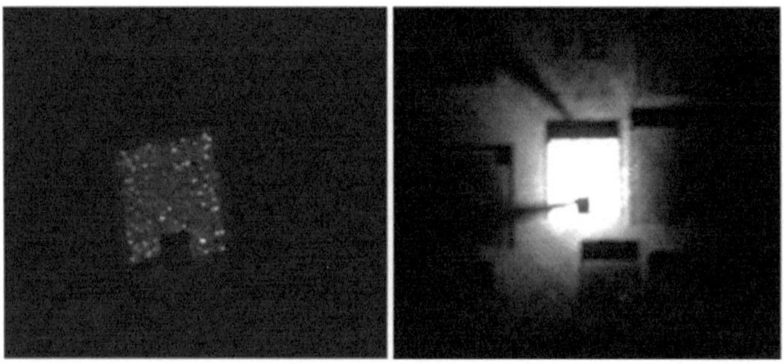

Fig. 7.24 The luminescence observed under a microscope of the LED grown on the c-plane GaN (left) and the LED grown on the a-plane GaN using lateral overgrowth (right) when driven by a current of 20 mA (400 × 400 μm junction area) [112]

based on nonpolar m-plane GaN created a historic breakthrough and became an important direction for the future development of GaN blue LEDs [87].

In addition, there are currently not many methods to improve the external quantum efficiency of nonpolar and semipolar LED devices. Since the external quantum efficiency of nonpolar and semipolar LED devices is still very low, improving the external quantum efficiency of nonpolar and semipolar LED devices is also an urgent problem to be solved.

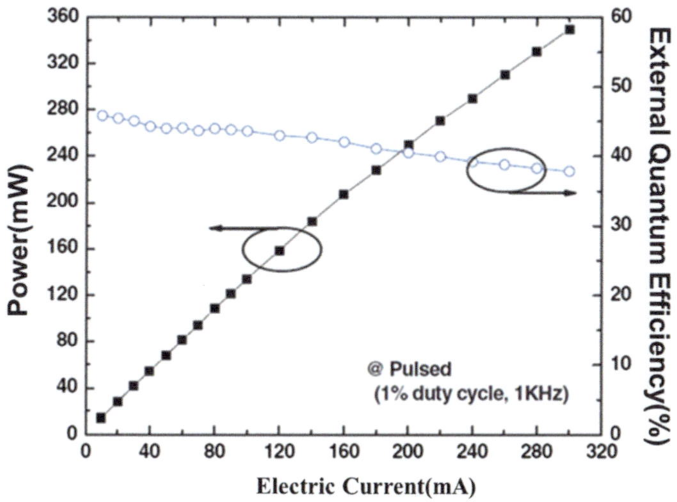

Fig. 7.25 Output power of m-plane LED with a luminescence wavelength of 402 nm. (Measured in an integrating sphere) and the relationship between external quantum efficiency and driving current

7.3 Micro Light Emitting Diode (Micro-LED)

Modern society has entered the information age and is moving towards intelligence, and display is the key link to realize information exchange and intelligence. Among many display technologies, Micro-LED display technology is considered to be a disruptive next-generation display technology. Micro-LED is a display technology that miniaturizes LED displays to the micron level. Compared with LCD and OLED, it has self-luminescence, high efficiency, low power consumption about 10% of LCD and 50% of OLED; high brightness 30 times higher than OLED ultra-high resolution, over 1500PPI, high color saturation, fast response speed, and also has extremely high contrast, wide viewing angle, seamless splicing can achieve any size display and a lifespan of more than 100,000 h.

The technology of Micro-LED originated from the microdisk laser technology of Bell Labs in the United States in 1992 [116]. In 2000, Jin et al. [117] first prepared a Micro-LED based on Group III nitrides. In 2001, Jiang et al. [118] reported a blue Micro-LED for display with a chip size of 12 μm and a 10 × 10 array. In 2004, Jeon et al. [119] reported a UV (ultraviolet) Micro-LED with a chip size of 20 μm and a 64 × 64 array. In 2008, Martin Dawson's team at the University of Strathclyde first inkjet printed organic luminescent materials on a 32 μm Micro-LED array and realized color display [120–122]. In 2012, the Kei May Lau research group at the Hong Kong University of Science and Technology reported a 360 PPI (Pixels Per Inch) Micro-LED display prototype, and upgraded it to 1700 PPI in 2014 [123, 124]. In 2015, Han et al. [125] reported the preparation of a full-color RGB Micro-LED with a chip size of 35 μm using a method of color conversion by UV light excitation of quantum dots; In 2017, Templier et al. [126] reported a blue Micro-LED with a

Fig. 7.26 Samsung micro-LED TV (the wall) (left) *Source* Samsung. Lehman optoelectronics micro-LED display (right). *Source* Lehman optoelectronics

chip size of about 2 μm based on a silicon substrate. In 2020, Bai et al. [127] reported a green Micro-LED with a chip size of 3.6 μm on a sapphire substrate prepared by a method that does not require etching. In the same year, the Hong Kong University of Science and Technology produced a silicon-based CMOS-driven Micro-LED array and combined it with red and green quantum dots to achieve a full-color display technology of 317 PPI [128].

The broad prospects of Micro-LED display technology have attracted major companies in the industry to invest a lot of resources in research and development and layout. In 2012, Sony first launched a 55-inch (139.7 cm) full HD Micro-LED TV, which includes 6 million independent Micro-LED devices, with a resolution of 1920 × 1080. In 2017, Sony exhibited a large-sized (970 cm × 270 cm) Micro-LED display with a resolution of 8 K × 2 K. In 2018, Samsung launched the world's first 146-inch (370.84 cm) Micro-LED TV (The Wall), see Fig. 7.26 left. At the same time, Lumens also exhibited two ultra-large size Micro-LED digital billboards of 130 inches and 139 inches, as well as a 0.57-inch Micro-LED display used as a head-up display.

Many domestic companies have achieved results in Micro-LED display applications. In July 2019, Lehman Optoelectronics launched a 324-inch 8 K micro-LED display, Fig. 7.26 right, at the end of October, Konka launched a 236-inch 8 K "Smart-Wall" micro-LED display, and in November, Leyard launched a 135-inch 8 K "The Great Space" micro-LED display.

LEDinside estimates that the Micro-LED market will reach 2.891 billion US dollars in 2025, with applications including mobile phones, wearable watches, car displays, virtual reality, TVs, etc. Although Micro-LED is a hot new technology in the current optoelectronic industry, Micro-LED is currently facing technical bottlenecks, including epitaxy and chip, transfer, full colorization, power drive, backplane and detection and repair technology, and it is difficult to achieve mass production in the short term. However, given that Micro-LED has multiple characteristics, it combines the advantages of LCD and OLED, but the current cost of LCD panels is already very competitive, and how new technologies find entry points will be a big challenge. With the gradual commercialization of 5G technology, the combination of 5G technology with VR, AR and 8 K ultra-high-definition video display technology will further promote the development of Micro-LED. This section will mainly introduce the current status and development trend of Micro-LED display in substrate technology, epitaxial technology, chip technology and packaging technology.

Since the chip size of Micro-LED has been miniaturized to < 50 μm, its yield and uniformity are more stringent than traditional LEDs. The choice of substrate material and epitaxial technology is crucial to the performance of Micro-LED devices. Therefore, higher requirements and challenges are put forward for the choice of substrate and epitaxial technology. When used in display technology, the injection current density of Micro-LED is very low, and the non-radiative recombination caused by defects is particularly prominent, which will greatly reduce the Micro-LED's light output efficiency. To obtain high-quality Micro-LEDs, the epitaxial wafer needs to achieve a lower defect density [129].

7.3.1 Substrate Material

Currently, the substrate materials for GaN-based blue-green LEDs mainly include sapphire substrates, silicon substrates, and GaN substrates [130, 131].

Sapphire is the most commonly used substrate for manufacturing GaN-based blue-green LEDs. The advantages of using sapphire as a substrate are: chemical stability at high temperatures (1000 °C), corrosion resistance, good light transmission, easy to obtain large sizes, and low cost [132, 133]. The disadvantages are: there is a large lattice mismatch (16%) and thermal expansion mismatch with GaN, and low thermal conductivity. Large lattice mismatch can lead to a high dislocation density in the GaN epitaxial layer, dislocations can reduce carrier mobility and minority carrier lifetime, thereby reducing thermal conductivity. Thermal mismatch can generate stress in the epitaxial layer during the cooling process, leading to the formation of cracks, ultimately reducing device performance. The thermal conductivity of sapphire is very low, and compared to other substrate materials, it is not easy to dissipate heat. In addition, sapphire substrates are non-conductive, making it difficult to manufacture device electrodes [134]. Currently, LEDs grown on sapphire substrates have the highest market penetration and the lowest cost.

The advantages of silicon (Si) substrates are: low cost, large area, high quality, good electrical and thermal conductivity. Since the thermal conductivity of Si is 5 times that of sapphire, good heat dissipation makes Si substrate LEDs have higher performance and longer life. At the same time, Si substrates can achieve damage-free peeling, eliminating the stress between the substrate and the GaN material layer. Silicon substrate LEDs are compatible with existing silicon-based integrated circuits, which gives them the advantage of developing towards miniaturization and high integration. Once the technology breaks through, the cost of epitaxial wafer growth and device processing will drop significantly. In addition, silicon substrate LEDs are expected to achieve higher wavelength yellow–red LED devices. The disadvantage is: there is a larger lattice mismatch and thermal stress mismatch between Si and GaN, which causes more defects to be generated in the epitaxial layer during the growth process, the epitaxial wafer is prone to warping, and it is easy to cause surface cracking [135–137]. In addition, silicon substrates will absorb visible light, causing the external quantum efficiency of LEDs to decrease.

For GaN-based LEDs, the most ideal substrate is GaN single crystal material, which can greatly improve the crystal quality of the epitaxial wafer, reduce the dislocation density, improve the device working life, luminous efficiency and device working current density. GaN substrates also have good electrical and thermal conductivity, and can be used in some high-performance devices, such as high-resolution AR/VR displays, etc.

In addition to the above three substrates, the substrate materials used for growing and preparing Micro-LEDs also include silicon carbide, zinc oxide, zinc germanium, etc.

7.3.2 Defect Control

As non-radiative recombination centers and leakage channels, dislocations significantly affect the performance of Micro-LED chips. Micro-LED chips require a low injection current density, and the requirements for dislocation density are more stringent. For example, for a 1 μm × 1 μm Micro-LED, the number of dislocations in a single LED chip is controlled to 1 or less, requiring a dislocation density of 10^7 cm^{-2} [138].

Currently, the main methods to reduce dislocation density and improve crystal quality for heteroepitaxial growth of GaN on sapphire or silicon substrates are substrate patterning technology and buffer layer technology [139]. The insertion of a buffer layer can provide nucleation centers for GaN growth, promote the transition of GaN's three-dimensional island growth to two-dimensional lateral growth, and reduce the dislocation density of GaN. For sapphire substrates, commonly used in LED epitaxial growth are GaN and AlN buffer insertion layers [140]. For silicon substrates, the buffer layers used in the LED preparation process usually include AlGaN/GaN, AlN/GaN superlattices, etc. [141, 142].

Figure 7.27 shows the Nanopatterned Sapphire Substrate (NPSS) prepared by the Wide Bandgap Semiconductor Research Center at Peking University [143] using a nanoimprint method. GaN is grown on the NPSS substrate. This method of epitaxial growth significantly reduces the time for nucleation islands to close, effectively relaxes the stress in the grown GaN thin film, reduces the dislocation density to 1.8×10^8 cm^{-2}, and reduces the surface roughness to below 0.1 nm.

By growing an AlN buffer layer on a silicon substrate and optimizing the superlattice insertion layer to filter dislocations, the quality of the silicon substrate LED epitaxial layer can be improved. Using defect introduction to create V-shaped pits, improving the In component to improve crystal quality, the active area changes

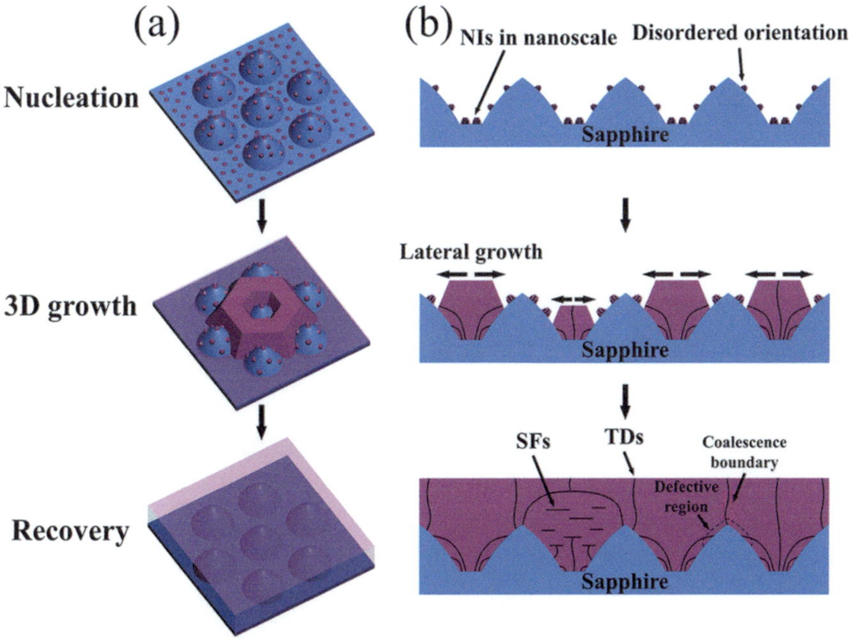

Fig. 7.27 Schematic diagram of NPSS substrate growth mode [143]

from a planar structure to a three-dimensional structure, forming vertical and horizontal p-n junctions, coordinating the two to control the carrier transport path and recombination position, thereby improving LED performance. The team at Nanchang University [141, 142, 144, 145] increased the external quantum efficiency (EQE) of the green-yellow LED with a wavelength of 551 nm to 37.7% ($4A/cm^2$) using this method. Its epitaxial structure and changes in external quantum efficiency are shown in Fig. 7.28a, b. Figure 7.28c shows a typical cross-section of a V-shaped pit and the carrier transport path and recombination position near the V-shaped pit.

High-quality GaN substrate homoepitaxy technology can obtain LED epitaxial wafers with low dislocation and high crystal quality, but due to substrate price costs, it is currently difficult to obtain large-size epitaxial wafers. In the future, with the continuous development and cost reduction of large-size high-quality GaN single crystal substrates, it is expected to make a big splash in the Micro-LED display field.

In addition to the defects existing in the LED epitaxial layer during the growth process, due to as the size of the device decreases, the proportion of the total area of the device occupied by the side walls increases, and the proportion of defects formed by etching damage increases. This leads to an increase in the proportion of non-radiative recombination, thereby reducing the luminous efficiency and also introducing new leakage channels to exacerbate the reverse leakage of the device [139]. In order to improve the performance of the device, it is necessary to effectively suppress surface defects, which can generally be achieved by optimizing etching conditions

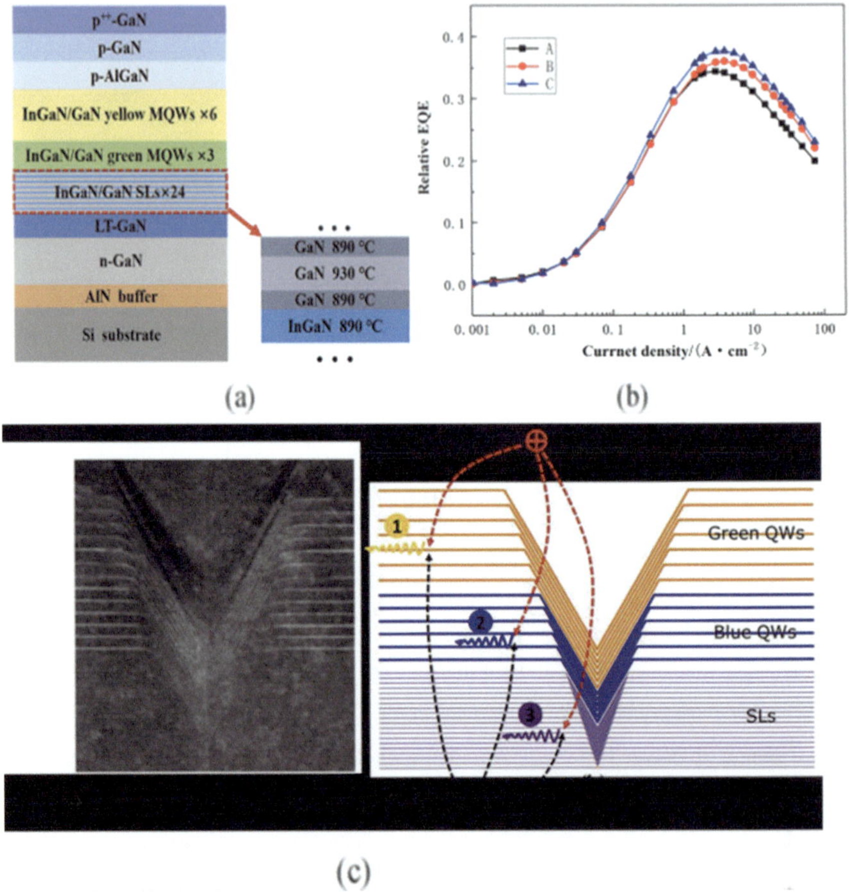

Fig. 7.28 **a** Schematic diagram of LED epitaxial layer structure with dislocation filtering layer; **b** Change curve of external quantum efficiency; **c** Typical cross-section of a V-shaped pit, diagram of carrier transport path and recombination position near the V-shaped pit [141, 142, 144, 145]

to achieve low-damage etching, etching the side walls after wet etching with solutions such as KOH, and growing passivation layers such as silicon dioxide, silicon nitride, and aluminum oxide on the etched side walls to partially repair etching damage, achieving the goal of reducing leakage and improving luminous efficiency [146].

7.3.3 Wavelength Uniformity

In high-resolution display applications, the color difference caused by the uneven emission wavelength of Micro-LEDs can greatly affect the display effect. To ensure the display effect, the standard deviation of the wavelength change of a single Micro-LED epitaxial wafer needs to be controlled within 0.8 nm. Currently, driven by market and process costs, 6-inch (15.24 cm) and larger LED wafers are gradually becoming mainstream. As the substrate size increases, controlling the wavelength uniformity during the epitaxial growth process becomes a challenge. The factors affecting uniformity mainly include: substrate thickness, size, and the control of airflow and temperature uniformity during the InGaN/GaN quantum well process in Metal–Organic Chemical Vapor Deposition (MOCVD). Generally speaking, the thicker the substrate, the easier it is to control uniformity; the larger the size, the harder it is to control uniformity. The key to improving the wavelength uniformity of LEDs is to optimize the airflow uniformity during the MOCVD epitaxial growth process.

Controlling Airflow Uniformity

In 2018, Veeco in the United States [147] reported that they used TurboDisc technology to improve the airflow uniformity above the sample during the epitaxial growth process and through a series of strain control engineering, achieved the control of the luminous wavelength uniformity within 1–2 nm in 95% of the area of the 6-inch sapphire substrate blue LED epitaxial wafer, and within 1–2 nm in 90% of the area of the 8-inch (20.32 cm) silicon substrate blue LED epitaxial wafer.

Controlling Temperature Uniformity

The In component in InGaN is very sensitive to temperature changes, usually, a temperature difference of 1°C will cause a shift of 1 ~ 3 nm in the emission wavelength [148], so improving the temperature uniformity during the epitaxial growth process is very important. The main methods include: increasing the substrate thickness, optimizing the MOCVD tray design, and adding insertion layers [149, 150]. Increasing the substrate thickness can effectively improve the warping of the epitaxial layer during growth, thereby improving the surface temperature field distribution. Optimizing the graphite tray design to give it a certain curvature can better match the warping of the epitaxial wafer during growth to achieve further improvement in temperature uniformity control. During epitaxial growth, introducing an insertion layer can effectively improve the wavelength uniformity of the epitaxial wafer.

In 2013, the Nishikawa research group [151] achieved a crack-free blue LED epitaxial layer with a thickness of 5.2 μm on an 8-inch silicon substrate by introducing an AlGaN/AlN buffer layer, with a standard deviation of wavelength change of 2.53 nm and a peak emission wavelength of 445.2 nm. Through further optimization of growth conditions, multi-insertion layer technology, and real-time strain adjustment during growth. In 2019 The research group [152] prepared an 8-inch (20.32 cm) silicon substrate LED with a wavelength standard deviation of 0.854 nm, and over 85% of the area had a peak wavelength offset of less than 2.5 nm. The corresponding PL mapping and luminous wavelength statistics are shown in Fig. 7.29, with a peak wavelength of 465.7 nm.

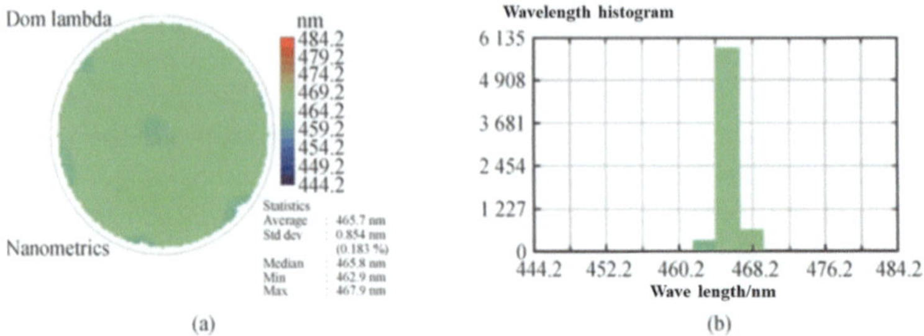

Fig. 7.29 PL mapping (**a**) and luminous wavelength statistics (**b**) of an 8-inch GaN-on-Si blue LED chip

7.3.4 *Micro-LED Device Preparation and Efficiency Degradation*

Currently, in high PPI resolution micro-display applications, the size of Micro-LED devices needs to be reduced to less than 10 μm. For example, in augmented reality/virtual reality (AR/VR) applications, the shorter viewing distance (~60 mm) requires a resolution of 1455 PPI, corresponding to a Micro-LED device size of less than 5 μm [153]. Existing research shows that as the size of Micro-LED devices decreases, the external quantum efficiency significantly decays, and the peak efficiency shifts towards high current density [154]. As shown in Fig. 7.30, as the Micro-LED device size decreases from 500 μm to 10 μm, its external quantum efficiency decays from 10 to 5%, and the current density corresponding to the peak efficiency shifts from 1 A/cm^2 to 30 A/cm^2 [154]. However, for Micro-LEDs used in micro-displays, the working current density is generally less than 5 A/cm^2, so the shift of the peak current further reduces the working efficiency of the Micro-LED device [155].

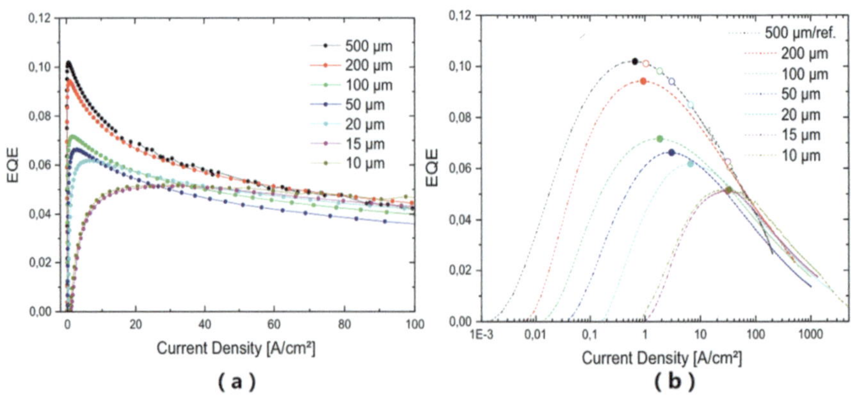

Fig. 7.30 Relationship between the external quantum efficiency of Micro-LED devices and size [154]

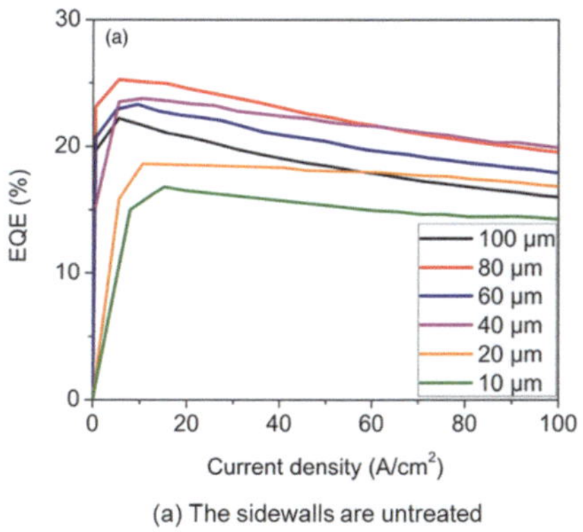

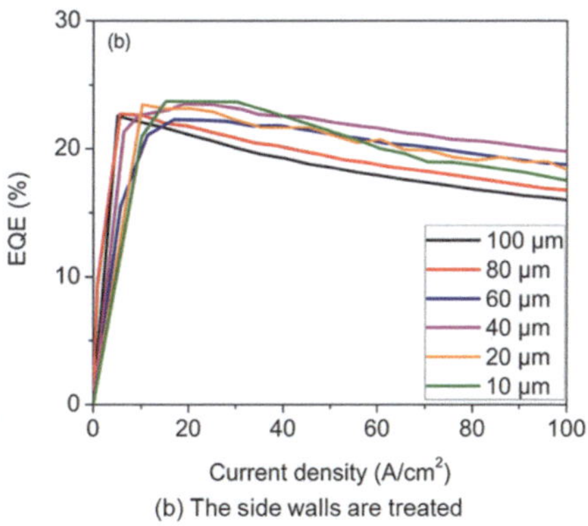

Fig. 7.31 Change in external quantum efficiency of Micro-LED devices after surface treatment with size [160]

As the size of the Micro-LED decreases, the aspect ratio of the device increases, so the efficiency decay caused by surface-related non-radiative recombination is more significant in small-size Micro-LEDs (<10 μm) [156]. Currently, to address surface non-radiative recombination, surface chemical treatment and passivation processes are introduced in the preparation of Micro-LED devices, which can effectively improve the quantum efficiency of the device [157–163]. Research shows that after ICP etching, chemical treatment of the etched sidewalls with KOH solution to remove the surface damage area introduced by the ICP process can effectively improve the device performance [160]. After KOH treatment, due to the anisotropy of KOH etching on GaN, the etched sidewalls expose the intact GaN non-polar face (m-plane), indicating that the etched damage area has been removed, thus reducing the density of surface non-radiative recombination centers. Further passivation of the sidewalls reduces the density of surface dangling bonds, thereby improving the working efficiency of the Micro-LED device. As shown in Fig. 7.31a, for the Micro-LED device without surface chemical treatment, as the size decreases from 100 to 10 μm, its external quantum efficiency decays from 23 to 15%. However, after chemical treatment of the Micro-LED sidewalls, the external quantum efficiency of the 10 μm device has significantly improved (~23%). In addition, the peak current density of the Micro-LED device did not show a significant shift, showing size-dependent characteristics [160].

As the size of the Micro-LED The development of device fabrication technology has successfully produced blue and green Micro-LED devices ranging from 1 to 10 μm using self-alignment technology. Figure 7.32 is a SEM photo of a 1 μm diameter Micro-LED structure prepared using self-alignment technology. Figure 7.33 shows the relationship between the external quantum efficiency of blue and green Micro-LED devices within the size range of 1 to 30 μm. Studies have found that as the device size decreases from 10 μm to 1 μm, the quantum efficiency of the blue Micro-LED device significantly decreases, accompanied by a shift of the efficiency

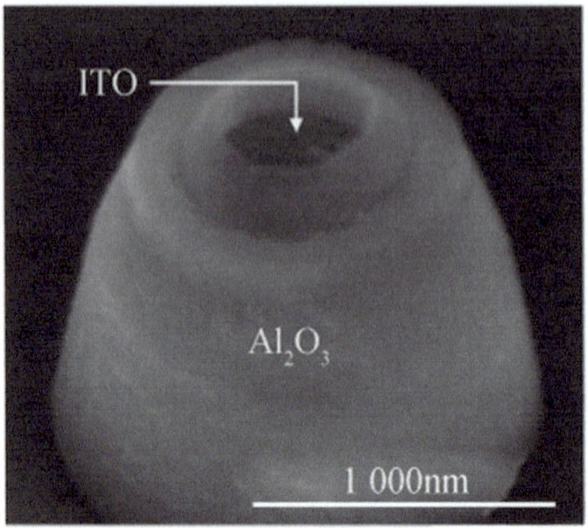

Fig. 7.32 SEM photo of the Al2O3 passivated Micro-LED structure with a diameter of 1 μm prepared using self-alignment technology [164]

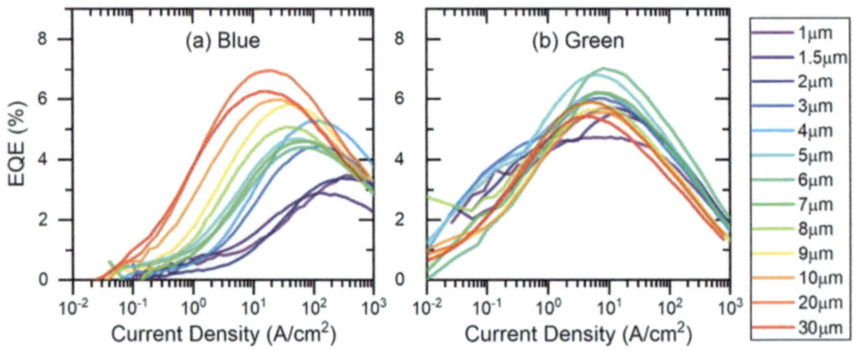

Fig. 7.33 Relationship between the external quantum efficiency of Micro-LED devices with sizes from 1 to 30 μm and the change in the blue-green wavelength range [164]

peak towards high current density. However, the quantum efficiency and peak current density of the green Micro-LED device did not change significantly. It should also be noted that within the size range of 1 to 10 μm, the external quantum efficiency of the green Micro-LED is higher than that of the blue Micro-LED. For GaN-based LEDs, the light-emitting core is the InGaN/GaN quantum well. Due to the polarization characteristics of Group III nitrides, a strong polarization electric field is generated in the InGaN/GaN quantum well, causing spatial separation of the electron and hole wave functions, thereby reducing the recombination rate of carriers in the quantum well. In green LEDs, as the In component increases, the quantum-confined Stark effect becomes more severe. Therefore, for conventional illumination LEDs, the quantum efficiency of green LEDs is lower than that of blue LEDs. However, the research results show that when the LED device size is less than 10 μm, the external quantum efficiency of the green Micro-LED is higher than that of the blue. At present, there is still no clear and definite physical explanation for this phenomenon, and further in-depth research is needed[129].

The c-plane grown GaN has a spontaneous polarization field, and the lattice mismatch between the source area InGaN and GaN will cause a piezoelectric polarization field [165]. The existence of the polarization field causes the energy band of the source area to tilt, the overlap of the electron and hole wave functions in space decreases, the radiative recombination probability decreases, thereby reducing the internal quantum efficiency of the Micro-LED, and at the same time due to the existence of the polarization field, when the current changes, the peak wavelength will shift, causing display color difference [166, 167]. In order to reduce the influence of the polarization field in the active area and increase the overlap of the electron and hole wave functions, the active area of the polar GaN can be structurally designed [168–180], or semi-polar GaN materials can be used to replace c-plane GaN [181–184].

The active area structural design methods include using the gradient growth temperature method to adjust the band shape of the InGaN/GaN quantum well [168–173], or using AlGaN [174–178], AlInN [179] and InN [180] insertion layers and other methods for polarization matching, to reduce the impact of the polarization field on the internal quantum efficiency of the Micro-LED. The gradient growth temperature method can control the In component in InGaN, grow multiple layers of InGaN potential wells with different In components, thereby forming staggered band shape InGaN/GaN quantum wells, as shown in Fig. 7.34a. In 2011, Zhao Deng et al. [168] systematically reported the role of quantum wells with staggered shapes in improving the luminous efficiency of LEDs. For the three-layer staggered $In_y Ga_{1-y} N/In_x Ga_{1-x} N/In_y Ga_{1-y} N$ quantum wells with a luminous wavelength of 520–525 nm, the carrier radiation recombination efficiency is 1.38–3.72 times higher than that of traditional InGaN/GaN quantum wells, which has increased the light output

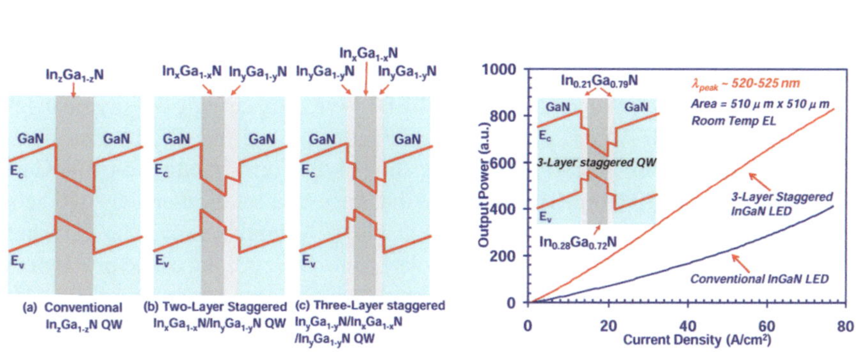

Fig. 7.34 **a** Schematic diagram of the conventional $In_zGa_{1-z}N$/GaN quantum well, two-layer staggered $In_xGa_{1-x}N/In_yGa_{1-y}N$ quantum well, and three-layer staggered $In_yGa_{1-y}N/In_xGa_{1-x}N/In_yGa_{1-y}N$ quantum well; **b** The relationship between the light output power and current density of LEDs based on the conventional InGaN quantum well and the three-layer staggered InGaN quantum well in the wavelength range of 520–525 nm, the inset is the band diagram of the three-layer staggered InGaN quantum well [168]

power of LEDs by 2.0–3.5 times [170] as shown in Fig. 7.34b. The AlGaN insertion layer can adjust the band structure of the quantum well, prevent the outward diffusion of In in InGaN, and the presence of the AlGaN layer can relax the compressive stress between InGaN/GaN and is conducive to the increase of the growth temperature of the GaN barrier layer, all of which are beneficial to improve the carrier radiation recombination efficiency [174–177]. In 2014, Hwang et al. [178] reported that when the luminous wavelength of the AlGaN/InGaN/GaN quantum well with the AlGaN insertion layer is 629 nm, the corresponding EQE is 2.9%, which is significantly higher than that of the conventional InGaN/GaN quantum well, indicating that the introduction of the AlGaN insertion layer can improve the luminous efficiency of LEDs in the long wavelength range such as red light, which is beneficial to the full-colorization of micro-LEDs.

In addition, the use of semi-polar GaN instead of c-plane GaN can effectively reduce the polarization electric field, reduce the quantum-confined Stark effect (QCSE) in the active area of Micro-LED, reduce the peak wavelength shift when the current changes, and improve the droop effect of GaN-based LEDs [181]. Semi-polar GaN has a higher In doping ability compared to polar and non-polar GaN. It can achieve longer wavelength light emission, so it has certain advantages for full-color display of micro-LEDs [182]. In 2015, Bai et al. [159] reported that LEDs based on $(11\bar{2}2)$ semi-polar GaN can achieve light emission from green to amber over a wide band, and the peak wavelength shift when the injection current changes is smaller than that of polar GaN, which also effectively alleviates the droop effect of GaN-based LEDs. In 2019, Li et al. [183] epitaxially grew $(11\bar{2}2)$ on a 2-inch patterned sapphire substrate. Semi-polar GaN has been used to prepare a series of green semi-polar Micro-LEDs with sizes ranging from 20 μm to 100 μm. Although the EQE is around 2%, far lower than the current c-plane GaN-based Micro-LEDs, it has advantages over c-plane GaN Micro-LEDs in terms of peak shift effect, droop effect, and leakage current control. In 2020, Huang-Chen et al. [184] grew semi-polar GaN on a 4-inch graphitized sapphire substrate, and prepared full-color RGB semi-polar micro-LEDs using quantum dot technology, as shown in Fig. 7.35. Its peak EQE can reach 60% of the corresponding c-plane GaN Micro-LED. When the current density changes from 1 A/cm^2 to 200 A/cm^2, the peak wavelength offset of the $(20\bar{2}1)$ semi-polar GaN Micro-LED is 3.2 nm, while the peak shift of the corresponding c-plane GaN Micro-LED reaches 13.6 nm. This indicates that the use of semi-polar GaN will help improve the color accuracy of Micro-LED displays. Although semi-polar GaN has advantages over c-plane GaN in terms of polarization effects, it faces the problem of difficult control of dislocation and stacking fault density. Due to the effect of lattice mismatch, semi-polar GaN has a higher dislocation density compared to c-plane GaN. Stacking faults, which are blocked at the growth interface and perpendicular to the growth direction of c-plane GaN, can penetrate to the surface of the device in semi-polar GaN materials, affecting the luminescent performance. Currently, the dislocation density can be reduced to the order of 10^8 cm^{-2} through buffer layer technology and lateral epitaxy, and the stacking fault density can be controlled at the

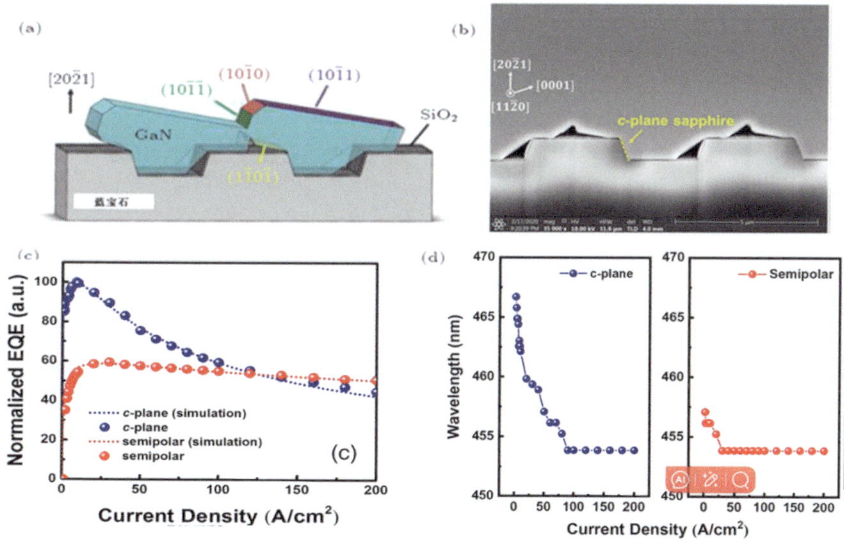

Fig. 7.35 Structure and optoelectronic characteristics of semi-polar Micro-LEDs grown on graphitized sapphire substrates. **a** Schematic of semi-polar GaN grown on graphitized sapphire substrates; **b** Scanning electron microscope image of the cross-section of (20$\bar{2}$1) GaN grown on graphitized sapphire substrates by orientation controlling epitaxy; **c** Experimental data and simulation curves of normalized EQE of c-plane and semi-polar Micro-LEDs; **d** Peak wavelength of c-plane and semi-polar Micro-LEDs when the current density changes in the range of 1–200 A/cm^2 [184]

order of 10^4 cm^{-2} [185, 186]. Further use of orientation controlling epitaxy (OCE) can almost eliminate stacking faults [187, 188], but the size, cost, and crystal quality of semi-polar GaN are still difficult to reach the level of device application. The traditional method of obtaining semi-polar GaN by cutting polar GaN at an angle is too costly, so there is still a lack of effective methods for obtaining high-quality semi-polar GaN at low cost [165].

7.3.5 Micro-LED Full-Color Display Technology

To use micro-LEDs for full-color display, the first problem to solve is the issue of the three primary colors R, G, B. Currently, InGaN/GaN LEDs can achieve efficient luminescence in blue and green light bands, but the luminescence efficiency of red light bands with higher indium(In) content is lower. Therefore, red LEDs generally use AlGaInP quaternary material. The main reasons for the difficulty in preparing red InGaN LEDs include the QCSE effect caused by the polarization field [189, 190], the carrier localization problem caused by In segregation [191], the more severe lattice mismatch and the high defect density problem caused by the lower InGaN growth temperature. Secondly, to achieve full-color display with R, G, B three-color

Micro-LEDs, a large number of three types of discrete Micro-LED devices need to be transferred to the same drive substrate. The transfer rate and yield during the transfer process are the technical bottlenecks for the industrialization of Micro-LEDs. The problems with this technical solution include the following points [129]: (1) For R, G, B three-color Micro-LED devices, blue and green Micro-LEDs belong to the InGaN material system, with a device turn-on voltage of 2.5–3 V; red Micro-LEDs belong to the AlGaInP material system, with a device turn-on voltage of about 1.7 V [192]. Therefore, the different electrical characteristics of R, G, B bring more challenges to the design of the driving circuit; (2) The electro-optical conversion efficiency of the three-color LEDs varies with the current density and the aging rate of the devices is also different, which can easily cause visual color difference in the display [192]; (3) Red AlGaInP material has a larger diffusion coefficient and surface recombination rate, causing the external quantum efficiency of red Micro-LEDs to show more significant attenuation as the size decreases, making it difficult to meet the high brightness requirements in AR/VR displays [193]; (4) There are still technical obstacles for the massive transfer of Micro-LED chips with a size less than 20 μm [192]. Therefore, for high PPI, high brightness full-color display applications, a single-chip integrated full-color display technology needs to be adopted. Single-chip integrated full-color display is to use GaN-based blue LED devices, excite red, green quantum dots (QD) or phosphors and other color conversion media, and achieve electrical interconnection between the display substrate and the drive substrate through wire bonding or flip-chip bonding, finally obtaining a single-chip full-color display [194–200]. There have been reports on the use of quantum dot single-chip full-color display technology. In 2015, Han et al. first used aerosol jet printing technology to spray RGB quantum dots on the surface of the UV Micro-LED array, achieving a single-chip full-color display [201]. The UV Micro-LED excitation light source has a wavelength of 395 nm, and the chip size is 35 μm × 35 μm. This research laid a good foundation for subsequent single-chip full-color displays, but the problem is the optical crosstalk (~32.8%) between pixels, causing the boundaries between adjacent pixels to blur. In 2017, Lin et al. successfully eliminated optical crosstalk by preparing a black photoresist mold (black matrix) between pixels [202]. In 2020, the Hong Kong University of Science and Technology used Si-based blue Micro-LEDs, combined with photolithography to realize quantum dot pixelation, and finally prepared an actively addressed single-chip full-color display prototype, as shown in Fig. 7.36 [195].

The problem with single-chip full-color display technology is how to eliminate the blue excitation light that is not absorbed by the red and green subpixels. One method is to introduce a distributed Bragg reflector (DBR) or color filter [203] above the QD layer. However, this method increases the complexity of the full-color display process and increases production costs. Another method is to increase the absorption rate of QD for blue light. Since QD has a high absorption rate in the wavelength region below 450 nm, it can achieve complete absorption of blue light. Simulation results show that a CdSe/CdS quantum dot thin film with a thickness of 5 μm and a volume fraction of 20–30% can absorb 99% of blue light [204]. But the increase in the thickness of the quantum dot thin film also increases the production cost.

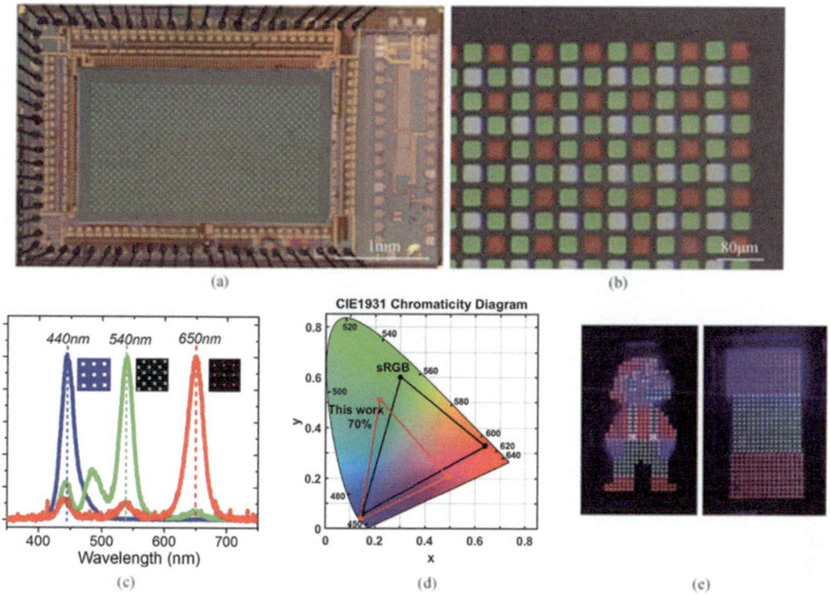

Fig. 7.36 Full-color display technology realized by combining silicon substrate Micro-LED chips with red and green quantum dots [195]

This technology route first electrically pumps blue light Micro-LED, the emitted blue light is absorbed by red and green quantum dots and converted into red and green light, in this process there are many energy loss channels, thereby reducing the color conversion efficiency. To solve this problem, the non-radiative resonant energy transfer (NRET) method can be used to transfer the energy of excitons in the blue light Micro-LED quantum well to the QD through coupling, thereby generating red and green light [205]. NRET is to use the resonance of excitons in the quantum well and QD to achieve energy transfer, avoiding the process of light emission and reabsorption, thereby greatly improving the color conversion efficiency. However, the energy transfer rate of NRET is proportional to R^{-c}, where R is the distance between the energy donor and the acceptor, and c is a constant [205]. In order to increase the energy transfer rate of NRET, it is necessary to reduce the distance between the quantum well and QD (<10 nm), so it is necessary to prepare nanopores in the Micro-LED by etching, and then fill the quantum dots in the nanopores by spin coating to improve the color conversion efficiency [206, 207].

In 2016, Krishnan prepared periodic nanopores on the LED surface and filled them with quantum dots, allowing the QD to directly contact the MQW. Thanks to the efficient energy transfer of NRET, its effective quantum yield reached a record 123% [208]. The rapid decay of time-resolved PL also proved that energy was effectively transferred from the quantum well excitons to the QD. In the same year, Nanjing University also reported a single-chip white light LED device with a high color rendering index using NERT [207]. By soft UV nanoimprinting, periodic nanopores

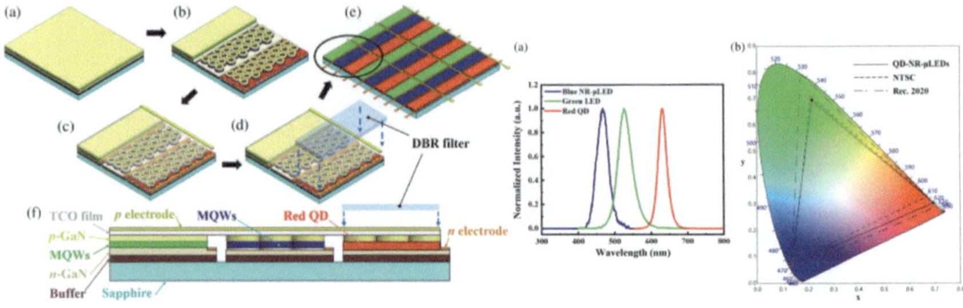

Fig. 7.37 Combining green LED nanoring structure with red quantum dots to achieve a single ring full-color display technology scheme [210]

were introduced into the LED and filled with quantum dots, finally achieving a color conversion efficiency of 69% and an effective quantum yield of 93%, and it was calculated that the efficiency of NRET between the quantum well excitons and QD reached 80%. The color rendering index of the white light LED device reached 82, and the correlated color temperature (CCT) could be adjusted within the range of 2629–6636 K.

Based on the high color conversion efficiency of NRET, a method combining LED nanorings and QD to achieve single-chip full-color display technology was reported in 2017 [209]. In this study, a nanoring structure was etched on the green LED epitaxial wafer. By controlling the wall thickness of the ring, the quantum-confined Stark effect introduced by stress in the green quantum well can be released to varying degrees, thereby achieving the transition of the emission wavelength from green to blue. Further, the improved inkjet printing technology was used to fill red quantum dots in some quantum rings to achieve the color conversion from blue to red [210]. The preparation process is shown in Fig. 7.37a. The emission wavelength of the green LED is 525 nm, the emission wavelength of the nanoring structure obtained by etching is 467 nm, and the emission wavelength of the red quantum dot is 630 nm. With this scheme, a full-color display is produced, with a single pixel size of 10 μm × 10 μm, and a single RGB The sub-pixel size is 3 μm × 10 μm. It should be noted that the traditional inkjet printing process used for spraying quantum dots is difficult to achieve such high PPI line width control. In this study, the improved inkjet printing process significantly improved the printing accuracy, achieving a line width of 1.65 μm. The full-color display scheme obtained in this study, its RGB spectrum and CIE color gamut are shown in Fig. 7.37b, c, and its color gamut area can reach 104.8% of the NTSC color gamut and 78.2% of Rec. 2020, which is sufficient to support full-color performance in practical applications.

Micro-LED display is a new type of self-emitting display technology and is considered an important choice for the next generation of mainstream display technologies. Because Micro-LED chips are much smaller than traditional lighting chips, and their stringent requirements for device yield and wavelength uniformity, they pose new challenges for substrate selection and foreign research technology. In addition, the luminous efficiency of Micro-LED devices significantly decays as the size decreases. Currently, through optimized surface treatment and passivation processes, the efficiency decay of devices above 10 μm can be alleviated, but there is still no effective way to improve the luminous efficiency of devices below 10 μm. Current research shows that using red and green quantum dots as color conversion media can achieve single-chip full-color display. However, in this technology, the pixelation process of quantum dots and color conversion efficiency still need further research and optimization.

Micro-LED, as a new generation of display technology, has advantages in display effects that LCD and OLED cannot match. With the promotion of commercial 5G technology, the demand for highly miniaturized and integrated display technologies is becoming stronger. At present, Micro-LED display technology has accumulated technology for as long as twenty years, related technical problems are gradually being overcome, and some Micro-LED display products are beginning to enter the public's view. Micro-LED display technology will play an irreplaceable role in various fields such as VR/AR, smartphones, tablets, high-end TVs, and wearable devices in the future. At this stage, more and more enterprises around the world are joining the research and development of Micro-LED displays, and many universities and enterprises in China have made outstanding contributions in this field, it is believed that Micro-LED display technology will definitely enter the public life in the near future, and occupy a crucial position in the display field.

7.4 Lasers

7.4.1 Applications of GaN-Based Lasers

Semiconductor lasers have the advantages of small size, narrow spectral half-width, high optical power density, good spatial coherence, high efficiency, long life, and fast response speed. They have a wide range of applications in information technology, life sciences, etc., such as in fiber optic communication, optical disc access, spectral analysis, optical information processing, laser micro-processing, laser printers, high-definition laser TVs, and the "optical tweezers" used in life science research using semiconductor lasers, etc. The lasing wavelength range that GaN-based semiconductor lasers can achieve can cover the ultraviolet to visible light bands, and they are used as indispensable light sources in laser display, laser lighting, storage and undersea communication, atomic clocks and quantum sensors, etc., with important application prospects and a wide market demand.

1. Laser Display

Laser display has many technical advantages: the light source has a narrow spectrum (~5 nm), which can achieve 12-bit color grayscale encoding without overlap; the laser wavelength is controllable, and according to the color television broadcasting standard established by the National Television Standards Committee of the United

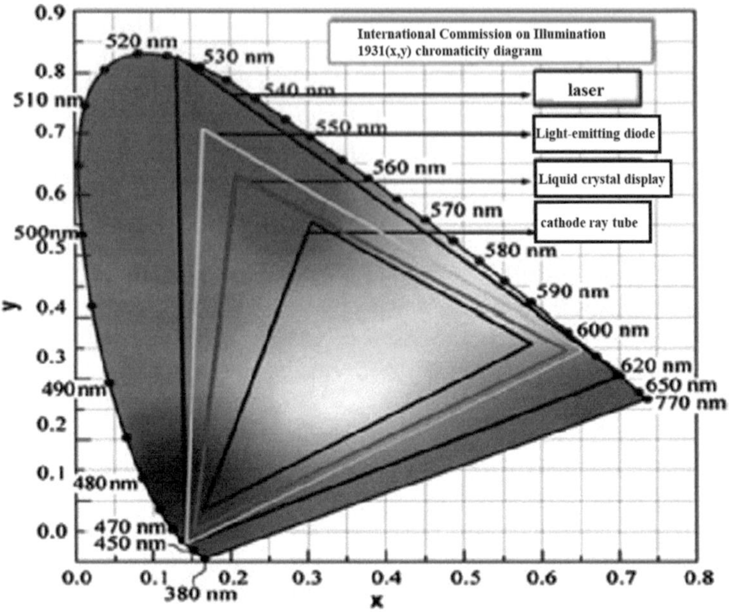

Fig. 7.38 Comparison of color gamut ranges that different display technologies can display [211]

States in 1952 (referred to as the "NTSC standard"), it can form a super-large color gamut of more than 150% NTSC (see Fig. 7.38); the laser brightness is high and can be accurately controlled in the optimal visual perception area of the human eye (8 K high-definition); the laser color temperature is precisely adjustable, making it easy to achieve seamless splicing display of super-large screens (hundreds of square meters); combined with holographic technology, it can reproduce the full information of the object's light wavelength (color), amplitude (intensity), and phase (stereo), realizing true three-dimensional display. Laser display is the only display technology that can achieve the BT.2020 ultra-high-definition international display standard, which can meet the ultimate demand for humans to pursue beautiful visual effects. Laser display is even called a "revolution in human visual history" and has a huge application prospect in the fields of television, home theater, and projectors [211].

Laser display has the characteristics of lightness, low cost, and green manufacturing. It consumes 50% less power than the same size LCD TV, has a lifespan of more than 2×10^4 h, is energy-saving and environmentally friendly; the product is small in size and light in weight, a 100-inch laser TV weighs about 20 kg (the same size LCD TV weighs about 150 kg), it is easy to be placed in elevators and enter ordinary people's homes; laser display uses reflective imaging, which is comfortable to watch; laser display belongs to the green manufacturing industry, does not require large-scale investment, compared to the huge investment in high-generation panels in traditional flat-panel displays (the display panel field has invested more than 1.2 trillion yuan), the laser display process is simple, in line with the development trend of new display flexibility, portability, low cost, high color gamut, and high light efficiency.

Laser display can take into account both the markets of viewing/entertainment and information, and will be the mainstream of next-generation TVs, movie machines, and super-large screen display products. As a representative of new display technology, laser display is highly aligned with our country's major national strategies, has become an advantageous industry that affects our country's national economy and people's livelihood and subsequent development, and is an important strategic industry for maintaining national industrial security, reflecting the level of national information technology intelligence, and promoting industrial transformation and upgrading.

Before the successful development of GaN-based blue and green lasers, the laser source used in laser displays was all-solid-state lasers. With the development and maturity of GaN-based blue and green lasers, due to their inherent advantages, the laser source of laser display is being replaced by direct semiconductor lasers. The red laser in the three primary colors uses an AlGaInP red laser, while the blue (445–450 nm) and green (520–530 nm) lasers are GaN-based lasers. In 2005, Sony developed a rear-projection TV with a GaN-based blue laser as the light source. In 2007, Shanghai Sanxin Technology Development Co., Ltd. developed a micro laser projector using a GaN-based blue laser. In 2014, Hisense Group launched a laser cinema system using a GaN-based blue laser [211].

In recent years, the development of laser display technology and industry in China has been rapid, from the "Tenth Five-Year Plan", "Eleventh Five-Year Plan", and "Twelfth Five-Year Plan" of the 863 From the planning, technology support plan, to the "Thirteenth Five-Year Plan" key research and development plan, key materials, devices and application technologies of laser display have gradually made breakthroughs, especially in the core of the laser display industry—three primary colors LD and the overall design and manufacturing, the progress is significant. Currently, the power of a single red light LD tube in China can reach 2W (with a lifespan exceeding 10,000 h), the maximum output power of a single blue light LD tube is 2.8 W (with a lifespan exceeding 5000 h), and the maximum output power of a green light LD tube reaches 500 mW. In terms of the whole machine, China has formed a number of core patented technologies in high-power laser modules, speckle suppression, and integrated manufacturing, overall reaching an international leading level. The state has issued a series of related policies and project support, greatly promoting the investment of backbone enterprises in laser display and the rapid development of the industry. Dozens of traditional home appliance companies and Internet companies in China, including Hisense Visual Technology Co., Ltd., Sichuan Changhong Electric Co., Ltd., Hangzhou Zhongke Aurora Technology Co., Ltd., Shenzhen Light Peak Technology Co., Ltd., TCL Technology Group Co., Ltd., etc., have carried out industrial layout around the entire chain of laser display. In 2019, the domestic laser display output value has exceeded 12.5 billion yuan, with a compound annual growth rate close to 100% in recent years, and the industry scale has reached an international leading level. 2. Optical Disc Storage GaN-based violet lasers have short wavelengths and can be used in the field of optical disc storage to increase the storage capacity of optical discs. Taking the common 5-inch double-sided optical disc as an example, using a red semiconductor laser with a wavelength of 670–690 nm, the storage capacity of the optical disc is 2.6 GB. If a GaN laser with a wavelength of 405 nm is used, the capacity of the same size optical disc can reach more than 20 GB. As the output power of the GaN-based laser increases, the number of data layers on the disc can also increase correspondingly, and its data capacity can also increase exponentially. Sanyo Electric's laser diode, with a pulse output power of 450 mW, can write 4 layers of data at 12 × speed, and the disc capacity can reach 100 GB [211]. 3. Laser Printing and Plate Making GaN-based violet lasers can also be used in high-resolution laser printing and plate making. Laser printers can directly receive text or images from computers, thus saving a lot of time, speeding up the process, and significantly reducing the pollution caused by printing. Currently, red light lasers (808 nm) are commonly used as the laser source in laser printers. Compared with red light lasers, GaN-based violet lasers (405 nm) have a shorter working wavelength and can obtain smaller diffraction spots. Therefore, using GaN-based violet lasers (405 nm) as the laser source can improve the resolution of laser printing and achieve higher print quality [211]. 4. Laser Communication GaN-based lasers with wavelengths of 470–540 nm have a lower absorption coefficient in seawater and therefore have stronger penetration. Their propagation distance can reach 600 m and can be used for deep-sea detection and anti-submarine communication. In addition, research shows that the scattering ratio of purple lasers and red lasers in smoke is different, so GaN-based violet lasers can also be used as laser fuses with anti-smoke interference capabilities [211]. With the increasing maturity of GaN-based laser technology, the continuous expansion of the wavelength range, and the gradual increase in output power, it is believed that more and more application fields will be opened up, and its market size will further expand.

7.4.2 Current Status of GaN-Based Laser Research

Internationally, since the successful development of the world's first room-temperature continuous lasing GaN-based violet laser by Nichia Corporation of Japan in December 1996, many research institutions have invested heavily in the research of GaN lasers. In recent years, driven by the huge market demand for laser displays, the research focus of GaN-based lasers has shifted to blue and green lasers [211]. The following introduces the development and research progress of GaN-based violet lasers, blue lasers, green lasers, and ultraviolet lasers.

1. Violet Lasers

The violet laser with a wavelength of about 405 nm is the earliest successful GaN-based laser, and its biggest application is high-density optical storage. In 1995, Nichia Corporation of Japan developed the world's first GaN-based violet laser. On January 12, 1999, Nichia Corporation announced the commercialization of the GaN-based violet laser, with a wavelength of 400 nm, a working current of 40 mA, a working voltage of 5 V, an output power of 5 mW, and a continuous working life at room temperature of more than 10,000 h [211].

Although the technology of violet lasers is mature, the output power of the laser is relatively small and the efficiency is low. At present, many studies are focused on improving the output power and efficiency of the laser. In 2003, Sony Corporation reported that the continuous working output power of a single-tube GaN-based laser reached 0.94 W, and the array power reached 6.1 W. Nichia Corporation's website also sells GaN-based laser arrays with a continuous working output power of up to 10 W. Another development direction of GaN-based violet lasers is picosecond lasers, which have important uses in medicine. In 2012, Sony Corporation achieved a pulsed laser with a wavelength of 405 nm, 300 W, and a repetition frequency of 1 GHz through laser mode locking and optical amplifiers. Domestically, in 2004, the Institute of Semiconductors of the Chinese Academy of Sciences and Peking University developed a GaN-based violet laser. In 2010, the Institute of Semiconductors of the Chinese Academy of Sciences used a self-supporting GaN substrate to further improve the performance of the laser, achieving a GaN-based violet laser with a threshold current density of $2.4 kA/cm^2$, a threshold voltage of 6.8 V, and a lasing wavelength of 413.7 nm [211].

2. Blue Lasers

Due to the huge application prospects of GaN-based blue and green lasers in the field of laser display, many research institutions are committed to extending the lasing wavelength of the laser to the blue and green range.

The epitaxial structure of the GaN-based blue laser is grown on the c-plane GaN substrate using the Metal–Organic Chemical Vapor Deposition (MOCVD) method, and then processed into a ridge waveguide structure GaN-based laser. The composition of the epitaxial structure is shown in Fig. 7.39, from bottom to top are Si-doped n-type GaN substrate, Si-doped n-type AlGaN lower cladding layer, Si-doped n-type GaN lower waveguide layer, Si-doped n The structure includes an n-type InGaN lower waveguide layer, a blue light InGaN/GaN quantum well active region, an unintentionally doped InGaN upper waveguide layer, a Mg-doped p-type AlGaN electron blocking layer, a Mg-doped p-type AlGaN/GaN superlattice upper confinement layer, a heavily doped p-type InGaN contact layer, and a p-electrode (Pd/Pt/Au). Ti/Au is deposited at the bottom as the n-electrode [212].

The growth temperature required for the blue light InGaN quantum well is 700 °C, and the growth temperature for the p-type AlGaN upper confinement layer is 950–1000 °C. High-temperature growth can improve the material quality and conductivity of the AlGaN confinement layer. However, high temperatures can cause thermal degradation of the grown InGaN quantum well. Thermal degradation can lead to the appearance of non-radiative recombination centers in the InGaN quantum well, which can be observed as dark spots under a confocal laser microscope (405 nm laser source) [213]. Figure 7.40 shows fluorescence microscope images of blue laser epitaxial wafers at different confinement layer growth temperatures. When the growth temperature is reduced from 960 to 940 °C, the dark spots disappear. This indicates that appropriately reducing the growth temperature of the p-type AlGaN confinement

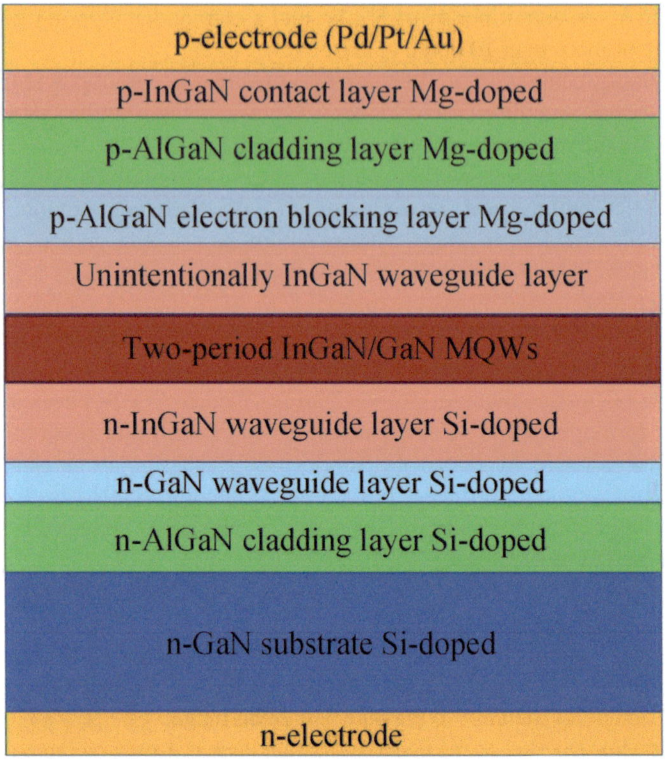

Fig. 7.39 GaN-based blue laser epitaxial structure [212]

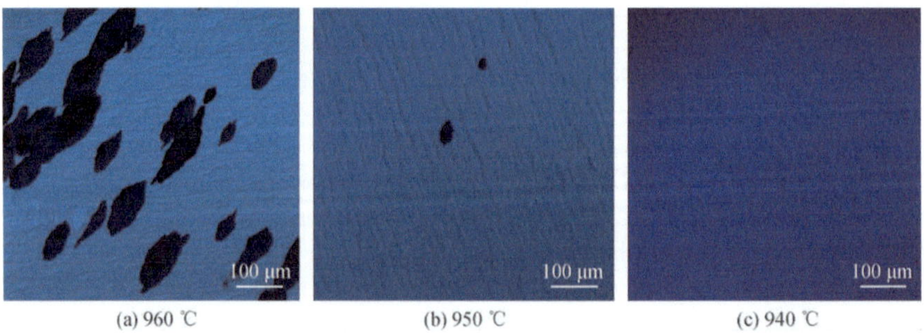

Fig. 7.40 Fluorescence images of blue laser epitaxial wafers at different p-AlGaN confinement layer growth temperatures [213]

layer can effectively suppress the thermal degradation of the quantum well, and the resistivity of the p-type AlGaN confinement layer does not significantly increase.

Reducing the internal optical loss of the laser can improve the slope efficiency of the laser and reduce the threshold current. The ionization rate of Mg doped in the p-type layer is very low, and a large amount of un-ionized acceptor impurity Mg is the main source of internal optical loss [214, 215]. To address this issue, increasing the thickness of the unintentionally doped InGaN upper waveguide layer can distribute more internal light field in the upper waveguide layer, thereby reducing the overlap of the light field with the p-type AlGaN electron blocking layer above the waveguide layer, and thus reducing the loss of the un-ionized acceptor impurity Mg to the light field [216]. Experiments have shown that increasing the thickness of the unintentionally doped InGaN upper waveguide layer from 20 to 110 nm can reduce the internal optical loss from 60 to 10 cm^{-1}.

Figures 7.41a, b show the corresponding slope efficiency and internal optical loss, and the P-I curve describing the current-output power relationship, under different unintentionally doped InGaN upper waveguide layer thicknesses. It can be seen that as the thickness of the InGaN upper waveguide layer increases, the slope efficiency increases and the internal loss gradually decreases. When the thickness of the InGaN upper waveguide layer is 20 nm, 70 nm, 90 nm, and 110 nm, the corresponding slope efficiencies are 0.36 W/A, 0.58 W/A, 1.1 W/A, and 1.5 W/A, respectively. The laser used to make the ridge waveguide structure has a cavity length of 400 μm, and the reflectivity of the front and rear cavity surfaces is 34% and 95%. The slope efficiency test is carried out under the condition of continuous operation (CW) of the laser at room temperature. Eight lasers were selected as samples for the test, and the slope efficiency is the average of 8 sets of data [217].

Due to the large effective mass and low mobility of holes in GaN base materials, it becomes difficult to inject holes into the quantum wells on the N side [218–220]. The necessary condition for lasing is to form a particle inversion in the active area, which requires a sufficient number of holes to be injected into the quantum wells to meet this condition, so improving the injection efficiency of holes is an important

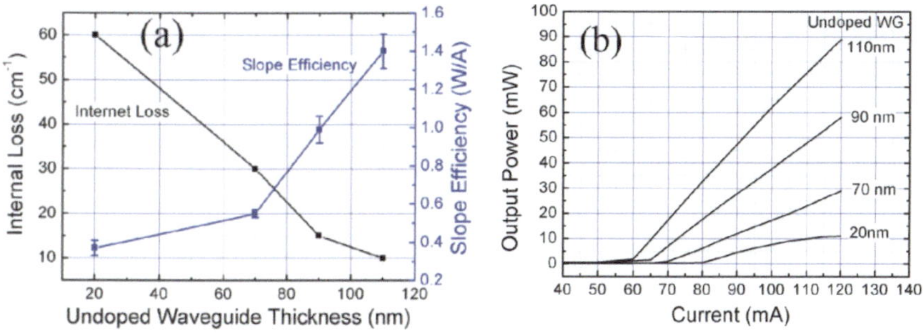

Fig. 7.41 **a** Slope efficiency and internal optical loss corresponding to different unintentionally doped InGaN upper waveguide layer thicknesses [217]; **b** Relationship between current and output power under different unintentional doping InGaN waveguide layer thicknesses [217]

Fig. 7.42 Output optical power-current (P-I) curves of blue lasers with different quantum barrier thicknesses [213]

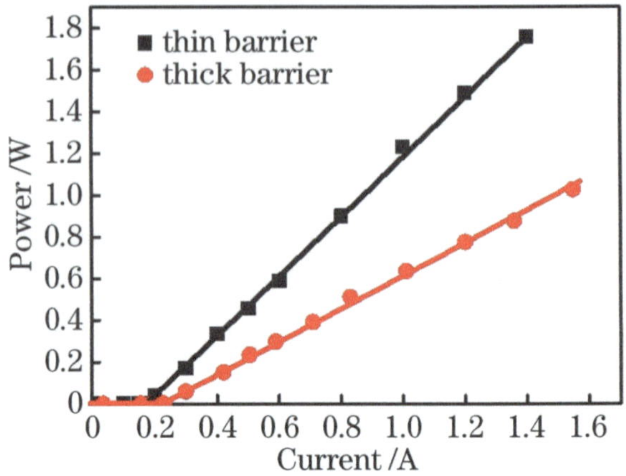

task. Studies have found that in GaN blue LEDs, when the thickness of the quantum barrier reaches 13 nm, the electrically injected holes are almost only distributed in the quantum wells near the p side, and almost no holes are injected into the quantum wells near the n side. Reducing the thickness of the quantum barrier and lowering the potential barrier height of the quantum barrier can increase the injection efficiency of holes [212]. Figure 5 shows two blue LDs with different quantum barrier thicknesses, testing their output optical power-current curves (P-I).

In Fig. 7.42, the threshold current density of the blue laser with a thinner quantum barrier is lower, at 1 kA/cm^2, and the slope efficiency is higher, at 1.5 W/A. The threshold current density of the blue laser with a thicker quantum barrier is larger, at 1.4 kA/cm^2, and the slope efficiency is lower, at 0.8 W/A. This is because the injection efficiency of carriers in the blue laser with a thinner quantum well is higher, and the distribution of holes in the quantum well is more uniform [212].

(1) Development of blue lasers

In September 1999, Nichia Corporation first reported a single quantum well blue laser grown on a lateral epitaxial GaN substrate, with a lasing wavelength of 450 nm, threshold current density and voltage of 4.6kA/cm^2 and 6.1 V, respectively, and a lifetime of 200 h at room temperature output of 5mW. In March 2001, Nichia Corporation used InGaN material as the waveguide layer to enhance optical confinement; at the same time, it improved the crystal quality of the active area, and the device's threshold current density dropped to 3.3 kA/cm^2, the threshold voltage dropped to 4.6 V, and the device life reached 3000 h when outputting 5mW at 50°C. With the continuous improvement of epitaxy, chip and heat dissipation packaging technology, the output power and life of the laser are constantly increasing. In 2013, Nichia Corporation reported a blue laser with a continuous output of 3.75 W, with a laser threshold current of 225 mA and a threshold current density of less than 1 kA/cm^2. In the market, Nichia Corporation has launched a 3.2 W blue laser. Osram's research on blue lasers started relatively late, but it has developed rapidly. In 2013,

Osram reported a blue laser with a maximum output power of up to 4 W. Currently, Osran has launched a continuous output 1.6W blue laser product in the market [211]. Nippon Nitto Chemical Industry Co., Ltd. [221] reported a blue laser with a slope efficiency of 1.8W/A and an optical power of about 5W; Sony Limited [222] reported The blue laser with a slope efficiency of 1.8W/A and optical power of approximately 5.2 W has been reported; the German Osram Group [223] also reported a blue laser with a slope efficiency of 1.6 W/A and an optical power of 4.5 W.

Domestically, the Suzhou Institute of Nanotechnology and Nanobionics of the Chinese Academy of Sciences took the lead in developing GaN blue lasers, and prepared and packaged high-power GaN-based blue edge-emitting lasers [217]. The laser cavity length is 1200 μm, and the ridge waveguide width is 45 μm. The laser characteristics are measured under pulsed current at room temperature, with a pulse width of 0.4 μs and a frequency of 10 kHz. The lasing wavelength is 445 nm, the threshold current density is 1 kA/cm^2, and the slope efficiency is 1.65 W/A. Under continuous working conditions, due to poor heat dissipation, the slope efficiency drops to 1 W/A, and the output optical power is 2.2 W at a current density of 6 kA/cm^2.

(2) Key issues in the preparation process of edge-emitting GaN-based blue-green lasers

The preparation of GaN-based blue-green lasers mainly solves the following key issues: how to prepare high-quality InGaN/GaN multiple quantum wells (MQWs), how to reduce internal optical loss, how to increase hole injection efficiency, and how to effectively eliminate the impact of quantum-confined Stark effect (QCSE) [212].

(1) Quality issue of InGaN/GaN multiple quantum wells (MQWs)

As the active region of GaN-based LD, the growth quality of InGaN/GaN multiple quantum wells is very important for laser performance. In recent years, the lasing wavelength range of GaN-based lasers has gradually increased. To achieve green LD, the In component in InGaN quantum wells is about 30%, but it is difficult to epitaxially grow high-quality high-In component InGaN quantum wells.

First, because the bond energy of In-N is low, the growth process of InGaN requires a reduction in growth temperature to ensure sufficient In is incorporated. However, at lower growth temperatures, the migration rate of surface atoms is low, the migration length is short, and the cracking efficiency of NH_3 at low temperatures is also low. In addition, the large miscibility gap between InN and GaN can easily cause crystal defects, component fluctuations, surface roughness, etc. in InGaN quantum wells [224]. In segregation and surface roughness of InGaN quantum wells can also cause non-uniform broadening of the photoluminescence (PL) spectrum.

Secondly, to obtain high conductivity p-type AlGaN, the growth temperature of its electron barrier layer and confinement layer needs to be about 1000 °C. However, when the p-type AlGaN layer is grown at high temperature, the temperature will cause thermal degradation of the InGaN/GaN quantum well active region, produce crystal defects, and the thermal degradation phenomenon is more significant in the green light InGaN quantum well with high In component [225–229].

Finally, there is a lattice mismatch between InGaN and GaN, and this lattice mismatch becomes more significant as the In component increases. In blue lasers, the lattice mismatch between InGaN quantum wells and GaN reaches 1.6%, and in green lasers, it reaches 3.3%. Severe lattice mismatch can cause crystal defects in InGaN quantum wells, which will reduce the radiation efficiency and shorten the life of the laser [230, 231].

(2) Internal optical loss

Reducing internal optical losses can lower the threshold current and increase the slope efficiency. The ionization energy of Mg doped in p-type GaN is relatively large ~ 160 meV, which will further increase with the increase of aluminum components [232, 233]. In reality, less than 10% of the acceptor impurities Mg ionize to form free holes, and the remaining 90% of Mg that has not been ionized becomes the main source of internal optical losses in GaN lasers [234, 235]. Larger internal optical losses will lead to an increase in the laser's threshold current and a decrease in slope efficiencies.

(3) Hole injection efficiency

The hole injection in the multi-quantum well active region is often uneven, which is not only due to the large effective mass of the hole, but also due to the potential barriers hidden in the multi-layer structure of the laser [236–238]. The number of hole injections in the quantum wells near the n side in the multi-quantum well active region is often insufficient. This is because the polarization field causes the energy band of the multi-quantum well in the active region to tilt, increasing the potential barrier of hole injection, thus making it difficult for holes to inject from the p side of the quantum well region to the n side of the quantum well region [239]. This uneven hole injection will also lead to an increase in the laser threshold current and a decrease in slope efficiency [240, 241]. In addition to the unevenness of hole injection, there is also the problem of hole overflow from the active region, that is, holes overflow from the active region to the lower waveguide layer to recombine with electrons [242–244].

(4) Quantum-confined Stark effect

GaN-based lasers are usually grown on c-plane GaN substrates. There is a certain lattice mismatch between InGaN quantum wells and GaN quantum barriers, so there is a certain stress at the interface, which leads to piezoelectric polarization at the interface. The polarization field in InGaN quantum wells caused by spontaneous polarization and piezoelectric polarization leads to the Quantum-Confined Stark Effect (QCSE) [245–247]. The polarization field causes the energy band of the quantum well region to tilt, which reduces the overlap of the wave functions of electrons and holes, ultimately leading to a redshift of the emission wavelength of the quantum

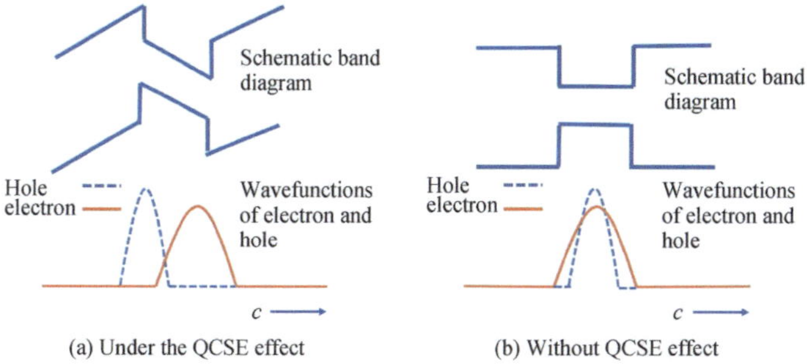

Fig. 7.43 Schematic diagram of the energy band in the InGaN/GaN quantum well region

well, a decrease in the radiative recombination rate, and a decrease in luminous efficiency. The change in the band structure is shown in Fig. 7.43a, b. The common solution is to use semi-polar or non-polar substrates to grow the laser structure.

3. Green lasers

Green laser is one of the three primary colors of laser display. With the development of laser display technology and the successful application of GaN-based blue laser in laser display technology, the demand for GaN-based green lasers has become more urgent. Gallium nitride green lasers are also a hot topic in the field of gallium nitride device research. There are two major difficulties in realizing gallium nitride green lasers. The first difficulty is to grow a high quantum efficiency green light InGaN quantum well active area. To achieve green light lasing, the In component of the InGaN quantum well needs to reach about 33%. It is very difficult to improve the luminous efficiency of high In component InGaN quantum wells, especially in the laser structure. The second difficulty is that the InGaN quantum well components and interfaces are uneven, leading to uneven light emission broadening and a decrease in the optical peak gain of the laser. Therefore, the preparation of GaN-based green lasers is more challenging than blue lasers [211].

The growth temperature of high In component green light InGaN quantum wells is usually below 700 °C, because the bond energy of In-N is small and the decomposition temperature is low. If the growth temperature is too high, the incorporation efficiency of In in InGaN is low, but the quality of the material grown at low temperature is very poor, often showing two-dimensional island-like morphology and three-dimensional island-like morphology [248–254]. This will increase the defects in the quantum well, cause In segregation, and is also one of the reasons for the roughness at the interface between layers, which will reduce the internal quantum efficiency (IQE) [250, 255–257]. Step-flow growth of InGaN quantum wells can obtain good morphology and steep interfaces [248]. Therefore, to suppress the formation of defects at the interface and improve the quality of InGaN/GaN multi-quantum wells, it is necessary to study the method of realizing high In component InGaN quantum well step-flow growth [212].

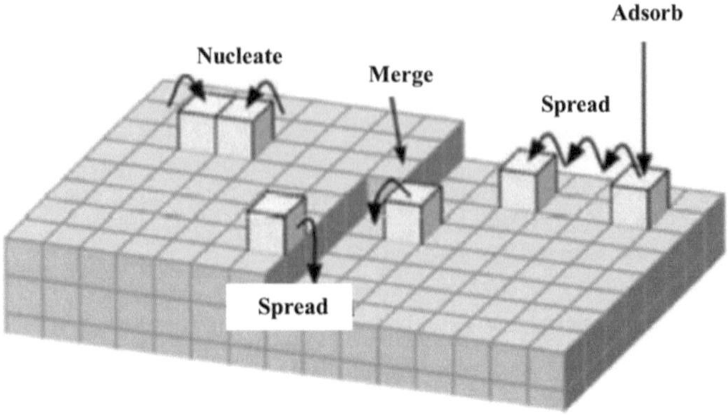

Fig. 7.44 Schematic diagram of atomic diffusion during epitaxial growth [258]

Step-flow growth

If the substrate used has a certain cut angle along a certain crystal face, steps will form on the cut surface, and the width and density of the steps are related to the cut angle. When the cut angle is large, the width of the atomic terrace is small, and when the cut angle is small, the width of the atomic terrace is large. When InGaN is grown on GaN with a step structure, the morphology of InGaN is affected by the growth temperature and atomic migration ability. Figure 7.44 shows a schematic diagram of atomic diffusion during epitaxial growth.

After the atoms are adsorbed on the terrace, they begin to diffuse and migrate in all directions. The atoms that fall on the terrace tend to grow at the edge of the step [259–263]. When InGaN is grown on the adjacent crystal face with a step morphology, assuming the terrace width is constant, if the atomic migration ability is sufficient to make its diffusion distance reach the terrace width, the atom can reach the edge of the upper step, forming step-flow growth. But if the atomic migration ability is insufficient, it cannot diffuse to the edge of the upper step, it is adsorbed on the step surface, gradually forming a nucleus, and the surrounding atoms tend to accumulate around the nucleus, so an island growth mode is formed. At low growth temperatures, atomic migration ability is often insufficient, and island growth is easy to form.

By adjusting the growth temperature and growth rate, high In component InGaN with step-flow morphology can be obtained. The composition of InGaN quantum wells is subject to many constraints. To increase the atomic migration distance, the temperature must be raised, but the incorporation efficiency of In at high temperatures is low. To incorporate enough In, the growth rate must also be increased. However, experiments have shown that even at a high growth temperature of 710 °C and a high growth rate of 0.06 nm/s, the growth morphology is still two-dimensional island-like [258]. Therefore, adjusting the growth temperature and growth rate to obtain a step-flow morphology is not an effective method [212].

The following describes an experiment using a GaN single crystal substrate [212], where the step-flow morphology is obtained by changing the substrate miscut angle.

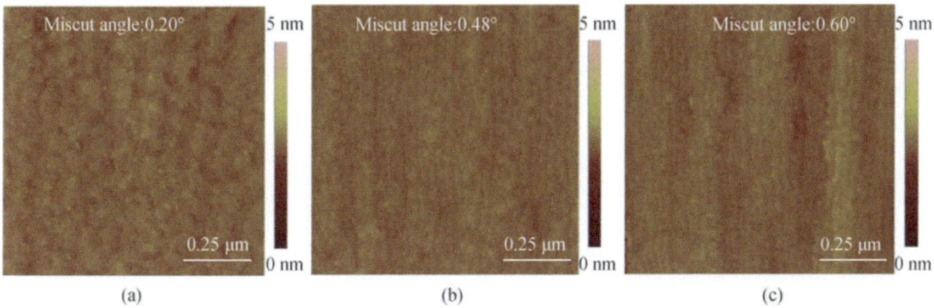

Fig. 7.45 InGaN green light quantum wells grown on substrates with different miscut angles (AFM photos) [258], **a** miscut angle 0.20°, terrace width 83.3 nm; **b** miscut angle 0.48°, terrace width 35.7 nm; **c** miscut angle 0.60°, terrace width 30.3 nm

GaN single crystal substrates have a low dislocation density, good thermal conductivity, easy cleavage, and are conducive to the fabrication of vertical structures. When the substrate's miscut angle is greater than 0.48°, a step-flow grown InGaN green light quantum well can be obtained [258]. Figure 7.45 shows a picture under an atomic force microscope (AFM). A single layer of InGaN quantum well with a thickness of 2.5 nm was grown on self-supporting GaN substrates with different miscut angles. It can be seen that when the miscut angle increases from 0.20° to 0.48°, the step width decreases from 83.3 nm to 35.7 nm. When the miscut angle increases to 0.60°, the step width decreases to 30.3 nm. As shown in Fig. 7.48b, a step-flow morphology is formed when the miscut angle increases to 0.48°. Since the step width of the step-flow morphology decreases with the increase of the substrate miscut angle, when the substrate miscut angle increases to a certain extent, the steps will merge and cause surface roughness. The experiment found that when the miscut angle increased to 0.60°, the phenomenon of step merging had already occurred, so the self-supporting GaN substrate with a miscut angle of 0.48° was chosen for the growth of InGaN green light quantum wells based on the experiment. The internal quantum efficiency (IQE) of the step-flow morphology and two-dimensional island-like morphology of InGaN/GaN green light quantum wells was measured by variable temperature photoluminescence spectroscopy, and it was found that the former was twice as high as the latter. From the full width at half maximum of the photoluminescence spectrum, the two-dimensional island-like morphology quantum well is 166 meV, which is larger than the 131 meV of the step-flow morphology quantum well.

The In component of the green light InGaN quantum well is high, and the lattice mismatch with GaN is large, resulting in a large stress, which produces a high density of interface defects, as shown in Fig. 7.46a. These defects seriously affect the luminescence efficiency of the InGaN/GaN green light quantum well. The main reason for the formation of interface defects is usually considered to be the In clusters on the surface of the InGaN quantum well. Heat treatment and H_2 treatment are common methods to remove In-rich clusters [265–271]. Both methods will effectively remove In-rich clusters, thereby reducing InGaN Density of interface defects in quantum wells. Adding a thin low-temperature GaN cap layer can effectively

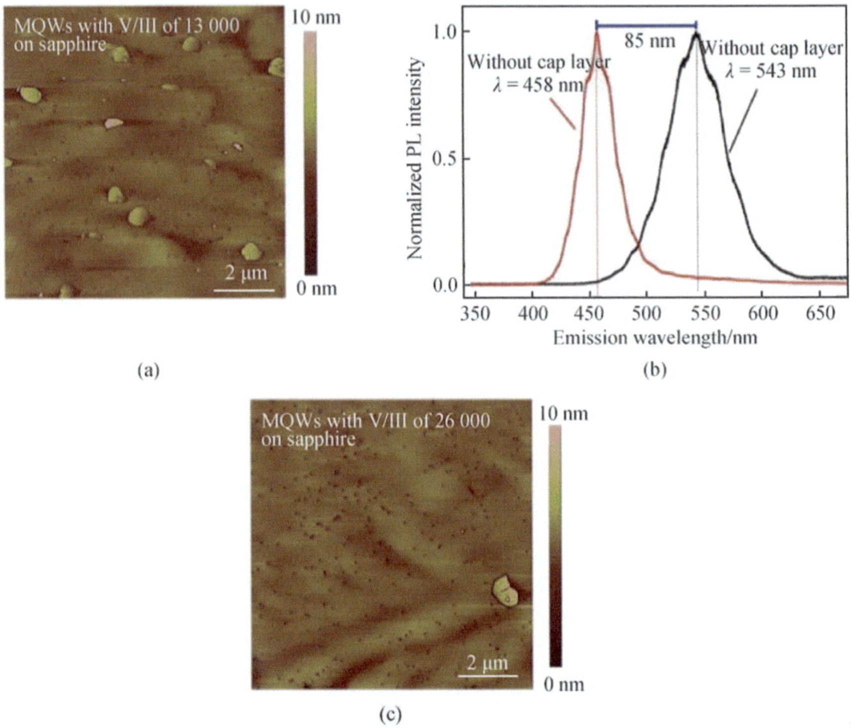

Fig. 7.46 **a** AFM photo of InGaN/GaN quantum well with V/III ratio of 13,000 [258]; **b** Photoluminescence (PL) spectra of quantum wells with and without low-temperature GaN cap layer [258]; **c** AFM photo of InGaN/GaN quantum well with V/III ratio of 26,000 [264]

protect the already grown InGaN quantum wells, as shown in Fig. 7.46b. Another method is to increase the V/III ratio from 13,000 to 26,000, so that high-quality InGaN/GaN green quantum wells with low interface defect density can be obtained even under single-temperature growth (the growth temperature of quantum wells and quantum barriers are the same) [264]. As shown in Fig. 7.46a–c, increasing the V/III ratio improves the crystal quality of single-temperature grown InGaN/GaN quantum wells [212].

The growth temperature of the p-type AlGaN confinement layer is high, and high-temperature growth of AlGaN easily causes thermal degradation of the active region InGaN quantum wells, severely affecting the performance of the laser [212]. As shown in Fig. 7.47a–c, there are obvious black areas in the fluorescence microscope image, indicating that the InGaN quantum wells here have undergone thermal degradation, thereby destroying the original crystal structure. When the p-type AlGaN confinement layer is grown at a lower temperature (900 °C), as shown in Fig. 7.47b–d, the fluorescence microscope image shows uniform luminescence everywhere, indicating that the crystal quality of the InGaN quantum wells is good at low temperatures. The growth of AlGaN layers at low temperatures will inevitably increase the

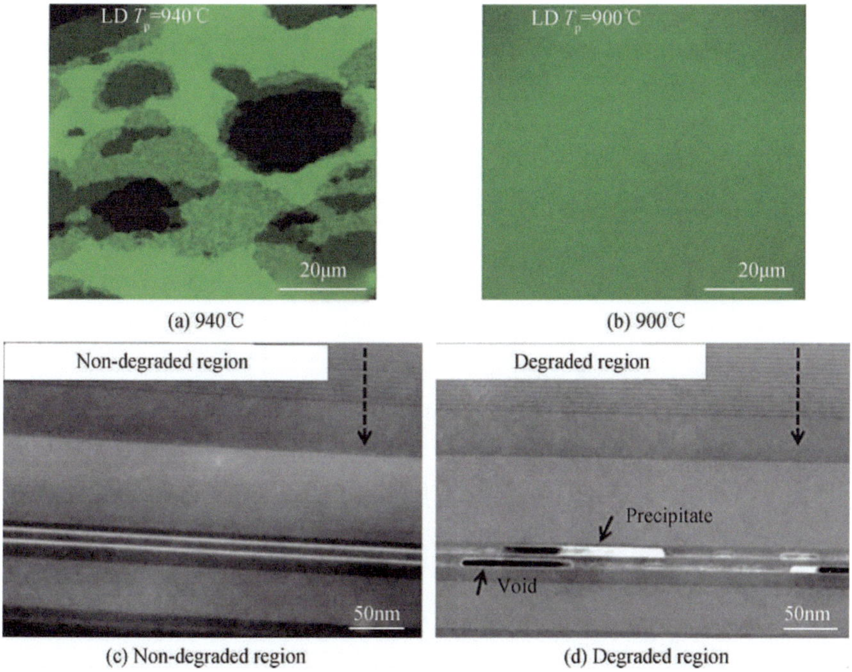

Fig. 7.47 Micro-PL images (**a**, **b**) and STEM photos (**c**, **d**) of p-type AlGaN grown at different temperatures in the structure of green lasers [228]

resistivity of the AlGaN layer and introduce more C impurities. By reducing the growth rate and increasing the growth pressure, the incorporation of C impurities can be effectively suppressed to an impurity concentration below 1×10^{17} cm^{-3}.

In addition, the thickness of the p-AlGaN confinement layer can be reduced by designing an ITO upper confinement layer. Figure 7.48a shows the structure of a green laser with a composite confinement layer with an ITO layer [272]. ITO has good conductivity and transparency, and has been widely used as an electrode material in GaN-based LEDs. The absorption coefficient of ITO is about two orders of magnitude lower than that of commonly used metal materials for electrodes, so inserting an ITO layer and thinning the p-type AlGaN layer will not introduce additional internal optical losses, as shown in Fig. 7.48b. The refractive index of ITO is about 2, which is less than that of the AlGaN layer, and can provide effective optical field confinement. The growth temperature of ITO is as low as 300°C, and the introduction of an ITO layer can reduce the high-temperature growth time of the AlGaN layer, effectively reducing the impact of high temperature on InGaN quantum wells and alleviating the thermal degradation of InGaN quantum wells at high temperatures. According to the results of pulse current tests at room temperature [273], compared with traditional green lasers, the threshold current density of composite green lasers with an ITO layer has dropped from 5–1.6 kA/cm^2, and the slope efficiency has increased from

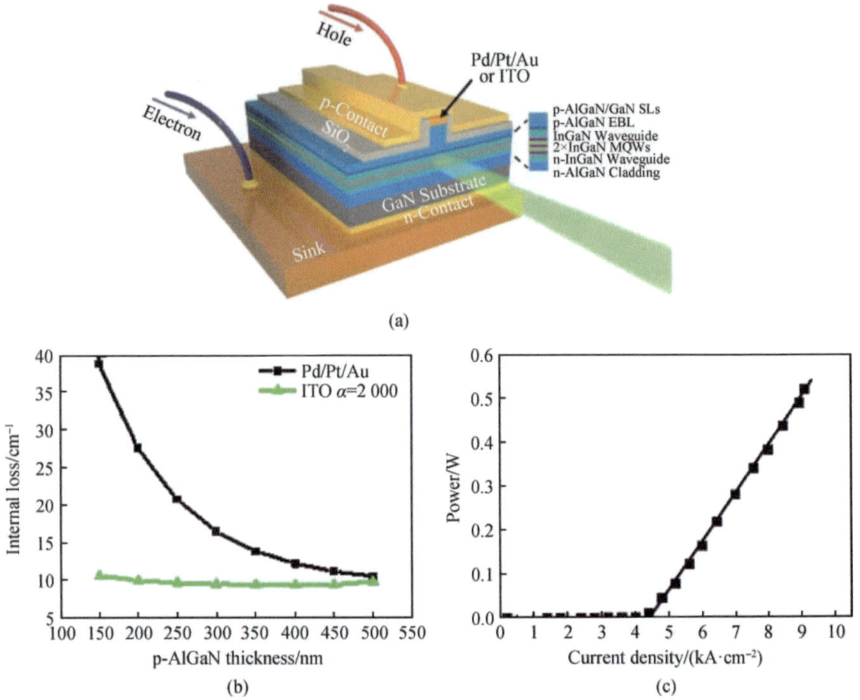

Fig. 7.48 Composite green laser and photoelectric characteristic curve. **a** Incorporating ITO as a limiting layer in the composite green laser and traditional green laser structure diagram [272]; **b** Calculated comparison of internal losses between the green laser using the traditional Pd/Pt/Au structure and the composite green laser using the ITO structure [272]; **c** Output power-current curve of the green laser with ITO as the limiting layer, p-type $Al_{0.035}$ $Ga_{0.965}$ N limiting layer thickness 300 nm, front cavity reflectivity 40% [272]

0.16–0.20 W/A. As shown in Fig. 7.48c, the output power-current (P-I) curve of the composite confinement layer green laser with an ITO layer was tested under the condition that the front cavity reflectance is 40%, and an output optical power of 500 mW was achieved under the condition of a working current of 1.6 A (current density of 9 kA/cm^2). Through further optimization, by reducing the absorption coefficient and contact loss of ITO, it is expected to obtain a composite structure green laser with higher performance [212].

Due to the difficulty in preparing green lasers, relevant international research did not achieve a breakthrough until 2009, when Nichia Corporation in Japan and Osram in Germany first achieved continuous room temperature lasing of green lasers.

After successfully developing blue lasers, companies like Nichia and Osram began to focus on the research of GaN-based green lasers. In 2009, Nichia and Osram made a breakthrough in the research of c-plane GaN green lasers. In January 2009, Osram was the first to achieve a green laser with a lasing wavelength greater than 500 nm on a c-plane GaN substrate, with a threshold current density of 6.2 kA/cm^2 and a slope efficiency of 0.65 W/A. In May 2009, Nichia reported a green laser with a lasing wavelength of 510–515 nm, a threshold current density of 4.4 kA/cm^2, a threshold voltage of 5.2 V, and a predicted lifespan of over 5000 h when continuously outputting 5 mW at 25°C. In 2010, Osram reported a c-plane room temperature continuous operation green laser with a lasing wavelength of 524 nm, a threshold current of 97 mA, a slope efficiency of 0.336 W/A, and an electro-optical conversion efficiency of 2.3% when outputting 50 mW. In February 2012, Osram reported a long-life green laser with a lasing wavelength of 519 nm, a continuous output power exceeding 100 mW, an electro-optical conversion efficiency of 6%, and a lifespan of 10,000 h when outputting 50 mW at 40 °C. In 2013, Osram reported a 250 mW green laser with a lasing wavelength of 520 nm. Nichia also made rapid progress in the research of green lasers, reporting a green laser with an output power of up to 1.01 W in 2013. In China, the Suzhou Institute of Nanotech and Nanobionics, Chinese Academy of Sciences, was the first to develop a GaN green laser in 2014, with a power-current curve as shown in Fig. 7.49a, a threshold current density of 8.5 kA/cm^2, and a slope efficiency of 0.22 W/A. Figure 7.49b shows its lasing spectrum, with a lasing wavelength of 508 nm [211]. The team successfully developed a GaN-based green laser in 2021, demonstrating a slope efficiency of 0.8 W/A and an output power of 1.7 W.

4. Semi-polar and non-polar GaN-based green lasers

Since the c-plane is the polar face of GaN, the polarization field in InGaN quantum wells can reach several megavolts per centimeter. As the lasing wavelength increases, the In component in the quantum well further increases, and the polarization field also increases. In the green light band, the In component in the quantum well is more than 30%, and the Quantum Confined Stark Effect (QCSE) has severely affected the recombination probability of carriers, limiting the development of GaN-based green lasers [211]. The polarization field on the semi-polar or non-polar face is smaller, and the impact of QCSE is weaker. According to the report by Park and others, the heavy hole band and light hole band of some quantum wells on the semi-polar or non-polar face are separated, making the quantum well have a higher material gain. Therefore, institutions such as the University of California, Santa Barbara (UCSB), Rohm Company of Japan, Sumitomo Electric, and Soraa Inc. of the United States are committed to the research of semi-polar and non-polar face GaN-based green lasers. In 2009, Rohm Company first reported a non-polar face GaN laser with a lasing wavelength greater than 500 nm, a threshold current density of 3.1 kA/cm^2, and a lasing wavelength of 500.2 nm when continuously outputting 15 mW. Sumitomo Electric later reported a green laser with a lasing wavelength of 531 nm in July 2009, using a self-supporting GaN substrate on the $(20\bar{2}1)$ face, with a threshold current density of 15.4 kA/cm^2. In July 2010, the company achieved a laser with a pulsed lasing wavelength of 533.6 nm and a continuous lasing wavelength of 523.3 nm on the semi-polar $(20\bar{2}1)$ face. In June 2012, the company achieved a long-life green

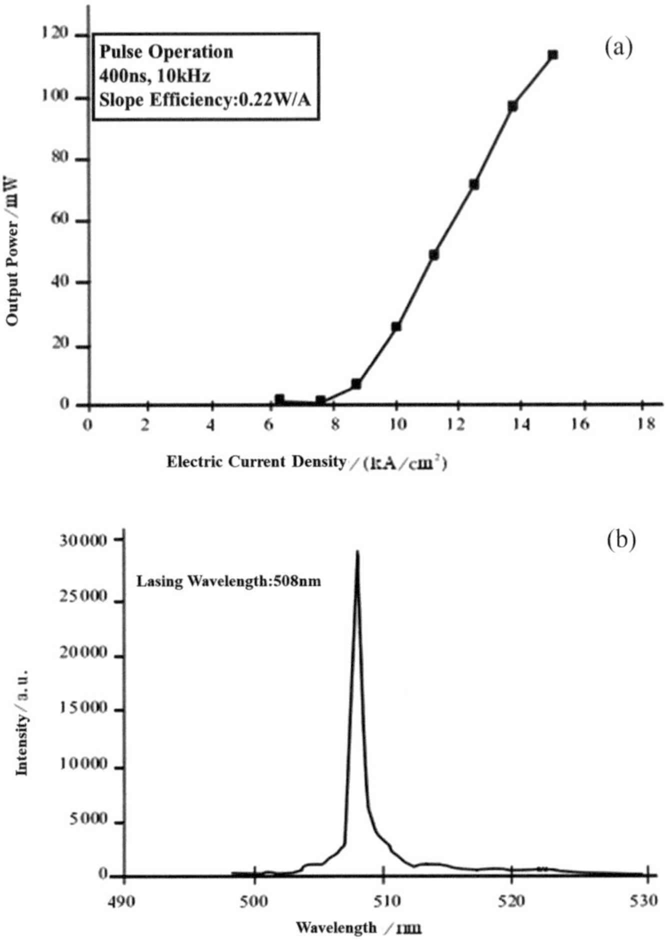

Fig. 7.49 Power-current curve (**a**) and lasing spectrum (**b**) of the green laser developed by the Suzhou Institute of Nanotech and Nanobionics [211]

laser with a continuous lasing wavelength of 525 nm and an output power of 50 mW on the semi-polar (1) face GaN substrate. In November 2010, Soraa Inc. reported a green laser with a lasing wavelength of 521 nm, and when continuously operating at an output of 60mW, the electro-optical conversion efficiency of the laser was 1.9%. Sony Corporation reported in 2018 [274] a green laser grown on the (20$\bar{2}$1) face, reaching an output power of nearly 1 W under a current of 1.2 A. Currently, the quality of GaN-based materials on semi-polar and non-polar faces is not as good as that on the c-plane GaN, and the size of the GaN substrate on the semi-polar and non-polar faces is smaller and more expensive, all of which hinder the development of lasers on the semi-polar and non-polar faces. Therefore, lasers on the semi-polar and non-polar faces have not yet been commercialized.

5. Ultraviolet Lasers

Ultraviolet lasers have many applications in biology and medicine. Compared to violet lasers, the research on ultraviolet lasers is more challenging. As the wavelength of the laser moves towards the short-wave direction, the laser quantum well needs to use AlGaN material, and the Al component in the optical limiting layer needs to increase accordingly. High-quality AlGaN materials are difficult to grow, so the efficiency of AlGaN quantum wells is low. Moreover, the ionization energy of Mg acceptors in p-AlGaN increases with the increase of Al components, so the resistivity of p-AlGaN is very high, and the working voltage of the laser is very high. All these make it difficult for ultraviolet lasers to achieve lasing [211]. In June 1996, I.Akasaki et al.in Japan successfully achieved room temperature pulsed lasing of a 376 nm single quantum well laser using a separately confined structure, with a threshold current density of 3.0kA/cm^2 and a threshold voltage of 16 V. In May 2001, Nichia Corporation reported the first continuously operating GaN-based ultraviolet laser. The active area of the laser is an AlInGaN single quantum well, the pulsed lasing wavelength is 366.4 nm, the threshold current density for continuous operation at room temperature is 3.5 kA/cm^2, and the threshold voltage is 4.8 V. When the output is 2 mW, the estimated life of the laser is 500 h. In June 2001, Nichia Corporation reported a single quantum well ultraviolet laser with a quantum well of GaN material, with a continuous working wavelength at room temperature of 369 nm, a threshold current density of 3.5 kA/c ^{2}, a threshold voltage of 4.6 V, and an output of 2 mW, the estimated life of the laser is 2000 h. To continue to expand towards the short-wave direction (<361 nm), some scholars use AlGaN materials as quantum wells. In December 2008, Hamamatsu Photonics Corporation reported an electrically injected pulsed working wavelength of 336 nm, a threshold current density of 17.6 kA/cm^2, and a threshold voltage of 34 V. Another group of scholars used narrow GaN quantum wells to reduce the polarization effect in the quantum wells, reduce band bending, and thus shorten the working wavelength of the laser. In 2013, Hamamatsu Photonics Corporation used 1–1.5 nm GaN multiple quantum wells to achieve room temperature pulsed lasing of a 340 nm laser, with a threshold current density of 15.4 kA/cm^2 and a threshold voltage of 27.9 V. In 2016, the State Key Laboratory of Integrated Optoelectronics of the Institute of Semiconductors of the Chinese Academy of Sciences, in cooperation with the Suzhou Institute of Nanotechnology of the Chinese Academy of Sciences, achieved room temperature electric injection lasing of a GaN ultraviolet laser. The laser with a stripe width of 10 μm and a cavity length of 600 μm has a threshold current density of 1.6–2.0kA/cm^2, a lasing wavelength of 392–395 nm, and a continuous lasing output light power of up to 80 mW. Figure 7.50a is the lasing spectrum of the ultraviolet laser, Fig. 7.50b is the P-I curve, and Fig. 7.51 is the blue fluorescent spot formed when the ultraviolet laser lasing is shone on copy paper (ultraviolet light is invisible to the human eye, and blue fluorescence is emitted when the ultraviolet laser shines on copy paper).

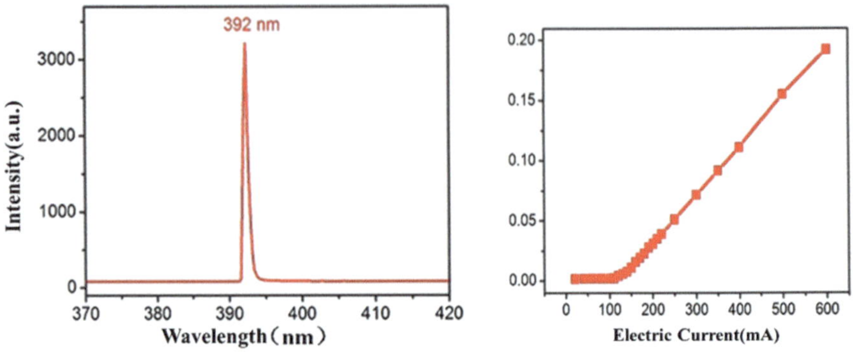

Fig. 7.50 **a** Lasing spectrum of GaN-based ultraviolet laser. **b** P-I curve of GaN-based ultraviolet laser

Fig. 7.51 The blue fluorescent spot formed on the photocopy paper when irradiated by an ultraviolet laser

7.5 GaN Vertical Cavity Surface Emitting Laser

Traditional edge-emitting lasers (EEL) emit lasers from the edge of the chip, while vertical-cavity surface-emitting lasers (VCSEL) emit lasers from the surface of the chip. The light emission window of EEL is a micrometer-level rectangle, and due to diffraction effects, the emitted light beam is elliptical; while the light emission window of VCSEL is circular, the emitted light beam has a circular divergence angle, which is higher than EEL when used for optical fiber coupling. VCSEL can simply achieve two-dimensional integration without segmentation, which can not only be used for wafer-level testing, but also provide higher output power. Thanks to the short cavity length and small active area, VCSEL has a lower threshold, better temperature stability of the lasing wavelength, and much higher modulation speed than EEL [275, 276]. Because the bandgap width of $In_x Ga_{1-x} N$ is continuously adjustable in the range of 0.7–3.5 eV [277], GaN-based VCSEL can cover the ultraviolet to visible light band.

7.5.1 GaN VCSEL Applications

GaN-based VCSELs have huge application prospects and market demand in high-resolution printing, solid-state lighting, visible light communication, miniature atomic clocks and sensors, etc. [278].

1. High-resolution printing

VCSEL has uniform performance, stable divergence angle and output power. The emission wavelength of GaN-based materials is shorter, so a smaller diffraction point can be obtained, which can improve the resolution of the printer. Combining infrared VCSEL with GaN-based blue and green VCSEL can achieve full-color printing.

2. Solid-state lighting

In solid-state lighting (SSL), commercial visible light LEDs suffer from a serious efficiency drop [279]. However, laser-based light sources can operate at peak efficiency under higher current densities and can avoid efficiency drop limitations [280]. At the same time, VCSEL is small in size, can provide a symmetrical circular beam directionally, which is easy to capture and focus, and is conducive to the miniaturization of lighting equipment. In addition, the elements in the two-dimensional VCSEL array can be individually addressed, which makes it possible to develop dynamic light emission modes in intelligent lighting systems [281].

3. Visible light communication

Although the development of LEDs has paved the way for the commercialization of visible light communication (VLC) systems, due to the longer carrier lifetime associated with the spontaneous emission process, the performance of the transmitter is limited by the spectral bandwidth and slow frequency response [282]. Compared with LEDs, VCSEL can provide a narrower spectrum and faster modulated light source, thereby improving the response speed of visible light communication and reducing energy consumption [278].

4. Atomic clocks

Miniature atomic clocks can be divided into microwave atomic clocks and optical atomic clocks. Theoretically, optical atomic clocks have higher accuracy because their frequency is four orders of magnitude higher than that of microwave atomic clocks [283]. Chip-level optical atomic clocks need to be modular and miniaturized, so they require small size and low power consumption light sources. In 2000, Kitching et al. [284] reported a full-optical atomic clock using modulated VCSEL, paving the way for significantly reducing the size and power consumption of atomic clocks. Currently, low-power and stable VCSELs play an irreplaceable role in chip-level optical atomic clocks [285].

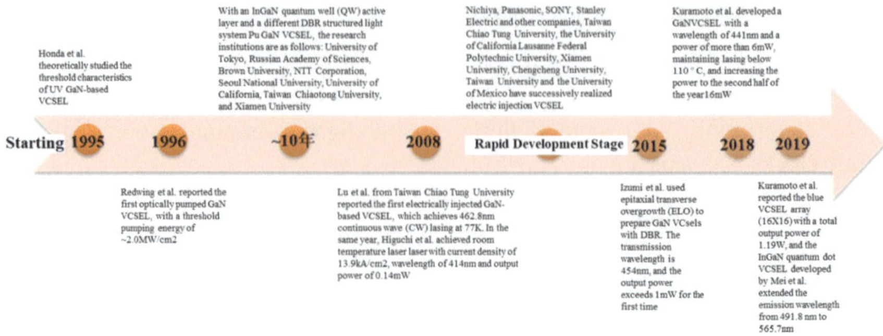

Fig. 7.52 Development history of GaN-based VCSELs

5. Sensors

VCSELs can conveniently achieve two-dimensional integration, have high temperature stability and low power consumption, making them very suitable for use in small sensors that need to work for a long time. In 2017, Apple first introduced an infrared sensor based on VCSEL arrays in its product iPhone X, which implemented facial recognition on mobile phones and led the 3D sensing market. In addition, larger high-power VCSEL arrays can also be used as lidar for cars [286]. Compared with infrared VCSELs, sensors using shorter wavelength visible light VCSELs have higher spatial resolution.

7.5.2 History of GaN VCSEL

Development Compared to the development process of GaAs-based VCSELs, the development of GaN-based VCSELs is slower, mainly because it was difficult to grow high-quality GaN films at the time. In 1989, Amano et al. [287] introduced Mg doping in GaN and successfully realized the epitaxial growth of p-GaN thin films for the first time. In 1991, Nakamura et al. [288] proposed a two-step growth method, which solved the p-doping problem in GaN and laid the foundation for the research of GaN-based VCSELs. The development history of GaN-based VCSELs is shown in the following Fig. 7.52.

7.5.3 The Present Research Status and the Existing Issues of Blue-Violet, Green, and Ultraviolet GaN VCSELs

1. Blue-Violet GaN VCSEL

In 1999, Krestnikov et al. from the Russian Academy of Sciences[289] grew 37 cycles of $Al_{0.15}Ga_{0.85}N$/GaN and 2λ vertical cavities. The active area in the middle of the 2λ cavity is composed of ultra-thin InGaN interlayers and InGaN/GaN multilayer structures (a total of 12 layers). The other side of the cavity is air, without growing DBR. Under optical pumping, a 401 nm laser was observed in the vertical cavity. In the same year, Someya et al. from the University of Tokyo [290] reported the first blue double DBR GaN VCSEL. Under the irradiation of a dye laser ($\lambda = 367$ nm), the VCSEL lased at 399 nm at room temperature, with a lasing linewidth less than 0.1 nm. This marked the practical blue VCSEL can be realized in the GaN material system [278].

The blue-violet GaN VCSEL did not achieve electric pumping until 2008 mainly because of two difficulties [291]:

(1) $Al_xGa_{1-x}N$ and DBRs composed of GaN have large lattice mismatches and thermal expansion coefficient differences, making the epitaxially grown DBR structure prone to cracking. Cracks can reduce its light reflectivity and increase scattering loss, and may also become a leakage current path; dielectric DBRs are non-conductive, so how to effectively inject current into GaN VCSELs is a challenge.

(2) Effective current injection requires low resistance and low light loss carrier injection paths, and p-type doping of GaN is very difficult to achieve, so how to effectively inject current into the quantum well active area is another challenge to consider.

Based on many years of optically pumped GaN VCSEL research, researchers from the National Chiao Tung University in Taiwan [291] made the first electrically injected GaN VCSEL in 2008; this device uses a hybrid structure of epitaxially grown AlGaN DBR and Ta_2O_5 /SiO_2 dielectric DBR, with a cavity length of 5λ, and 10 pairs of InGaN/GaN multi-quantum well active layers in the middle; at the same time, the device uses highly transparent and conductive ITO as the current spreading layer; at 77 K, the device achieved continuous lasing at 462.8 nm, with a threshold current of 1.4 mA.

In 2008, Nichia Corporation [292] reported the RT-CW performance of GaN VCSEL: lasing at 414 nm, with a threshold current of 7 mA and a threshold voltage of 4.3 V. This device uses a double dielectric SiO_2 /Nb_2O_5 DBR, and the active area consists of two pairs of InGaN (9 nm) /GaN (13 nm) multi-quantum wells. The device can achieve RT-CW operation mainly due to two points: one is to improve heat dissipation by mounting the device on a high thermal conductivity Si substrate through laser lift-off and wafer bonding; the second is to control the thickness of the cavity by growing Nb_2O_5 adjustment layer, so that the ITO layer is located at the node of the light field, and the active area is located at the antinode of the light field, which increases the gain and reduces the loss. Although the threshold current density of this device is large, its threshold voltage is small, which is because the device's non-p-type mesa vertical structure avoids the current crowding effect. In 2009, Omae et al. from Nichia Corporation [293] replaced the sapphire substrate with a GaN substrate, grew high-quality epitaxial layers, and increased the output power of the device from 140 to 620 μW. Later, after two years of optimization of the epitaxial structure and process conditions, Kasahara et al. from Nichia Corporation [294] made a blue optically pumped GaN VCSEL with a threshold current of 1.5 mA, a lasing wavelength of 451 nm, and a maximum output power of up to 700 μW.

In 2012 and 2014, the University of California developed two non-polar VCSEL devices [295, 296], the difference in their preparation methods is: epitaxial growth on non-polar m-plane rather than the usual c-plane GaN substrate, and then using bandgap selective photoelectrochemical etching was used to replace laser lift-off technology to remove the substrate. Epitaxial growth on the m-plane substrate can make the light polarization degrees of the two devices reach 72 and 100% respectively. High polarization degree is beneficial for the manufacture of polarization-locked GaN VCSEL arrays, which have certain application value in the development of displays, projectors, sensors, etc. [278].

The Micro/Nano Optoelectronics Research Lab at Xiamen University has studied electrically injected blue GaN VCSEL has been thoroughly researched. In 2014, Liu et al. [297] manufactured a high-quality VCSEL device with a Q value of up to 3570 using a two-step substrate transfer technique. This device has a coupled quantum well active area composed of 12 pairs of top and 17.5 pairs of bottom ZrO_2/SiO_2 DBRs, 5 pairs of InGaN quantum wells, and a GaN barrier layer. However, the heat dissipation ability of the device using adhesive bonding is poor, and no lasing is produced; after switching to metal bonding, a GaN VCSEL with a threshold current as low as 0.93 mA and a lasing wavelength of 424 nm was obtained.

In 2016, Izumi et al. from Sony Corporation [298] reported that when GaN VCSEL works under RT-CW conditions, the lasing wavelength is 453.9 nm, and the output power has reached 1.1 mW, which is the first time that the output power of GaN VCSEL has reached the mW level. The main feature of this device is that there are dielectric DBRs on both sides of the cavity, and one of the DBRs is embedded in the n-type GaN grown by lateral overgrowth technology. This structure has several outstanding advantages: ① High reflectivity dielectric DBRs can be formed on both sides of the cavity; ② It is easy to obtain a shorter cavity length through overgrowth; ③ High heat dissipation efficiency.

In 2019, blue-violet GaN VCSEL made breakthroughs in increasing output power and reducing threshold current. Stanley Electric Company [299] manufactured a 16 × 16 two-dimensional GaN VCSEL array, as shown in Fig. 7.53. Thanks to the high-quality AlIn/GaN DBR and SiO_2 buried structure, the total output power of this array reached 1.19 W. Sony Corporation [300] introduced boron ion injection for lateral current limitation and used curved DBR for lateral optical limitation, and obtained lasing at 445.3 nm. Directly growing a curved DBR under the substrate can simplify the process and eliminate the steps of laser stripping and substrate transfer, but it lengthens the cavity and causes the lasing wavelength to drift.

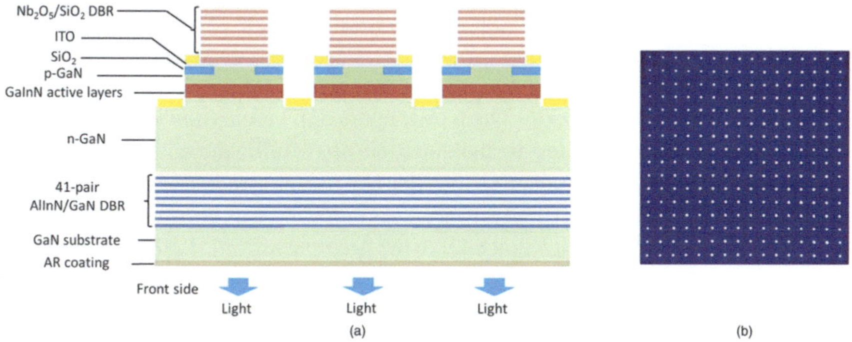

Fig. 7.53 Blue VCSEL array [299]. **a** Schematic diagram of the blue VCSEL array; **b** Emission image of the blue VCSEL array working under conditions below the threshold

2. Green GaN VCSEL

InGaN QW has been successfully used in blue LEDs and LDs, and there are many reports on blue-violet GaN VCSELs with InGaN QW as the active layer. Theoretically, InGaN QW can cover green and longer wavelengths, but within the green wavelength range, the QW structure usually suffers from low emission efficiency, which is commonly referred to as the "green gap" [301]. To extend the emission wavelength to the green band, it is necessary to increase the In content in the InGaN QW layer. For this, quantum wells usually need to be grown at low temperatures, but this will lead to a decrease in crystal quality. At the same time, due to the large lattice constant difference between GaN and InN, an increase in In content will lead to more defects and a larger strain-induced electric field [278].

Defects are non-radiative recombination centers, and the built-in electric field will cause the quantum confined Stark effect (QCSE), which pulls electrons and holes to different sides of the quantum well, thereby reducing the recombination efficiency [302].

In addition, the effective mass of carriers in the GaN-based material system is relatively large, leading to a higher transparent carrier density [303], which is another obstacle to the preparation of high-performance green lasers with InGaN QW structures. However, the situation is different in quantum dot structures. It is well known that using quantum dots as the active region can effectively reduce the threshold current in laser devices [304]. GaN-based quantum dots are zero-dimensional nanocrystals, where electrons and holes are confined in a very small space, thus producing discrete states with δ function state density. This state density has the potential to reach high gain peaks in quantum dot structures. If the quantum dots are small enough, the energy difference between discrete energy levels is large, electrons tend to concentrate at the ground state energy level, therefore, carriers in higher sub-bands can be ignored, and the behavior of internal carriers is similar to the two-level system in a single atom. When quantum dots are used as the gain medium for semiconductor lasers, the lasing performance is ultimately independent of the effective mass of the carriers. In addition, the localized state of carriers in quantum dots may prevent them from being captured by defects and dislocations.

Quantum dots used as the active region for light-emitting devices are usually grown by strain-driven SK mode, which can significantly reduce the strain remaining in the quantum dots compared to the situation of two-dimensional QW epitaxial layers. Strain relaxation can reduce QCSE, thereby reducing radiative recombination. These characteristics suggest that InGaN quantum dots can be used as the active region to manufacture VCSELs that emit in the "green gap" [305]. In 2008, Cai et al. [306] achieved the first optically pumped blue-green GaN VCSEL by optimizing the growth of InGaN/GaN QW; they observed lasing at room temperature, with a lasing wavelength of 498.8 nm and a linewidth of 0.15 nm. In 2016, Weng et al. [307] achieved the first continuous lasing of an electrically injected green GaN VCSEL; this GaN VCSEL device uses InGaN quantum dots as the gain medium, with a lasing wavelength of 560.4 nm under RT-CW operation, and a low current threshold of 0.61 mA. In 2017, Mei et al. [308] also used InGaN quantum dot active region devices, and by adjusting the cavity length, the wavelength of the device was extended from 491.8 nm (blue-green) to 565.7 nm (yellow-green), covering most of the "green gap", as shown in Fig. 7.53. In the same year, Xu et al. [309] achieved a GaN VCSEL that emits both blue and green light under electrical injection by using quantum dots in the quantum well active region (QD-in-QW); the device first lased at a threshold current of about 2 μA, emitting green light with a wavelength of 545 nm, and then another lasing peak appeared at 430 nm with a further increase in current, with a threshold current of about 5 mA; Fig. 6 shows the emission spectra at different injection currents. In 2018, Xu et al. achieved a green VCSEL based on localized state emission in quantum wells by combining the usually blue-emitting InGaN/GaN quantum well with a microcavity. Due to the resonant cavity effect, the emission efficiency of the localized states resonating with the cavity mode is greatly improved, resulting in a rapid increase in green (peak near 493 nm) emission at high injection currents, achieving lasing; Fig. 7.54 shows the emission spectra at different injection

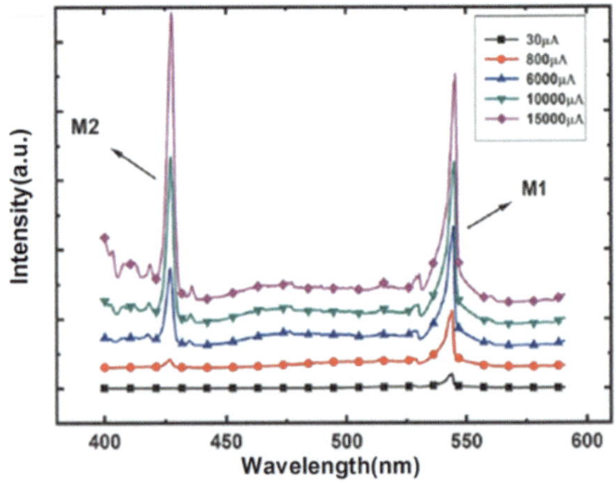

Fig. 7.54 Laser emission spectra under various injection currents measured at room temperature [309]

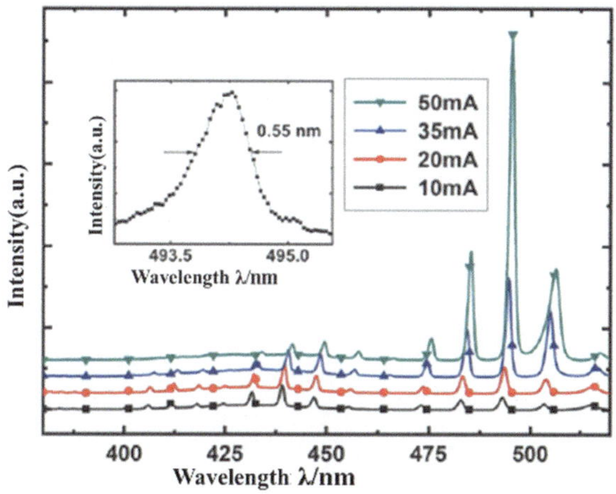

Fig. 7.55 EL spectra of VCSEL measured at different currents. (The insert is the linewidth of the emission peak measured at high resolution)

currents. The above research provides ideas for the design and manufacture of green VCSELs (Fig. 7.55).

3. Ultraviolet GaN VCSEL

The UV GaN VCSEL was the earliest reported GaN VCSEL. In 1996, Redwing et al. reported the first optically pumped GaN VCSEL with a lasing wavelength of 363.5 nm and a lasing threshold of about 2 MW/cm^2. Since then, GaN VCSELs have developed rapidly. Over the past 20 years, UV, blue, and green GaN VCSELs have all achieved optical and electrical pumping, but the development of UV GaN VCSELs has encountered a bottleneck, and few people have reported electrically pumped UV GaN VCSELs. Compared with other GaN VCSELs, the technical difficulty of realizing UV VCSELs is great, facing many difficulties. The two main difficulties are: First, the UV emission band requires $Al_xGa_{1-x}N$ with a high aluminum content. When the aluminum content is high, it is difficult to obtain high conductivity $Al_x Ga_{1-x} N$; Second, the reflectivity of III-N type DBR in the UV light band is relatively low, and there are problems such as lattice mismatch, thermal expansion coefficient mismatch, in-plane component changes, and low refractive index difference, so it is difficult to obtain DBR with high reflectivity. In 2010, Chen et al. from Nanyang Technological University realized a vertical cavity structure composed of in-situ grown bottom $Al_xGa_{1-x} N/Al_yGa_{1-y}N$ distributed Bragg reflectors and top SiO_2/HfO_2 dielectric mirrors; this structure lased at 343.7 nm with a lasing threshold of 0.52 MW/cm^2. They used nanosphere lithography (NSL) to prepare nanopillars with an average diameter of about 500 nm, which can provide excellent optical confinement, so the lasing threshold is low. In 2015, Liu et al. from Georgia Institute of Technology realized a GaN VCSEL with conductive UV n-DBR, its lasing wavelength is 367.5 nm, and the lasing threshold is about 1 MW/cm^2; the following year, they achieved lasing at 374.9 nm, with a threshold power density of 1.64 MW/cm^2. In 2018, Chang et al. from National Chiao Tung University demonstrated the first optically pumped GaN

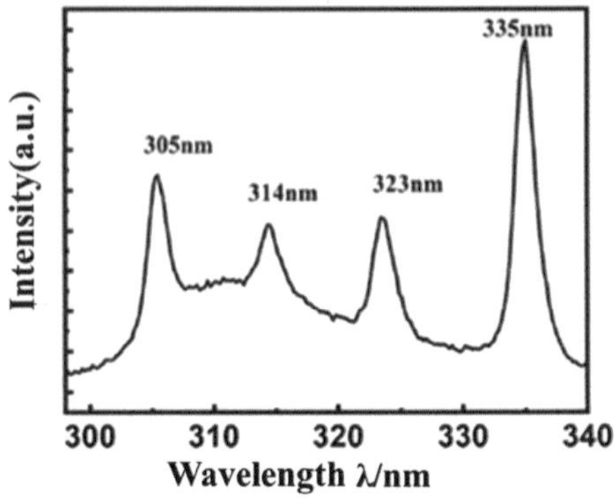

Fig. 7.56 Emission spectrum of the microcavity at room temperature

VCSEL using a high contrast grating (HCG) as the top reflector, its lasing wavelength is 369.1 nm, and it has a low threshold of 0.69 MW/cm^2. Currently, HCG has been widely studied in long-wavelength VCSELs. It has advantages such as small thickness, high reflection bandwidth, and allowing the setting of resonant wavelength through grating parameters, but its manufacturing process is particularly complex. In 2019, Park et al. from Georgia Institute of Technology introduced an air gap/ $Al_{0.05}GaN_{0.95}$ N DBR to increase the refractive index difference, thereby achieving a DBR with high reflectivity; this device lased at 375 nm under optical pumping, with a threshold power density as low as 270 kW/cm^2. When the band is extended to the deep ultraviolet (DUV) band, the lasing of GaN VCSEL becomes more difficult, with high optical losses in its cavity. Zheng et al. analyzed the optical losses in the nitride DUV microcavity in detail and successfully fabricated a DUV nitride Fabry–Perot (F-P) microcavity with an AlGaN quantum dot active layer and double-sided HfO_2 / SiO_2 DBR. The emission spectrum of this microcavity at room temperature is shown in Fig. 7.56, and the optical loss in the cavity can be inferred from the Q value of the cavity mode to be 10^3 cm^{-1}. Through calculations, Zheng et al. believe that the main source of optical loss is interface scattering loss, and the two key parameters determining interface loss are the interface roughness after laser stripping and the overlap of the rough interface with the standing wave. This research provides useful information for improving DUV VCSEL devices and reducing their cavity losses.

VCSEL based on the GaAs material system has matured and entered the market, while VCSEL based on the GaN material system is rapidly developing. It is reported that the output power of a single-tube blue VCSEL has reached tens of milliwatts, and the output power of an array of blue VCSEL has reached the watt level. It can be expected that there will be blue VCSEL products in the near future. In addition, the use of quantum dots can largely eliminate the effects caused by electric fields and defects in quantum well structures, and eliminate the problem of higher transparent carrier concentration caused by larger effective carrier mass. At the same time, quantum dots have a strong carrier localization effect, which can break through the traditional "green gap" limitation and achieve green VCSELs with very low threshold currents [278].

Another method to make green VCSELs is to use localized states in blue quantum wells, combined with the resonance effect of the resonant cavity to improve their radiation efficiency, and achieve green lasers under high injection. This method is conducive to obtaining high output power. However, to achieve commercialization, the performance of green VCSELs needs to be further improved, especially to increase their output power. Possible approaches include arraying, using localized states in quantum wells, using QD-in-QW active region structures, optimizing DBR structures, etc. Extending the emission wavelength range covered by GaN VCSELs to longer wavelengths (such as red light) and shorter wavelengths (such as deep ultraviolet) is a future research direction. This requires growing active regions of different compositions and structures according to the design wavelength, as well as high-reflectivity, high-quality mirrors that match the emission wavelength. Whether from an industrial perspective or a scientific research perspective, GaN-based VCSELs pose greater challenges than GaAs-based VCSELs. One of the technical bottlenecks to achieving the former is material issues [278]:

(1) To reduce the defect density in the material, high-quality GaN substrates are needed, but their growth is still very difficult and the price is still high;
(2) The technology for epitaxial growth on sapphire or silicon substrates also needs to be improved to obtain high-quality epitaxial materials and devices;
(3) In In-rich InGaN, there are high-density defects and large polarization fields. QCSE spatially separates holes from electrons, leading to reduced internal quantum efficiency and increased threshold current;
(4) In the deep ultraviolet band, the doping of materials and the absorption loss of DBR are still key issues to be solved.

Another bottleneck lies in the design of the device structure[278]:

(1) Well-designed current spreading and contact structures to reduce losses in the carrier injection process;
(2) Well-designed optical confinement structures to reduce light loss;
(3) Well-designed heat dissipation structures, such as elongating the cavity length, transferring the substrate, while balancing the loss of wavelength thermal stability brought about by this and the difficulty of process implementation.

Although there are many technical difficulties, GaN-based VCSEL still has a lot of opportunities in the future. In the fields of data transmission and intelligent sensing, it will become an essential key component to support the future information society.

7.6 Development Trends

7.6.1 LED Development Trends

1. Smart Lighting

Future LED lighting will continue to mature, constantly reducing costs and improving performance, with increasing penetration rates in lighting, display, and other fields; and as performance improves and understanding of LEDs deepens, new application fields will continue to emerge. New smart lighting technologies with intelligent control interfaces or functions are constantly developing, such as indoor and outdoor smart lighting applications in smart homes and smart buildings. On the other hand, Nobel laureate Shuji Nakamura believes that lasers are the future of lighting. He predicted in 2016 that laser lighting would replace LED lighting in the next 10 years. Laser lighting is smaller and more compact. In addition to laser lighting, laser displays in televisions, projectors, cars, and other fields will have broader applications. Some high-end cars have already used laser lighting technology for headlights. Nakamura pointed out that the current price of laser lighting is still too high, but with the gradual reduction in cost, laser lighting is a major trend in the future.

LED light sources in agriculture can completely replace natural lighting and traditional artificial lighting, which is of great significance in promoting the decentralization, urbanization, scaling, and automated operation of modern agriculture. LED lighting conditions have a significant impact on the growth of crops, leafy vegetables, fruits and vegetables, fungi and algae, etc. LEDs can also increase yield in livestock breeding and breeding, aquaculture, etc. At the same time, it can improve pest control, disinfection and sterilization LED equipment and combine with new smart lighting technologies such as artificial intelligence to form an integrated smart lighting solution. The bottleneck of this technology is the lack of systematic research on lighting conditions, which requires a systematic and clear understanding of the suitable spectral conditions, cultivation environment, etc. for various crops, poultry, and aquatic products. In the design of supplementary lamps, these knowledge should be used for special optimization. Whether from the perspective of promoting the development of modern agriculture or from the perspective of energy conservation and environmental protection, the application of LEDs in agriculture is a trend in the future. With the continuous improvement of China's agricultural lighting technology, its cost will gradually decrease, and China, as a large agricultural country, also provides a broad development space for agricultural lighting.

2. Visible Light Communication

With the current popularity and significant cost reduction of LED lights, visible light communication is gradually emerging. LED lights can achieve extremely fast flickering through chip control, which is sufficient to support faster switching speeds than traditional light sources. The switching of the switch means the switching of logical information of 0 and 1, that is, real-time data transmission can be achieved. This communication method eliminates the need for traditional physical fiber optic laying, which can save costs; at the same time, it also avoids electromagnetic pollution such as radio waves and microwaves, which is more energy-saving and environmentally friendly; and as the transmission rate continues to increase, future application experience can exceed WiFi signals. There are still some technical problems in visible light communication, such as upload data bottlenecks, LED light signal modulation frequency, high-speed white light transceiver modules, reducing long-distance communication error rates, etc., all need to be further improved.

3. Ultraviolet Detection

GaN is also a good material for preparing ultraviolet optoelectronic devices, and ultraviolet optoelectronic chips have a wide range of application scenarios in both military and civilian fields. Typical military applications include fire suppression systems (ground tanks and armored vehicles, ships and aircraft), aircraft landing (land) guidance, space exploration, nuclear radiation and biological warfare agent monitoring, explosive detection, etc.; typical civilian applications include: flame detection, corona discharge detection, medical monitoring and diagnosis, water quality monitoring, atmospheric monitoring, criminal biological detection, etc.

4. Ultraviolet Sterilization

Deep ultraviolet light-emitting diodes (LEDs) based on gallium nitride semiconductors are the mainstream development direction of ultraviolet disinfection light sources. Their light sources are small in size, high in efficiency, and long in life. They have all the potential advantages of LED cold light sources and are recognized as future green energy-saving and environmentally friendly products to replace ultraviolet mercury lamps. However, the threshold of deep ultraviolet LED technology is very high, and it is currently only used in some high-end fields in large quantities. The market is mainly dominated by Japanese and Korean manufacturers, but more and more domestic semiconductor companies are entering the field.

Deep ultraviolet light-emitting diodes (LEDs) based on gallium nitride semiconductors are the mainstream development direction of ultraviolet disinfection light sources. Their light sources are small in size, high in efficiency, and long in life. A chip module the size of a thumb cap can emit ultraviolet light stronger than a mercury lamp. Because it has all the potential advantages of LED cold light sources, deep ultraviolet LEDs are recognized as future green energy-saving and environmentally friendly products to replace ultraviolet mercury lamps. However, the threshold of deep ultraviolet LED technology is very high, and it is still in the development stage. There is still room for improvement in terms of light power, luminous efficiency, life, and cost. In recent years, the technical level and chip performance of deep ultraviolet LEDs have improved rapidly, and they have been used in large quantities in some high-end fields, and are expected to be more widely used in the future.

At present, high-end deep ultraviolet LED products on the market are still mainly Japanese and Korean manufacturers, but more and more domestic semiconductor companies have begun to pay attention to the deep ultraviolet industry and have made deep layouts. Such as Qingdao Jiesheng (Yuanrong Optoelectronics) which has laid out the deep ultraviolet chip-packaging-module industry chain, Sanan Optoelectronics, Hubei Deep Purple, Zhongke Lu'an, Huacan Optoelectronics, Hongli Bingyi, which are deep ultraviolet LED chips, and Gaomin Optoelectronics with high-performance ultraviolet sensing chips. At present, Gaomin Optoelectronics is the only company in China that has ultraviolet sensing chip technology. The high-end gallium nitride and silicon carbide ultraviolet sensing chips it has developed have been put into mass production and have been used on a large scale in ultraviolet purification fields such as drinking water, air, food, clothing, and medical equipment.

7.6.2 Micro-LED Display Technology Development

Micro-LED, as a new generation of display technology, is based on forming a new pattern that complements the current LCD and OLED, fully leveraging its irreplaceable advantages, focusing on high-end display applications, and exploring its commercialization process in both ultra-large size and micro size directions, promoting the industrial development of Micro-LED display technology.

Regarding the ultra-large size Micro-LED display, in recent years, with the continuous progress of Micro-LED large size display preparation technology, the commercialization process of ultra-large size Micro-LED display has accelerated, gradually entering the market, becoming an important direction for the current Micro-LED industrial development. Our country already has a relatively complete industrial chain, and at the same time, there is a huge market demand, which is expected to build a pattern of ultra-large size Micro-LED display industry development based on domestic circulation and dual circulation at home and abroad.

The application of Micro-LED in micro size is currently mainly focused on AR glasses display, because AR glasses can best leverage the irreplaceable advantages of Micro-LED, which can support the requirements of AR glasses near-eye display for high brightness, high contrast, high resolution, high refresh rate, and extremely low latency, promoting the high-quality development of the AR glasses industry.

In the 5G information era, AR is the new generation of portable smart terminal devices to be developed after smartphones, with a wide range and large quantity, broad application prospects, and huge market space, such as applications in healthcare, architecture, education, and product design fields.

7.6.3 The Trend of Semiconductor Lasers Development

Semiconductor lasers are the core components of optical communication, laser sensing, laser processing, and laser pumping, and can also be directly applied to fields such as lidar, laser ranging, laser weapons, missile guidance, and photoelectric countermeasures. They are currently widely used in the fields of material processing and lithography, communication and optical storage, scientific research and military, medical and beauty, instruments and sensors, and entertainment, display and printing.

The rise of the Internet of Things and artificial intelligence has prompted the industrial field to begin to appear new pattern changes. The natural advantages of laser processing technology, such as the integration of numerical control technology and remote processing, along with its tool-free operation, position it at the forefront of the next generation of intelligent manufacturing. The rise of high-power blue semiconductor lasers brings another surprise to laser technology. 2 kW blue semiconductor lasers have shown their advantages in metal processing, especially in the processing of high-reflective metal materials. The brightness and power of blue semiconductor lasers are constantly increasing to new limits, which will lead to a wider range of applications. For example, the additive manufacturing capability of blue lasers is still being explored. In addition, apart from efficient metal material processing, blue semiconductor lasers are expected to have cross-departmental applications, especially the mechanical engineering department will be able to perform laser material processing underwater with blue light. For the manufacturing industry, this is a huge advantage. Although the processing applications based on high-power blue semiconductor lasers have just begun, with the development and progress of future technology and processes, it may become one of the core tools of the next generation of cutting-edge intelligent manufacturing.

Semiconductor laser monitoring technology can significantly improve the accuracy and efficiency of medical monitoring technology, and also promote the development of medical monitoring equipment towards precision and intelligence. Laser medicine is developing into a large industry, and with the improvement of people's quality of life, the scale of the industry will continue to expand.

Semiconductor laser technology can also be applied to ranging, imaging, pointing, guidance, communication, and countermeasures, etc., changing the face of warfare in a certain sense while improving equipment performance. In particular, high-energy lasers are entering the field of equipment research and development. Their inherent nature of military-civilian integration can be used in public security, disaster relief, such as laser locust extermination, laser de-icing, and other fields.

The research on the integration of military and civilian lasers is vigorously driving the development of many branches of optics, such as quantum optics, strong light optics, aspherical optics, atmospheric optics, adaptive optics, optical materials, transient optics, nano-guide star technology, etc. Various applications have given birth to a variety of laser industries, bringing huge economic and social benefits. The development of laser technology is global and strategic, and it is an indispensable core key technology field and strategic emerging industry for a strong science and technology country.

References

1. Popular Information on the Noble Prize in Physics 2014 [EB/OL]. [2017-06-26]. http://www.nobleprize.org
2. Weikun G (2014) The enlightenment of the 2014 nobel prize in physics. Phys Eng 24(6):3–8
3. Xin R, Shunfeng L, Weikun G (2017) Physics and engineering of the third generation semiconductor III nitrides. Phys Eng 27(6)
4. Eliashevich I, Osinsky A, Karlicek RF (1999) InGaN blue light-emitting diodes with optimized n-GaN layer]. In: Proceedings of SPIE—the international society for optical engineering, 3621
5. Amano H, Kito M, Hiramatsu K et al (2014) P-type conduction in Mg-doped GaN treated with low-energy electron beam irradiation (LEEBI)]. Jap J Appl Phys 28(Part2,No.12):L2112-L2114
6. Qi M (2016) Research on the photoelectric characteristics of GaN-based LEDs [Master's Thesis]. Shandong University, Shandong
7. Ryou J-H, Lee W, Limb J, D. Yoo, Liu JP, Dupuis RD, Wu ZH, Fischer AM, Ponce FA (2008) Effect of silicon doping in the quantum-well barriers on the electrical and optical properties of visible green light-emitting diodes. Appl Phys Lett 92:101113
8. Jia CY, Yu TJ, Lu HM, Zhong CT, Sun YJ, Tong YZ, Zhang GY (2013) Performance improvement of GaN-based LEDs with step stage InGaN/GaN strain relief layers in GaN-based blue LEDs. Opt Express 21:8444
9. Son JH, Lee J-L (2010) Strain engineering for the solution of efficiency droop in InGaN/GaN light-emitting diodes. Opt Express 18:5466
10. ShihH Y, Che YF, Lin TY (2012) Symmetrically tunable optical properties of InGaN/GaN multiple quantum disks by an external stress. Appl Phys Lett 100:171916
11. Nakamura S (1998) The roles of structural imperfections in InGaN-based blue light-emitting diodes and laser diodes. Science 281:956
12. Niu NH, Wang HB, Liu JP, Liu NX, Xing YH, Han J, Deng J, Shen GD (2007) Enhanced luminescence of InGaN/GaN multiple quantum wells by strain reduction. Solid State Electron 51:860
13. Leem SJ, Shin YC, Kim KC, Kim EH, Sung YM, Moon Y, Hwang SM, Kim TG (2008) The effect of the low-mole InGaN structure and InGaN/GaN strained layer superlattices on optical performance of multiple quantum well active layers. J Cryst Growth 311:103
14. Nakamura S (1999) InGaN-based violet laser diodes. Semicond Sci Technol 14:R27
15. Lu CM, Wang L, Lu JN, Li R, Liu L, Li D, Liu NY, Li L, Cao WY, Yang W, Chen WH, Du WM, Lee C, Hu XD (2013) Investigation of the electroluminescence spectrum shift of InGaN/GaN multiple quantum well light-emitting diodes under direct and pulsed currents. J Appl Phys 113:013102
16. Wang HN, Ji ZW, Qu S, Wang G, Jiang YZ, Liu BL, Xu XG, Mino H (2012) Influence of excitation power and temperature on photoluminescence in InGaN/GaN multiple quantum wells. Opt Express 20:3932
17. Ma ZX, Pierz K (2002) Carrier thermalization and activation within self-assembled InAs/AlAs quantum dot states. Surf Sci 511:57
18. Skromme BJ, Zhao H, Wang D, Kong HS, Leonard MT, Bulman GE, Molnar RJ (1997) Strain determination in heteroepitaxial GaN. Appl Phys Lett 71:829
19. Jr Watson GH, Daniels WB, Wang CS (1981) Measurements of Raman intensities and pressure dependence of phonon frequencies in sapphire. J Appl Phys 52:956
20. Ji LW, Fang TH, Meen TH (2006) Effects of strain on the characteristics of InGaN-GaN multiple quantum-dot blue light emitting diodes. Phys Lett A 355:118
21. Shengjun Z (2011) Research on the design and manufacturing technology of high-power GaN-based LED chips [PhD thesis]. Shanghai: Shanghai Jiaotong University
22. Yoshikawa A, Ohshima E, Fukuda T et al (2001) GaN and related materials II]. Philadephia: Gordon and Breach Sci Publ 143(14):234–236
23. Han R, Xu X, Hu X et al (2003) Development of bulk SiC single crystal grown by physical vapor transport method. Opt Mater 23(2):415–420

24. News on CREE website [EB/OL]. [2017–06–26]. http://www.cree.com/news-media/news/article/cree-first-tobreak-300-lumens-per-watt-barrier
25. Chuanbing X, Fengyi J, Wenqing F et al (2008) Study on the stress change before and after the transfer of GaN blue light-emitting material on silicon substrate. J Lumin 57(5):3176–3180
26. Xu K et al (2015) Progress in bulk GaN growth. Chin Phys B 24:066105
27. Krames M (2013) GaN-on-GaN platform removes cost/performance tradeoffs in LED lighting. Laser Focus World, 49(9):37–40
28. Wierer JJ, Steigerwald DA, Krames MR et al (2001) High-power AlGaInN flip-chip light-emitting diode. Appl Phys Lett 78(22):3379–3381
29. Shchekin OB, Epler JE, Trottier TA (2006) High performance thin-film flip-chip InGaN-GaN light-emitting diodes. Appl Phys Lett 89(7):071109
30. Chu CF, Lai FI, Chu JT et al (2004) Study of GaN light-emitting diodes fabricated by laser lift-off technique. J Appl Phys 95(8):3916–3922
31. Wong WS, Sands T, Cheung NW (1997) Damage-free separation of Ga N thin films from sapphire substrates. Appl Phys Lett 72(5):599–601
32. Ryu JH, Chandramohan S, Kim HG (2010) Effect of ridge growth on wafer bowing and light extraction efficiency of vertical GaN-based light-emitting diodes. Jpn J Appl Phys 49:072102
33. Hsu SC, Liu CY (2005) Stress analysis of transferred thin-GaN LED by Au-Si wafer bonding. Proc SPIE 5941:594116
34. Tavernier PR, Clarke DR (2001) Mechanics of laser-assisted debonding of films. J Appl Phys 89(3):1527–1536
35. Cho J, Jung J, Chae JH et al (2007) Alternating-current light emitting diodes with a diode bridge circuitry. Jpn J Appl Phys 46(48):L1194–L1196
36. Onushkin GA, Lee YJ, Yang JJ et al (2009) Efficient alternating current operated white light-emitting diode chip. IEEE Photonics Technol Lett 21(1):33–35
37. Yen HH, Kuo HC, Yeh WY (2008) Characteristics of single-chip GaN-based alternating current light-emitting diode. Jpn J Appl Phys 47(12):8808–8810
38. Cheng JH, Wu YS, Liao WC (2010) Improved crystal quality and performance of GaN-based light-emitting diodes by decreasing the slanted angle of patterned sapphire. Appl Phys Lett 96:051109
39. Park EH, Jang J, Gupta S et al (2008) Air-voids embedded high efficiency In Ga N-light emitting diode. Appl Phys Lett 93:191103
40. Gao H, Yan F, Zhang Y (2008) Enhancement of the light output power of InGaN/Ga N light-emitting diodes grown on pyramidal patterned sapphire substrates in the micro-and nanoscale. J Appl Phys 103:014314
41. Chang CT, Hsiao SK, Chang EY (2009) 460-nm InGaN-based LEDs grown on fully inclined hemisphere-shape-patterned sapphire substrate with submicrometer spacing. IEEE Photon Technol Lett 21(19):1366–1368
42. Kao CC, Su YK, Lin CL (2010) The aspect ratio effects on the performances of GaN-based light-emitting diodes with nanopatterned sapphire substrates. Appl Phys Lett 97:023111
43. Hong EJ, Byeon KJ, Park H (2009) Effect of nano-patterning of p-GaN cladding layer on photon extraction efficiency. Solid-State Electron 53(10):1099–1102
44. Huang HW, Kao CC, Chu JT et al (2005) Improvement of InGaN-GaN light-emitting diode performance with a nano-roughened p-GaN surface. IEEE Photon Technol Lett 17(5):983–985
45. Lin CF, Zheng JH, Yang ZJ (2006) High-efficiency InGaN-based light-emitting diodes with nanoporous GaN: Mg structure. Appl Phys Lett 88(8):083121
46. Pan SM, Tu RC, Fan YM et al (2003) Improvement of In GaN-GaN light-emitting diodes with surface-textured indium- tin-oxide transparent ohmic contacts. IEEE Photon Technol Lett 15(5):649–651
47. Lin CL, Chen PH, Chan CH et al (2007) Light enhancement by the formation of an Al oxide honeycomb nanostructure on the n-GaN surface of thin-GaN light-emitting diodes. Appl Phys Lett 90:242106
48. Kao CC, Kuo HC, Yeh KF (2007) Light-output enhancement of nano-roughened Ga N laser lift-off light-emitting diodes formed by ICP dry etching. IEEE Photon Technol Lett 19(11):849–851

49. Fujii T, Gao Y, Sharma R (2004) Increase in the extraction efficiency of Ga N-based light-emitting diodes via surface roughening. Appl Phys Lett 84(6):855–857
50. Lee YJ, Kuo HC, Lu TC et al (2006) High Light-extraction GaN-based vertical LEDs with double diffuse surfaces. IEEE J Quantum Electron 42(12):1196–1201
51. Horng R H,Zheng X, Hsieh CY et al (2010) Light Extraction He Anhe, He Miao, Zhu Xuehui, Chen Xianwen, Fan Guanghan, Zhang Yong. Research on improving the light output of GaN-based LEDs using Ni nanoparticle masks. Optoelectronics·Laser 21(7):967–969
52. Jingmei F, Liangchen W, Zhiqiang L (2009) The effect of surface roughening on the light output efficiency of GaN-based vertical structure LEDs. Optoelectron·Laser. 20(8):995–996
53. China semiconductor lighting industry development yearbook. Science Press pp 152–156 (2006)
54. Wang P, Gan Z, Liu S (2009) Improved light extraction of GaN-based light-emitting diodes with surface-patterned ITO. Opt Laser Technol 41:823–826
55. Chang SJ, Shen CF, Chen WS et al (2007) Nitride-based light emitting diodes with indium tin oxide electrode patterned by imprint lithography. Appl Phys Lett 91:013504
56. Byeon KJ, Hong EJ, Park H et al (2009) Enhancement of the photon extraction of green and blue LEDs by patterning the indium tin oxide top layer. Semicond Sci Technol 24(10):105004
57. Oder TN, Kim KH, Lin JY et al (2004) III-nitride blue and ultraviolet photonic crystal light emitting diodes. Appl Phys Lett 84(4):466–468
58. Wierer JJ, Krames MR, Epler JE (2004) InGaN/GaN quantum-well heterostructure light-emitting diodes employing photonic crystal structures. Appl Phys Lett 84(19):3885–3887
59. Kim DH, Cho CO, Roh YG (2005) Enhanced light extraction from GaN-based light-emitting diodes with holographically generated two-dimensional photonic crystal patterns. Appl Phys Lett 87:203508
60. Truong TA, Campos LM, Matioli E et al (2009) Light extraction from GaN-based light emitting diode structures with a noninvasive two-dimensional photonic crystal. Appl Phys Lett 94:023101
61. Wierer JJ, Jr DA (2009) III-nitride photonic-crystal light-emitting diodes with high extraction efficiency. Nat Photon 3:163–169
62. Kang X, Zhang B, Dai T et al (2008) Improvement of light extraction from GaN-based LED with surface photonic lattices. In: 5th China international forum on solid state lighting, pp 195–200
63. Haiyang H, Xingsheng X, Lin L, Qian S, Wei D, Chunxia W, Hongda C (2008) Improving the light output efficiency of photonic crystal GaN LEDs using conventional processes. Optoelectron ·Laser 19(5):569–572
64. Wenfeng (2010) Theoretical and experimental study on white light emitting diodes [PhD thesis]. Huazhong University of Science and Technology, Wuhan
65. Xiaoli L (2009) Research on the fabrication of photonic crystal structures and their applications using nanoimprint technology [PhD thesis]. Shanghai Jiao Tong University, Shanghai
66. Kim JK, Gessmann T, Schubert EF (2006) GaInN light-emitting diode with conductive omni-directional reflector having a low-refractive-index indium-tin oxide layer. Appl Phys Lett 88:013501
67. Windisch R, Butendeich R, Illek S et al (2007) 100-lm/W InGaAlP thin-film light-emitting diodes with buried microreflectors. IEEE Photon Technol Lett 19(10):774–776
68. Jianming Z, Deshu Z, Liu Sinan X, Chen SG (2007) New omnidirectional reflective aluminum gallium indium phosphide thin film light emitting diode. Acta Physica Sinica 56(5):2905–2909
69. Krames M, Holcomb M, Collins D (1999) High-power truncated-inverted- pyramid (Al_x Ga_{1-x})0.5In0.5P/GaP light-emitting diodes exhibiting >50% external quantum efficiency. Appl Phys Lett 75:2365–2367
70. Hui KN, Hui KS, Lee H et al (2011) Enhanced light output of angled sidewall light-emitting diodes with reflective silver films. Thin Solid Films 519:2504–2507
71. Douglas WP, Michael DC, Gloria EH (2006) Forming an optical element on the surface of a light emitting device for improved light extraction. United States Patent. US 6987613B2, pp 1–18

72. Hsueh TH, Sheu JK, Huang HW et al (2005) Enhancement in light output of In Ga N-based microhole array light-emitting diodes. IEEE Photon Technol Lett 17(6):1163–1165
73. Lee JS, Lee J, Kim S et al (2007) Fabrication of reflective Ga N mesa sidewalls for the application to high extraction efficiency LEDs. Phys Status Solidi (c). 4(7):2625–2628
74. Lee JS, Lee J, Kim S et al (2008) GaN light-emitting diode with deep-angled mesa sidewalls for enhanced light emission in the surface-normal direction. IEEE Trans Electron Dev 55(2):523–526
75. Chang CS, Chang SJ, Lee CT et al (2004) Nitride-based LEDs with textured side walls. IEEE Photon Technol Lett 16(3):750–752
76. Kao CC, Kuo HC, Huang HW et al (2005) Light-output enhancement in a nitride-based light-emitting diode With 220 undercut sidewalls. IEEE Photon Technol Lett 17(1):19–21
77. Tan LX, Li J, Liu S (2008) A light emitting diode's chip structure with low stress and high light extraction efficiency. In: The 58th electronic components and technology conference, pp 783–788.
78. Xianpeng Z, Yanjun H, Yi L, Xiaolin X, Lai W, Yang J (2008) ICP Etching p-GaN surface microstructure GaN-based blue LED. Semiconductor Optoelectron 29(1):6–15
79. Alivov YI (2005) Forward-current electroluminescence from GaN/ZnO of InGaN-based light-emitting diodes with various well widths. Jpn J Appl Phys 44:7348–7351
80. Jinju C, Buran W, Hong D (2011) Progress in the preparation technology of non-polar GaN epitaxial films for LEDs. Semiconductor Optoelectronics 8(32):4
81. Yi W Epitaxial growth and characterization of non-polar and semi-polar GaN-based nitrides [Master's thesis]. Nanjing, Southeast University
82. Chichibu SF, Uedono A, Onuma T, Haskell BA, Chakraborty A, Koyama T, Fini PT, Keller S, DenBaars SP, Speck JS, Mishra UK, Nakamura S, Yamaguchi S, Kamiyama S, Amano H, Akasaki I, Han J, Sota T (2006) Nat Mater 5(10):810–816
83. Chichibu SF, Abare AC, Minsky MS et al (1998) Effective band gap inhomogeneity and piezoelectric field in InGaN/GaN multiquantum well structures]. Appl Phys Lett 73(14):2006
84. Choi CK, Kwon Y H, Little BD et al (2001) Time-resolved photoluminescence of In_x Ga_{1-x} N/GaN multiple quantum well structures: effect of Si doping in the barriers. Phys Rev B, 64(24):245339
85. Miller DAB, Chemla DS, Damen TC et al (1984) Band-edge electroabsorption in quantum well structures: the quantum-confined stark effect. Phys Rev Lett 53:2173–2176
86. Waltereit P, Brandt O, Tarampert A et al (2000) Nitride semicondutors free of electrostatic fields for efficient white light-emitting diodes. Nature, 406(6798):865–868
87. Haskell BA, Chakraborty A, Wu F et al (2005) Microstructure and enhanced morphology of planar nonpolar m-plane GaN grown by hydride vapor phase epitaxy. J Electron Mater 34(4):357–360
88. Baker TJ, Haskell BA, Wu F et al (2006) Characterization of planar semipolar gallium nitride films on sapphire substrates. Jap J Appl Phys 45(2L):L154
89. Koyama T, Onuma T, Masui H et al (2006) Prospective emission efficiency GaN substrates. Appl Phys Lett 89(9)
90. Yamada H, Iso K, Saito M, Hirasawa H, Fellows N, Masui H, Fujito K, Speck JS, DenBaars SP, Nakamura S (2008) Comparison of InGaN/GaN light emitting diodes grown on m-plane and a-plane bulk GaN substrates. Phys Stat Sol (RRL), 2(2):89–91
91. Liu JP, Limb JB, Ryou J-H, Yoo D, Horne CA, Dupuis RD, Wu ZH, Fischer AM, Ponce FA, Hanser AD, Liu L, Preble EA, Evans KR (2008) Blue light emitting diodes grown on freestanding (11–20)-plane GaN substrates. Appl Phys Lett 92:011123
92. Funato M, Kotani T, Kondou T, Kawakami Y, Narukawa Y, Mukai T (2006) Tailored emission color synthesis using microfacet quantum wells consisting of nitride semiconductors without phosphors. Appl Phys Lett 88(26):261920
93. Yamada H, Iso K, Saito M, Masui H, Fujito K, DenBaars SP, Nakamura S (2008) Compositional dependence of nonpolar m-plane InxGa1−xN/GaN light emitting diodes. Appl Phys Express 1(4):041101

94. Schmidt MC, Kim K-C, Sato H, Fellows N, Masui H, Nakamura S, DenBaars SP, Speck JS (2007) High power and high external efficiency m-Plane InGaN light emitting diodes. Jpn J Appl Phys 46(7):L126–L128
95. Kim K-C, Schmidt M C, Sato H et al (2007) Study of nonpolar m-plane InGaN/GaN multiquantum well light emitting diodes grown by homoepitaxial metal-organic chemical vapor deposition. Appl Phys Lett 91(18):181120
96. Chakraborty A, Haskell BA, Keller S et al (2004) Nonpolar InGaN/GaN emitters on reduced-defect lateral epitaxially overgrown a-plane GaN with drive-current-independent electroluminescence emission peak. Appl Phys Lett 85:5143
97. Iso K, Yamada H, Hirasawa H, Fellows N, Saito M, Fujito K, DenBaars SP, Speck JS, Nakamura S (2007) High brightness blue InGaN/GaN light emitting diode on nonpolar m-plane Bulk GaN substrate. Jpn J Appl Phys 46(40):L960–L962
98. Sato H, Chung RB, Hirasawa H, Fellows N et al (2008) Optical properties of yellow light-emitting diodes grown on semipolar (11–22)bulk GaN substrates Appl. Phys Lett 92(22):221110
99. Tyagi A, Zhong H, Fellows NN, Iza M, Speck JS, DenBaars SP, Nakamura S (2007) High brightness violet InGaN/GaN light emitting diodes on semipolar (10–1–1) bulk GaN substrates. Jpn J Appl Phys 46(4):L129–L131
100. Detchprohm T, Zhu M, Li Y, Xia Y, Wetzel C, Preble EA, Liu L, Paskova T, Hanser D (2008) Green light emitting diodes on a-plane GaN bulk substrates. Appl Phys Lett 92(24):241109
101. Funato M, Ueda M, Kawakami Y, Narukawa Y, Kosugi T, Takahashi M, Mukai T (2006) Blue, green, and amber InGaN/GaN lightemitting diodes on semipolar 11–22 GaN bulk substrates. Jpn J Appl Phys 45(24):L659–L662
102. Sato H, Tyagi A, Zhong H, Fellows N et al (2007) High power and high efficiency green light emitting diode on free-standing semipolar (11–22) bulk GaN substrate. Phys Stat Sol (RRL) 1(4):162–164
103. Masui H, Nakamura S, DenBaars SP, Mishra UK (2010) Nonpolar and semipolar III-nitride light-emitting diodes: achievements and challenges. IEEE Trans Electron Dev 57(1):88–100
104. Chen C-Q, Adivarahan V, Yang J-W, Shatalov M, Kuokstis E, Khan MA (2003) Ultraviolet light emitting diodes using non-polar a-plane GaN-AlGaN multiple quantum wells. Jpn J Appl Phys 42:L1039
105. Imer B, Schmidt M, Haskell B et al (2008) Improved quality nonpolar a-plane GaN/AlGaN UV LEDs grown with sidewall lateral epitaxial overgrowth (SLEO)]. Phys Stat Sol (A) 205(7):1705–1712
106. Saito Y, Okuno K, Boyama S et al (2009) m-plane GaNInN light emitting diodes grown on pattern a-plane sapphire substrates. Appl Phys Express 2:041001–041003
107. Liu B, Zhang R, Xie ZL et al (2007) Nonplane m-plane thin film GaN and InGaN/GaN light-emitting diode on LiAlO(100) substrates. Appl Phys Lett 91:253506
108. Kim KC, Schmidt MC, Sato H et al (2007) Improved electroluminescence on nonpolar m-plane InGaN/GaN quantum wells LEDs. Phys Status Solidi(RRL)-Rapid ResLett 1:125–127
109. McCall SL, Levi AFJ, Slusher RE, Pearton SJ, Logan RA (1992) Whispering-gallery mode microdisk lasers. Appl Phys Lett 60(289)
110. Jin SX, Li J, Li JZ, Lin JY, Jiang HX (2000) GaN microdisk light emitting diodes. Appl Phys Lett 76(631)
111. Jiang HX, Jin SX, Li J, Shakya J, Lin JY (2001) III-nitride blue microdisplays. Appl Phys Lett 78(1303)
112. Jeon CW, Choi HW, Gu E, Dawson MD (2004) High-density matrix-addressable AlInGaN-based 368-nm microarray light-emitting diodes. IEEE Photon Technol Lett 16(2421)
113. Wu M, Gong Z, Kuehne AJ et al (2009) Hybrid GaN/organic microstructured light-emitting devices via ink-jet printing. Opt Exp 17(19):16436–16443
114. Mckendry JJD, Green RP, Kelly AE et al (2010) High-speed visible light communications using individual pixels in a micro light-emitting diode array. IEEE Photon Technol Lett 22(18):1346–1348

115. Islim MS, Ferreira RX, He XY et al (2017) Towards 10 Gb/s orthogonal frequency division multiplexing-based visible light communication using a GaN violet micro-LED. Photon Res 5(2):A35
116. Chong WC, Cho WK, Liu ZJ et al (2014) 1700 pixels per inch (PPI) passive-matrix micro-LED display powered by ASIC. In: 2014 IEEE compound semiconductor integrated circuit symposium (CSICS), pp 1–4
117. Liu ZJ, Chong WC, Wong KM et al (2013) 360 PPI flip-chip mounted active matrix addressable light emitting diode on silicon (LEDoS) microdisplays. J Display Technol 9(8):678–682
118. Han HV, Lin HY, Lin CC, Chong WC, Li JR, Chen KJ, Yu P, Chen TM, Chen HM, Lau K M, Kuo HC (2015) Resonant-enhanced full-color emission of quantum-dot-based micro LED display technology. Opt Express 23(32504)
119. Templier F, Benaïssa L, Aventurier B, Nardo C D, Charles M, Daami A, Henry F, Dupré L (2017) A novel process for fabricating high-resolution and very small pixel-pitch GaN LED Microdisplays. SID Symp Dig Tech Pap 48(1):268–271
120. Bai J, Cai Y, Feng P, Fletcher P, Zhao X, Zhu C, Wang T (2020) A direct epitaxial approach to achieving ultrasmall and ultrabright InGaN micro light-emitting diodes (μLEDs)]. ACS Photon 7(411)
121. Zhang X, Qi LH, Chong WC et al (2020) Active matrix monolithic micro-LED full-color micro-display. J Soc Inf Display n /a
122. Fulong J, Feifan X, Zhaojun L, Bin L, Youyong Z (2020) Research progress on GaN-based Micro-LED display technology. J Synthet Cryst 49(11)
123. Li G, Wang W, Yang W et al (2016) GaN-based light-emitting diodes on various substrates: a critical review. Rep Prog Phys Phys Soc (Great Britain) 79(5):056501
124. Ponce FA, Bour DP (1997) Nitride-based semiconductors for blue and green light-emitting devices. Nature 386(6623):351–359
125. Egawa T, Ohmura H, Ishikawa H et al (2002) Demonstration of an InGaN-based light-emitting diode on an AlN/sapphire template by metalorganic chemical vapor deposition. Appl Phys Lett 81(2):292–294
126. Zhang B, Egawa T, Liu Y et al (2003) InGaN multiple-quantum-well light-emitting diodes on an AlN/sapphire template by metalorganic chemical vapor deposition. Physica Status Solidi (c) 7:2244–2247
127. Chang SJ, Lin YC, Su YK et al (2003) Nitride-based LEDs fabricated on patterned sapphire substrates. Solid-State Electron 47(9):1539–1542
128. Dadgar A, Bläsing J, Diez A et al (2000) Metalorganic chemical vapor phase epitaxy of crack-free GaN on Si (111) exceeding 1μm in thickness. Jap Jo Appl Phys 39:L1183–L1185.
129. Kikuchi A, Kawai M, Tada M et al (2004) InGaN/GaN multiple quantum disk nanocolumn light-emitting diodes grown on (111) Si substrate. Jap J Appl Phys 43(12A):L1524-L1526
130. Zhao DG, Xu SJ, Xie MH et al (2003) Stress and its effect on optical properties of GaN epilayers grown on Si(111), 6H-SiC(0001), and c-plane sapphire. Appl Phys Lett 83(4):677–679
131. Hwang D, Mughal A, Pynn CD et al (2017) Sustained high external quantum efficiency in ultrasmall blue III-nitride micro-LEDs. Appl Phys Exp 10(3):032101
132. Liu B, Chen DJ, Lu H et al (2019) Hybrid light emitters and UV solar-blind avalanche photodiodes based on III-nitride semiconductors. Adv Mater 2019:1904354
133. Amano H, Sawaki N, Akasaki I et al (1986) Metalorganic vapor phase epitaxial growth of a high quality GaN film using an AlN buffer layer. Appl Phys Lett 48(5):353–355
134. Yang XD, Zhang JL, Wang XL et al (2020) Enhance the efficiency of green-yellow LED by optimizing the growth condition of preparation layer. Superlattices Microstruct 141:106459
135. Zhang JL, Wang XL, Liu JL et al (2018) Study on carrier transportation in InGaN based green LEDs with V-pits structure in the active region. Opt Mater 86:46–50
136. Chen YF, Chen ZZ, Li JZ et al (2018) A study of GaN nucleation and coalescence in the initial growth stages on nanoscale patterned sapphire substrates via MOCVD. CrystEngComm 20(42):6811–6820

137. Mo CL, Fang WQ, Pu Y et al (2005) Growth and characterization of InGaN blue LED structure on Si(111) by MOCVD]. J Cryst Growth 285(3):312–317
138. Quan ZJ, Wang L, Zheng CD, Liu LJ, Jiang FY (2014) Roles of V-shaped pits on the improvement of quantum efficiency in InGaN/GaN multiple quantum well light-emitting diodes. J Appl Phys 116:183107
139. Tian PF, Mckendry JJD, Gong Z et al (2012) Size-dependent efficiency and efficiency droop of blue InGaN micro-light emitting diodes. Appl Phys Lett 101(23):231110
140. Moustakas TD, Paiella R (2017) Optoelectronic device physics and technology of nitride semiconductors from the UV to the terahertz. Rep Progress Phys Phys Soc (Great Britain) 80(10):106501
141. Dupré L, Marra M, Verney V et al (2017) Processing and characterization of high resolution GaN/InGaN LED arrays at 10 micron pitch for micro display applications. In: Gallium nitride materials and devices XII. San Francisco, California, USA. SPIE, 2017: n/a
142. Nishikawa A, Loesing A, Slischka B (2019) Achieving high uniformity and yield for micro LED applications with precise strain-engineered large-diameter epiwafers.In: SPIE OPTO. Proceeding SPIE 10940, light-emitting devices, materials, and applications, San Francisco, California, USA, vol 1094, pp 109400Z
143. Aida H, Aota N, Takeda H et al (2012) Control of initial bow of sapphire substrates for III-nitride epitaxy by internally focused laser processing. J Cryst Growth 361:135–141
144. Nishikawa A, Groh L, Solari W et al (2013) 200-mm GaN-on-Si based blue light-emitting diode wafer with high emission uniformity. Jap J Appl Phys 52(8S):08JB25
145. Liu XY, Wang XH, Zhang YG et al (2018) Insight into the near-conduction band states at the crystallized interface between GaN and SiNx grown by low-pressure chemical vapor deposition. ACS Appl Mater Interfaces 10(25):21721–21729
146. Jr WJJ, Tansu N (2019) III-nitride micro-LEDs for efficient emissive displays. Laser Photon Rev 13(9):1900141
147. Olivier F, Tirano S, Dupré L et al (2017) Influence of size-reduction on the performances of GaN-based micro-LEDs for display application. J Lumin 191:112–116
148. Jia XT, Zhou YG, Liu B et al (2019) A simulation study on the enhancement of the efficiency of GaN-based blue light-emitting diodes at low current density for micro-LED applications. Mater Res Express 6(10):105915
149. Olivier F, Daami A, Licitra C et al (2017) Shockley-read-hall and auger non-radiative recombination in GaN based LEDs: a size effect study. Appl Phys Lett 111(2):022104
150. Konoplev SS, Bulashevich KA, Karpov SY (2018) From large-size to micro-LEDs: scaling trends revealed by modeling]. Phys Status Solidi (a) 215(10):1700508
151. Wong MS, Hwang D, Alhassan AI et al (2018) High efficiency of III-nitride micro-light-emitting diodes by sidewall passivation using atomic layer deposition. Opt Exp 26(16):21324
152. Kou J, Shen CC, Shao H et al (2019) Impact of the surface recombination on InGaN/GaN-based blue micro-light emitting diodes. Opt Exp 27(12):A643–A653
153. Wong MS, Lee C, Myers DJ et al (2019) Size-independent peak efficiency of III-nitride micro-light-emitting-diodes using chemical treatment and sidewall passivation. Appl Phys Express 12(9):097004
154. Zhu J, Takahashi T, Ohori D et al (2019) Near-complete elimination of size-dependent efficiency decrease in GaN micro-light-emitting diodes. Phys Status Solidi (a) 216(22):1900380
155. Lee DH, Lee JH, Park JS et al (2020) Improving the leakage characteristics and efficiency of GaN-based micro-light-emitting diode with optimized passivation. ECS J Solid State Sci Technol 9(5):055001
156. Ley RT, Smith JM, Wong MS et al (2020) Revealing the importance of light extraction efficiency in InGaN/GaN microLEDs via chemical treatment and dielectric passivation. Appl Phys Lett 116(25):251104
157. Smith JM, Ley R, Wong MS et al (2020) Comparison of size-dependent characteristics of blue and green InGaN microLEDs down to 1 μm in diameter. Appl Phys Lett 116(7):071102
158. Zojian P, Zhizhong C, Fei J et al (2020) Review of key technologies for micro-light-emitting diode epitaxy and chips for display applications. Acta Phys Sinica 69(19):198501

159. Fiorentini V, Bernardini F, Della Sala F, Di Carlo A, Lugli P (1999) Effects of macroscopic polarization in III-V nitride multiple quantum wells. Phys Rev B 60:8849
160. Ambacher O, Majewski J, Miskys C, Link A, Hermann M, Eickhoff M, Stutzmann M, Bernardini F, Fiorentini V, Tilak V, Schaff B, Eastman LF (2002) Pyroelectric properties of Al(In)GaN/GaN hetero- and quantum well structures J Phys Condens Matter 143399
161. Zhao HP, Liu GY, Zhang J, Poplawsky JD, Dierolf V, Tansu N (2011) Approaches for high internal quantum efficiency green InGaN light-emitting diodes with large overlap quantum wells. Opt Express 19:A991
162. Zhao HP, Arif RA, Ee YK, Tansu N (2009) Design analysis of staggered InGaN quantum wells light-emitting diodes at 500–540 nm. IEEE J Quantum Electron 15:1104
163. Zhao HP, Liu GY, Li X, Arif R, Huang GS, Poplawsky JD, Penn S, Dierolf V, Tansu N (2009) Design and characteristics of staggered InGaN quantum-well light-emitting diodes in the green spectral regime. IET Optoelectron 3:283
164. Tsai MC, Yen SH, Kuo YK (2011) Investigation of blue InGaN light-emitting diodes with step-like quantum well. Appl Phys A 104:621
165. Arif RA, Ee YK, Tansu N (2007) Polarization engineering via staggered InGaN quantum wells for radiative efficiency enhancement of light emitting diodes. Appl Phys Lett 91:091110
166. Yang ZW, Li R, Wei QY, Yu T, Zhang YZ, Chen WH, Hu XD (2009) Analysis of optical gain property in the InGaN/GaN triangular shaped quantum well under the piezoelectric field. Appl Phys Lett 94:061120
167. Shioda T, Yoshida H, Tachibana K, Sugiyama N, Nunoue S (2012) Enhanced light output power of green LEDs employing AlGaN interlayer in InGaN/GaN MQW structure on sapphire (0001) substrate. Phys Status Solidi A 209:473
168. Kimura S, Yoshida H, Uesugi K, Ito T, Okada A, Nunoue S (2016) Performance enhancement of blue light-emitting diodes with InGaN/GaN multi-quantum wells grown on Si substrates by inserting thin AlGaN interlayers. J Appl Phys 120:113104
169. Alhassan AI, Farrell RM, Saifaddin B, Mughal A, Wu F, DenBaars SP, Nakamura S, Speck JS (2016) High luminous efficacy green light-emitting diodes with AlGaN cap layer. Opt Express 24:17868
170. Al Muyeed SA, Sun W, Wei X, Song R, Koleske DD, Tansu N, Wierer JJ (2017) Strain compensation in InGaN-based multiple quantum wells using AlGaN interlayers. AIP Adv 7:105312
171. Hwang JI, Hashimoto R, Saito S, Nunoue S (2014) Development of InGaN-based red LED grown on (0001) polar surface. Appl Phys Expres 7:071003
172. Sun W, Al Muyeed SA, Song R, Wierer JJ, Tansu N (2018) Integrating AlInN interlayers into InGaN/GaN multiple quantum wells for enhanced green emission. Appl Phys Lett 112:201106
173. Zhao HP, Liu GY, Tansu N (2010) Analysis of InGaN-delta-InN quantum wells for light-emitting diodes. Appl Phys Lett 97:131114
174. Wang T (2016) Topical review: development of overgrown semi-polar GaN for high efficiency green/yellow emission. Semi Sci Technol 31:093003
175. Poyiatzis N, Athanasiou M, Bai J, Gong Y, Wang T (2019) Monolithically integrated white light LEDs on (11–22) semi-polar GaN templates. Sci Rep 9:1383
176. Bai J, Xu B, Guzman FG, Xing K, Gong Y, Hou Y, Wang T (2015) (11–22) semipolar InGaN emitters from green to amber on overgrown GaN on micro-rod templates. Appl Phys Lett 107:261103
177. Li HJ, Wong MS, Khoury M, Bonef B, Zhang HJ, Chow YC, Li PP, Kearns J, Taylor AA, Mierry PD, Hassan Z, Nakamura S, DenBaars SP (2019) Study of efficient semipolar (11–22) InGaN green micro-light-emitting color micro-LED display. SID Sympos Digest Tech Papers 50(1):48–51
178. Huang Chen SW, Huang YM, Singh KJ, Hsu YC, Liou FJ, Song J, Choi J, Lee PT, Lin CC, Chen Z, Han J, Wu TZ, Kuo HC (2020) Full-color micro-LED display with high color stability using semipolar (20–21) InGaN LEDs and quantum-dot photoresist. Photon Res 8(5):630
179. Zhang Y, Bai J, Hou Y, Smith RM, Yu X, Gong Y, Wang T (2016) Defect reduction in overgrown semi-polar (11–22) GaN on a regularly arrayed micro-rod array template. AIP Adv 6:025201

180. Monavarian M, Rashidi A, Feezell D (2019) A decade of nonpolar and semipolar III-nitrides: a review of successes and challenges. Phys Status Solidi A 216:1800628
181. Song J, Choi J, Zhang C, Deng Z, Xie YJ, Han J (2019) Elimination of stacking faults in semipolar gan and light-emitting diodes grown on sapphire. ACS Appl Mater Interfaces 11:33140
182. Song J, Han J (2020) High quality, mass-producible semipolar GaN and InGaN light-emitting diodes grown on sapphire. Phys Status Solidi B 257:1900565
183. Ambacher O, Majewski J, Miskys C, Link A, Hermann M,Eickhoff M, Stutzmann M, Bernardini F, Fiorentini V, Tilak V, Schaff B, Eastman L (2002) Pyroelectric properties of Al(In)GaN/GaN hetero-and quantum well structures. J Phys Condens Matter 143399
184. Auf der Maur M, Pecchia A, Penazzi G, Rodrigues W, Di Carlo A (2016) Efficiency drop in Green InGaN/GaN light emitting diodes: the role of random alloy fluctuations. Phys Rev Lett 116:027401 (2016)
185. Liu Z, Lin CH, Hyun BR et al (2020) Micro-light-emitting diodes with quantum dots in display technology Light. Sci Appl 9:83
186. Boroditsky M, Gontijo I, Jackson M et al (2000) Surface recombination measurements on III-V candidate materials for nanostructure light-emitting diodes. J Appl Phys 87(7):3497–3504
187. Chong WC, Cho WK, Liu ZJ et al (2014) 1700 pixels per inch (PPI) passive-matrix micro-LED display powered by ASIC. In: 2014 IEEE compound semiconductor integrated circuit symposium (CSICS). October 19–2, La Jolla,CA, USA. IEEE, 2014:1–4
188. Zhang X, Qi LH, Chong WC et al (2020) Active matrix monolithic micro-LED full-color micro-display. J Soc Inf Display, n/a
189. Liou JC, Yang CF (2018) Design and fabrication of micro-LED array with application-specific integrated circuits (ASICs) light emitting display. Microsyst Technol 24(10):4089–4099
190. Gou FW, Hsiang EL, Tan GJ et al (2019) High performance color-converted micro-LED displays. J Soc Inf Display 27(4):199–206
191. Kim HM, Ryu M, Cha JHJ et al (2019) 4–3: distinguished paper: 10 μm pixel, quantum-dots color conversion layer for high resolution and full color active matrix micro-LED display. SID Symp Digest Tech Papers 50(1):26–29
192. Zhang X, Li PA, Zou XB et al (2019) Active matrix monolithic LED micro-display using GaN-on-Si epilayers. IEEE Photon Technol Lett 31(11):865–868
193. Kang JH, Li BJ, Zhao TS et al (2020) RGB arrays for micro-light-emitting diode applications using nanoporous GaN embedded with quantum dots. ACS Appl Mater Interfaces 12(27):30890–30895
194. Han HV, Lin HY, Lin CC et al (2015) Resonant-enhanced full-color emission of quantum-dot-based micro LED display technology. Opt Exp 23(25):32504–32515
195. Pai YM, Lin CH, Lin HY et al (2017) Optical cross talk reduction in a quantum-dot-based full-color micro-LED display by a lithographic-fabricated photoresist mold. Photonics Res 5(5):411–416
196. Sabnis RW (1999) Color filter technology for liquid crystal displays. Displays 20(3):119–129
197. Osinski J (2019) Palomaki P. 4–5: quantum dot design criteria for color conversion in MicroLED displays. SID Symposium Digest of Technical Papers, 50(1):34–37
198. Clapp AR, Medintz IL, Mattoussi H (2006) Förster resonance energy transfer investigations using quantum-dot fluorophores]. ChemPhysChem 7(1):47–57
199. Chanyawadee S, Lagoudakis PG, Harley RT et al (2010) Increased color-conversion efficiency in hybrid light-emitting diodes utilizing non-radiative energy transfer. Adv Mater 22(5):602–606
200. Zhuang Z, Guo X, Liu B et al (2016) High color rendering index hybrid III-nitride /nanocrystals white light-emitting diodes. Sci Found China 24(3):20
201. Krishnan C, Brossard M, Lee KY et al (2016) Hybrid photonic crystal light-emitting diode renders 123% color conversion effective quantum yield. Optica 3(5):503
202. Wang S W, Hong KB, Tsai YL et al (2017) Wavelength tunable InGaN/GaN nano-ring LEDs via nano-sphere lithography. Sci Rep 7:42962

203. Huang Chen SW, Shen CC, Wu TZ et al (2019) Full-color monolithic hybrid quantum dot nanoring micro light-emitting diodes with improved efficiency using atomic layer deposition and nonradiative resonant energy transfer. Photon Res 7(4):416
204. Jianping L, Hui Y (2015) Global research status of gallium nitride laser materials and devices. New Mater Ind 10(45)
205. Fangzhi L, Lei H, Aiqin T, Lingrong J, Liqun Z, Deyao L, Masao I, Jianping L, Hui Y (2020) Current status and future development trend of GaN-based blue-green laser diodes. J Synthet Crystals 49(11)
206. Lei H, Liqun Z, Jianping L et al (2020) High power gallium nitride-based blue laser diodes. China Laser 47:0701025
207. Kioupakis E (2010) Auger recombination and free-carrier absorption in nitrides from first principles. Am Phys Soc 81(24):775–780
208. Kioupakis E, Rinke P, Delaney KT et al (2011) Indirect Auger recombination as a cause of efficiency droop in nitride light-emitting diodes. Appl Phys Lett 98(16):161107
209. Feng MX, Liu JP, Zhang SM et al (2013) Design considerations for GaN-based blue laser diodes with InGaN upper waveguide layer. IEEE J Sel Top Quantum Electron 19(4):1500705–1500705
210. Liu J, Zhang L, Li D et al (2017) GaN-based blue laser diodes with 2.2 W of light output power under continuous-wave operation. IEEE Photon Technol Lett 24:1–1
211. Krames MR, Shchekin OB, Mueller-Mach R et al (2007) IEEE/OSA status and future of high power light emitting diodes for solid state lighting. J Display Technol 3(2):160–175
212. Verzellesi G, Saguatti D, Meneghini M et al (2013) Efficiency droop in InGaN/GaN blue light-emitting diodes: Physical mechanisms and remedies. J Appl Phys 114(7):071101
213. Wang CH, Ke CC, Lee CY et al (2010) Hole injection and efficiency droop improvement in InGaN/GaN light-emitting diodes by band-engineered electron blocking layer. Appl Phys Lett 97(26):261103
214. Masui S, Nakatsu Y, Kasahara D et al (2017) Recent improvement in nitride lasers. Proc SPIE 10104:101041H
215. Murayama M, Nakayama Y, Yamazaki K et al (2018) Watt-class green (530nm) and blue (465nm) laser diodes. Phys Status Solidi(a), 215(10):1700513
216. Strauβ U, Hager T, Brüderl G, Wurm T, Somers A, Eichler C, Vierheilig C, Löffler A, Ristic J, Avramescu A (2014) Recent advances in c-plane GaN visible lasers. In: Proceeding SPIE 8986, gallium nitride materials and devices IX, 89861L (8 March 2014)
217. Morishita Y, Nomura Y, Goto S et al (1995) Effect of hydrogen on the surface-diffusion length of Ga adatoms during molecular-beam epitaxy. Appl Phys Lett 67(17):2500–2502
218. Queren D, Schillgalies M, Avramescu A et al (2009) Quality and thermal stability of thin InGaN films. J Cryst Growth 311(10):2933–2936
219. Strauβ U, Avramescu A, Lermer T et al (2011) Pros and cons of green InGaN laser on c-plane GaN. Phys Status Solidi (b) 248(3):652–657
220. Queren D, Avramescu A, Schillgalies M et al (2010) Epitaxial design of 475 nm InGaN laser diodes with reduced wavelength shift. Phys Status Solidi 6(s2):S826-S829
221. Li Z, Liu J, Feng M et al (2013) Suppression of thermal degradation of InGaN/GaN quantum wells in green laser diode structures during the epitaxial growth. Appl Phys Lett 103(15):152109
222. Liu J, Li Z, Zhang L et al (2014) Realization of InGaN laser diodes the electroluminescence peak energy in green InGaN single quantum well LEDs. J Appl Phys 115:033112
223. Hager T, Lermer T, Bruederl G et al (2014) Carrier transport in InGaN-based light-emitting diodes. Phys Status Solidi A 211(1):22–29
224. Hader J, Moloney JV, Lermer T et al (2013) On the importance of radiative and Auger losses in GaN-based quantum wells. Appl Phys Lett 103:071101
225. Hader J, Moloney JV, Lermer T et al (2014) Temperature dependence of the radiative recombination coefficients in GaN-based quantum wells. Appl Phys Lett 104:231107
226. Hader J, Moloney JV, Lermer T et al (2014) On the importance of multi-particle effects in the temperature dependence of the radiative efficiency in InGaN quantum wells. Appl Phys Lett 105:071104

227. Hader J, Moloney JV, Lermer T et al (2015) On the importance of green gap in InGaN quantum wells. Appl Phys Lett 106:101104
228. Hader J, Moloney JV, Lermer T et al (2016) On the importance of high temperature carrier leakage in InGaN quantum wells. Appl Phys Lett 108:141104
229. Hader J, Moloney JV, Lermer T et al (2017) On the importance of Auger processes in InGaN quantum wells. Appl Phys Lett 110:071105
230. Hader J, Moloney JV, Lermer T et al (2018) On the importance of carrier localization and delocalization in InGaN quantum wells. Appl Phys Lett 112:071106
231. Hader J, Moloney JV, Lermer T, et al (2019) On the importance of carrier capture and escape processes in InGaN quantum wells. Appl Phys Lett 113:071107
232. Hader J, Moloney JV, Lermer T, et al (2020) On the importance of carrier recombination processes in InGaN quantum wells. Appl Phys Lett 114:071108
233. Hader J, Moloney JV, Lermer T et al (2021) On the importance of carrier injection processes in InGaN quantum wells. Appl Phys Lett 115:071109
234. Hader J, Moloney JV, Lermer T, et al (2022) On the importance of carrier transport processes in InGaN quantum wells. Appl Phys Lett 116:071110
235. Zhang L, Liu J, Tian A et al (2017) High-efficiency green InGaN/GaN light-emitting diodes with graded superlattices. Appl Phys Lett 110(14):141108
236. Liu J, Zhang L, Tian A et al (2017) High-efficiency InGaN-based green light-emitting diodes with quaternary InAlGaN superlattice electron blocking layer. Appl Phys Lett 110(13):131101
237. Zhang L, Liu J, Tian A et al (2017) High-efficiency InGaN-based green light-emitting diodes with quaternary InAlGaN superlattice electron blocking layer. Appl Phys Lett 110(12):121101
238. Liu J, Zhang L, Tian A et al (2017) High-efficiency InGaN-based green light-emitting diodes with quaternary InAlGaN superlattice electron blocking layer. Appl Phys Lett 110(11):111101
239. Zhang L, Liu J, Tian A et al (2017) High-efficiency InGaN-based green light-emitting diodes with quaternary InAlGaN superlattice electron blocking layer. Appl Phys Lett 110(10):101101
240. Liu J, Zhang L, Tian A et al (2017) High-efficiency InGaN-based green light-emitting diodes with quaternary InAlGaN superlattice electron blocking layer. Appl Phys Lett 110(9):091101
241. Zhang L, Liu J, Tian A et al (2017) High-efficiency InGaN-based green light-emitting diodes with quaternary InAlGaN superlattice electron blocking layer. Appl Phys Lett 110(8):081101
242. Liu J, Zhang L, Tian A et al (2017) High-efficiency InGaN-based green light-emitting diodes with quaternary InAlGaN superlattice electron blocking layer. Appl Phys Lett 110(7):071101
243. Zhang L, Liu J, Tian A et al (2017) High-efficiency InGaN-based green light-emitting diodes with quaternary InAlGaN superlattice electron blocking layer. Appl Phys Lett 110(6):061101
244. Liu J, Zhang L, Tian A et al (2017) High-efficiency InGaN-based green light-emitting diodes with quaternary InAlGaN superlattice electron blocking layer. Appl Phys Lett 110(5):051101
245. Zhang L, Liu J, Tian A et al (2017) High-efficiency InGaN-based green light-emitting diodes with quaternary InAlGaN superlattice electron blocking layer. Appl Phys Lett 110(4):041101
246. Massabuau FCP, Sahonta SL, Trinh-Xuan L et al (2012) Morphological, structural, and emission characterization of trench defects in InGaN/GaN quantum well structures. Appl Phys Lett 101:212107
247. Massabuau CP, Trinh-Xuan L, Lodie D et al (2013) Correlations between the morphology and emission properties of trench defects in InGaN/GaN quantum wells]. J Appl Phys 113(7):3675
248. Tian Ai Qin (2017) MOCVD growth and characterization of GaN-based green lasers. University of Science and Technology of China, Hefei
249. Pohl UW (2013) Epitaxy of semiconductors: introduction to physical principles]. Graduate Texts in Physics 31(1):45–50
250. Graham WR, Ehrlich G (1973) Surface self-diffusion of atoms and atom pairs]. Phys Rev Lett 31(23):1407–1408
251. Wang SC, Ehrlich G (1993) Adatom motion to lattice steps: a direct view]. Phys Rev Lett 70(1):41–44
252. Liu SJ, Wang EG, Woo CH et al (2000) Three-dimensional Schwoebel-Ehrlich barrier]. J Comput Aided Mater Des 7(3):195–201

253. Liu SJ, Huang H, Woo CH (2002) Schwoebel- Ehrlich barrier: from two to three dimensions. Appl Phys Lett 80:3295–3297
254. Suihkonen S, Svensk O, Lang T et al (2007) The effect of InGaN/GaN MQW hydrogen treatment and threading dislocation optimization on GaN LED efficiency. J Cryst Growth 298:740–743
255. Suihkonen S, Lang T, Svensk O et al (2007) Control of the morphology of InGaN/GaN quantum wells grown by metalorganic chemical vapor deposition. J Cryst Growth 300:324–329
256. Taylor EFang FOehler Fet al (2013) Composition and luminescence studies of InGaN epilayers grown at different hydrogen flow rates. Semiconductor Sci Technol 28:065011
257. Scholz F, Off J, Fehrenbacher E et al (2000) Investigations on structural properties of GaInN/GaN multi quantum well structures. PhysicsStatus Solisi (a) 180:315–320
258. Liu JP, Wang YT, Yang H et al (2004) Investigations on V-defects in quaternary AlInGaN epilayers. Appl Phys Lett 84:5449
259. Shiojiri M, Chuo CC, Hsu JT et al (2006) Structure and formation mechanism of V defects in multiple InGaN/GaN quantum well layers. J Appl Phys 99:073505
260. Florescu DI, Ting SM, Ramer JC et al (2003) Investigation of V-defects and embedded inclusions in InGaN/GaN multiple quantum wells grown by metalorganic chemical vapor deposition on (0001) sapphire. Appl Phys Lett 83:33
261. Tian A, Liu J, Zhou R et al (2019) Green laser diodes with constant temperature growth of InGaN/GaN multiple quantum well active region. Appl Phys Express, 12:064007
262. Hu L, Ren X, Liu J et al (2020) High-power hybrid GaN-based green laser diodes with ITO cladding layer. Photon Res 8:279
263. Tian A, Hu L, Zhang L et al (2020) Design and growth of GaN-based blue and green laser diodes. Sci China Mater 63(8):1348–1363
264. Haglund A, Hashemi E, Bengtsson J et al (2016) Progress and challenges in electrically pumped GaN-based VCSELs. Proc SPIE 9892:98920Y
265. Murayama M, Nakayama Y, Yamazaki K et al (2017) Watt-class green (530 nm) and blue (465 nm) laser diodes. Phys Status Solidi (A)Appl Mater 215(10):1700513.1–1700513.5
266. Tatum JA (2014) Evolution of VCSELs. Proc SPIE 9001:90010C
267. Feezell DF (2015) Status and future of GaN-based vertical-cavity surface-emitting lasers. Proc SPIE 9363:93631G
268. Guoqiang R, Jianfeng W, Zongliang L et al (2019) Progress in the study of GaN single crystal growth. J Synthet Crystals 48(9):1588–1598
269. Tianrui Y, Huan X, Yang M, Rongbin X, Baoping Z, Leiyin Y (2020) Progress in the research of GaN vertical cavity surface emitting lasers. Chinese Laser 47(7)2020
270. Kioupakis E, Rinke P, Delancy KT et al (2011) Indirect Auger recombination as a cause of efficiency droop in nitride light-emitting diodes. Appl Phys Lett 98(16):161107
271. Jr Wierer JJ, Tsao JY, Sizov DS (2013) Comparison between blue lasers and light-emitting diodes for future solid-state lighting. Laser Photonics Rev 7(6):963–993
272. McKendry JJD, Massoubre D, Zhang SL et al (2012) Visible-light communications using a CMOS-controlled micro-light-emitting-diode array. J Lightwave Technol 30(1):61–67
273. Fang F, Aimin Z, Tianchu L (2019) Time: from astronomical time to atomic seconds. Metrol Technol 5:7–10
274. Kitching J, Knappe S, Vukicevic M et al A microwave frequency reference based on VCSE-driven dark line line resonances in Cs vapor. IEEE Trans Inst Measur 49(6):1313–1317.
275. Miah MJ, Al-Samanch A, Kern A et al (2013) Fabrication and characterization of low-threshold polarization-stable VCSELs for CS-based miniaturized atomic clocks. IEEE J Selected Top Quantum Electron 19(4):17011410
276. Warren ME, Podva D, Dacha P et al (2018) Low-divergence high-power VCSEL arrays for lidar application. Proc SPIE 10552:105520E
277. Amano H, Kito M, Hiramatsu K et al (1989) P-type conduction in Mg-doped GaN treated with low-energy electron beam irradiation(LEEBI). Jpn J Appl Phys 28(12):L2112–L2114

278. Nakamura S (1991) GaN growth using GaN buffer layer. Jap J Appl Phys 30(10A):L1705–L1707
279. Krestnikov IL, Lundin WV, Sakharov AV et al (1999) Room-temperature photopumped InGaN/GaN/AIGaN vertical-cavity surface-emitting laser. Appl Phys Lett 75(9):1192–1194
280. Someya T, Werner R, Forchel A et al (1999) Room temperature lasing at blue wavelengths in gallium nitride microcavities. Science 285(5435):1905–1906
281. Lu T C, Kao CC, Kuo HC et al (2008) CW lasing of current injection blue GaN-based vertical cavity surface emitting laser. Appl Phys Lett 92(14):141102
282. Higuchi Y, Omae K, Matsumura H et al (2008) Room-temperature CW lasing of a GaN-based vertical-cavity surface-emitting laser by current injection. Appl Phys Exp 1:121102
283. Omae K, Higuchi Y, Nakagawa K et al (2009) Improvement in lasing characteristics of GaN-based vertical-cavity surface-emitting lasers fabricated using a GaN substrate. Appl Phys Exp 2:052101
284. Kasahara D, Morita D, Kosugi T et al (2011) Demonstration of blue and green GaN-based vertical-cavity surface-emitting lasers by current injection at room temperature. Appl Phys Exp 4(7):072103
285. Holder C, Speck JS, DenBaars SP et al (2012) Demonstration of nonpolar GaN-based vertical-cavity surface emitting lasers. Appl Phys Exp 5(9):092104
286. Holder CO, Leonard J T, Farrell RM et al (2014) Nonpolar III-nitride vertical-cavity surface emitting lasers with a polarization ratio of 100% fabricated using photoclectrochemical etching. Appl Phys Lett 105(3):031111
287. Liu WJ, Hu XL, Ying LY et al (2014) Room temperature continuous wave lasing of electrically injected GaN-based vertical cavity surface emitting lasers. Appl Phys Lett 104(25):251116
288. Izumi S, Fuutagawa N, Hamaguchi T et al (2015) Room-temperature continuous-wave operation of GaN-based vertical-cavity surface-emitting lasers fabricated using epitaxial lateral overgrowth. Appl Phys Exp 8(6):062702
289. Kuramoto M, Kobayashi S, Akagi T et al (2019) Watt-class blue vertical-cavity surface-emitting laser arrays. Appl Phys Express 12(9):091004
290. Hamaguchi T, Nakajima H, Tanaka M et al (2019) Sub-milliampere-threshold continuous wave operation of GaN-based vertical-cavity surface-emitting laser with lateral optical confinement by curved mirror. Appl Phys Exp 12(4):0440.
291. Langer T, Kruse A, Ketzer FA et al (2011) Origin of the "green gap": increasing nonradiative recombination in indium-rich GalnN/GaN quantum well structures. Phys Status Solidi (c) 8(7/8):2170–2172
292. Waltereit P, Brandt O, Trampert A et al (2000) Nitride semiconductors free of electrostatic fields for efficient white light emitting diodes. Nature 406(6798):865–868
293. Meney AT, O' Reilly EP, Adams AR (1996) Optical gain in wide bandgap GaN quantum well lasers. Semiconductor Sci Technol 11(6):897–903
294. Tao RC, Arakawa Y (2019) Impact of quantum dots on III-nitride lasers; a theoretical calculation of threshold current densities. Jap J Appl Phys 58(SC):SCCC31
295. Arakawa Y (2002) Progress in GaN-based quantum dots for optoelectronics applications. IEEE J Selected Top Quantum Electron 8(4):823–832
296. Cai LE, Zhang JY, Zhang BP et al (2008) Blue-green optically pumped GaN-based vertical cavity surface emitting laser. Electron Lett 44(16):972–974
297. Weng GE, Mei Y, Liu JP et al (2016) Low threshold continuous-wave lasing of yellow-green InGaN-QD vertical-cavity surface-emitting lasers. Opt Exp 24(14):15546–15553
298. Mei Y, Weng G E, Zhang BP et al (2017) Quantum dot vertical-cavity surface-emitting lasers covering the 'green gap'. Light: Sci Appl 6(1):e16199
299. Xu R B, Mei Y, Zhang B P, et al. Simultaneous blue and green lasing of GaN-based vertical-cavity surface-emitting lasers].Semiconductor Science and Technology, 2017,32(10):105012.
300. Xu RB, Mei Y, Xu H et al (2018) Green vertical-cavity surface-emitting lasers based on combination of blue-emitting quantum wells and cavity-enhanced recombination. IEEE Trans Electron Dev 65(10)4401–4406

301. Mei Y, Xu RB, Ying LY et al (2019) Room temperature continuous wave lasing of GaN-based green vertical-cavity surface-emitting lasers. Proc SPIE 10918:10918H
302. Redwing JM, Loeber DAS, Anderson NG et al (1996) An optically pumped GaN-AlGaN vertical cavity surface emitting laser. Appl Phys Lett 69(1):1–3
303. Detchprohm T, Li X, Shen SC et al (2017) III-N wide bandgap deep-ultraviolet lasers and photodetectors. Semiconduct Semimetals 96(4):121–166
304. Chen R, Sun HD, Wang T et al (2010) Optically pumped ultraviolet lasing from nitride nanopillars at room temperature. Appl Phys Lett 96(2):241101
305. Liu YS, Haq AF MS, Kao TT et al (2015) Development for ultraviolet vertical cavity surface emitting lasers. In: 2015 European conference on lasers and electro-optics-european quantum electronics conference, June 21–25, 2015, Munich, Germany. Washington DC; OSA, 2015:PD_A_2
306. Liu YS, Saniul Haq AFM, Mchta K et al (2016) Optically pumped vertical-cavity surface-emitting laser at 374.9 nm with an electrically conducting n-type distributed Bragg reflector. Appl Phys Exp 9(11):111002
307. Chang TC, Kuo SY, Hashemi E et al (2018) GaN vertical-cavity surface-emitting laser with a high-contrast grating reflector. Proc SPIE 10542:105420T
308. Park YJ, Detchprohm T, Mehta K et al (2019) Optically pumped vertical-cavity surface-emitting lasers at 375nm with air-gap/Al0.05Ga0.95N distributed Bragg reflectors].Proceedings of SPIE 10938:109380 A
309. Zheng ZM, Li YQ, Paul O (2018) Loss analysis in nitride deep ultraviolet planar cavity. J Nanophoton 12(4):043504

Chapter 8
Applications of Freestanding Gallium Nitride Substrates in Power Electronic and Microwave RF Devices

Power electronic technology is used in the transmission, processing, storage, and control of electrical energy, and its main function is realized by power electronic devices. The main requirements for power electronic devices are high breakdown voltage, low on-resistance, low switching loss, high switching speed, and high reliability. Si-based power electronic technology is the mainstream technology currently, but due to the inherent limitations of the Si material, Si-based power electronic devices face significant bottlenecks in breakdown voltage, on-resistance, switching loss, switching speed, and reliability.

Microwave radio frequency technology is applied in the generation, modulation, frequency mixing, drive amplification, power amplification, transmission, spatial propagation, reception, low-noise amplification, intermediate-frequency amplification, demodulation, detection, filtering, attenuation, phase shifting, and switching of microwave and radio frequency signals. Its core function is to achieve power amplification. The main requirements for microwave radio frequency devices are high efficiency, high power, and wide bandwidth. Currently, Si-based LDMOS and GaAs-based devices are the mainstream technologies, however, the operating frequency of Si-based LDMOS is limited (below 3 GHz), and the output power of GaAs-based devices is relatively low (below 50 W), which makes it difficult to meet the future requirements of microwave radio frequency devices.

GaN material possesses properties such as wide band gap, high electron mobility, high thermal conductivity, strong breakdown electric field, and strong radiation resistance [1]. It is suitable for operation under harsh conditions such as high temperature, high frequency, high voltage, high power, and strong radiation. GaN-based power electronic devices have advantages such as low on-state resistance, high switching speed, and low switching loss. GaN-based microwave radio frequency devices have advantages such as high efficiency, high power, and wide bandwidth. They can meet the requirements of high speed, high energy efficiency, high power, high frequency, high temperature, and radiation resistance in practical applications. GaN-based devices are the core of power electronics and microwave radio frequency

K. Xu et al., *GaN Single Crystal Growth and Application*, Wide Bandgap Semiconductors, https://doi.org/10.1007/978-981-96-7572-2_8

devices. They have broad application prospects in the fields of next-generation mobile communication, smart grid, electric vehicles, consumer electronics, etc. They are the new strategic high ground of the global semiconductor industry.

At present, the research on GaN electronic devices is mainly divided into two categories. The first category is GaN planar devices prepared on semi-insulating heterosubstrates such as Si, SiC, etc. The second category is vertical devices fabricated on GaN single crystal substrates. Planar devices have disadvantages such as high dislocation density, difficulty in fabricating normally-off devices, severe current collapse effect, low breakdown voltage, high leakage current, non-uniform electric field distribution and strong influence of surface states on device performance. Although it is feasible to fabricate vertical GaN devices on Si substrates, the heteroepitaxial GaN layer grown on Si substrates has a high dislocation density on the order of 10^8–10^{10} cm^{-2}, resulting in poor breakdown performance of GaN-on-Si vertical structure devices.

In order to fundamentally improve the breakdown voltage of the device and solve the current collapse effect, the vertical device based on GaN single crystal substrate is an important development direction. Vertical devices have many advantages over planar devices: (1) Vertical devices are fabricated with homogeneous epilayers on single crystal substrates, so the defect density in the vertical devices is lower than that in the planar devices. Moreover, the leakage current of the device is lower; (2) Because vertical structures are fabricated with single crystal substrates via homoepitaxy, they require no buffer layer. This allows for the growth of thicker homoepitaxial layers, which can withstand higher breakdown voltages. Furthermore, the peak electric field is shifted from the surface to the interior of the device, thereby improving its reliability; (3) Vertical structure devices can recover from avalanche breakdown, leading to higher device reliability; (4) The high electric field region in vertical structure devices is transferred to the internal of the device, which effectively alleviates the problem of current collapse in planar structure devices; (5) Vertical structure devices are conducive to the miniaturization and integration of devices.

As the performance of Si-based devices approaches the material limit, the search for new alternative materials has become an urgent task, leading to the rapid development of third-generation semiconductor materials such as SiC and GaN. Figure 8.1 summarizes the performance comparison of various devices, including vertical GaN devices and planar GaN devices, SiC devices, and Si power devices. As shown in the figure, the performance of vertical GaN devices has surpassed that of planar GaN devices and is approaching the theoretical limit of GaN.

Advances in GaN single-crystal substrate technology have led to significant improvements in crystal quality, electrical performance, and wafer size, along with a substantial cost reduction. These advancements have accelerated the development of GaN-based vertical devices, allowing their great potential to be progressively realized.

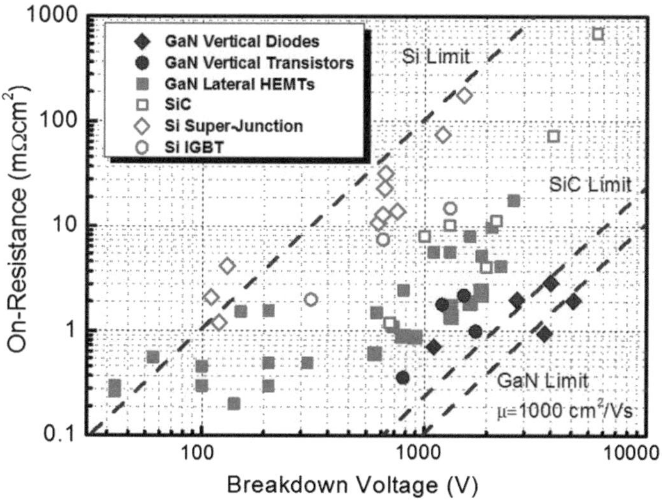

Fig. 8.1 Comparison of the performance of different types of devices

8.1 Market Analysis

8.1.1 Market Analysis of Power Electronic Device

Compared with traditional Si-based power electronic devices, GaN-based power electronic devices have the following advantages. First, the conversion efficiency of GaN-based power electronic devices is higher than traditional silicon devices. The bandgap of GaN is about three times that of silicon, and the critical breakdown field is 10 times that of silicon. Therefore, at the same rated voltage, the on-resistance of GaN power devices is roughly three orders of magnitude lower than that of Si devices, which greatly reduces the conduction loss of the switch. Second, the operating frequency of GaN-based power electronic devices is higher. The electron transit time in GaN is 10 times lower than that in silicon, and the electron velocity is more than 2 times higher than that in silicon. The reverse recovery time can basically be ignored, so the operating frequency of GaN switching power devices can be at least 20 times higher than that of silicon devices. This greatly reduces the volume of energy storage components such as capacitors and inductors in the circuit, leading to a multi-fold reduction in both overall device volume and the consumption of valuable raw materials like copper. In addition, the high switching frequency can also reduce switching losses, further reducing the overall energy consumption of the power supply. Third, the operating temperature of GaN-based power electronic devices is higher. GaN has a wide bandgap of 3.4 eV, resulting in an extremely low intrinsic carrier concentration and high carrier excitation energy. Consequently, GaN devices can theoretically operate at temperatures exceeding 800 °C.

GaN, SiC, and Si materials have their own distinct advantages, but there is also overlap. GaN material with the highest electron saturation drift velocity is suitable for high frequency application scenarios, but it is inferior to SiC in high voltage and high power scenarios. With the decrease in cost, GaN is expected to replace silicon-based power devices in the medium and low voltage applications such as diodes, IGBTs, MOSFETs, etc. In terms of operating voltage, Si dominates the 0–300 V range, while SiC is preferred for 600 V and above. The voltage range between 300 V and 600 V is the competitive domain for GaN technology. According to Yole's estimation, in the low-voltage market of 0 ~ 900 V, GaN has a large application potential, accounting for about 68% of the total power market.

The market for GaN-based power electronic devices has a significant growth potential, with rapidly increasing demand in smart grids, electric vehicles, power and photovoltaic inverters, power supplies, etc. Currently, GaN-based power electronic devices have made significant progress in the field of power supply devices. Due to their heterojunction two-dimensional electron gas structure, GaN devices have a higher operating frequency than that of SiC devices, and the breakdown voltage of GaN is lower than that of SiC, so GaN power electronic devices are more suitable for high-frequency, small volume, cost-sensitive and low-power requirements in the power supply field, such as light-weight power adapters for consumer electronics, ultra-light-weight power supplies for drones, wireless charging devices, etc.

8.1.2 Market Analysis of Microwave Radio Frequency Devices

Compared with traditional Si-based LDMOS and GaAs-based devices, GaN-based microwave radio frequency devices have the following advantages. First, the efficiency of GaN-based microwave radio frequency devices is higher. Higher efficiency can reduce power consumption, save electricity, reduce cooling costs, and lower total operating costs. Second, the bandwidth of GaN-based microwave radio frequency devices is wider. A larger bandwidth can increase information carrying capacity, cover multiple frequencies with fewer devices, and reduce product costs of customers. It is also suitable for spread spectrum communications, electronic countermeasures and other areas. Third, the power of GaN-based microwave radio frequency devices is higher. In the frequency band above 4 GHz, GaN-based devices can output much higher power than GaAs-based ones, making them especially suitable for high-power applications such as radar, satellite communication, relay communication, and other fields.

In the low-frequency range, Si-based LDMOS is currently the mainstream in the market. However, the bandwidth of LDMOS will decreases significantly with the increasing frequency (being effective only up to approximately 3.5 GHz). GaN devices with a 0.25 μm process can reach a frequency up to four times higher Fabricated on a 0.25-μm process, GaN devices can operate at frequencies four times higher, increase bandwidth by 20%, and achieve a power density of 6–8 W/mm (compared to 1–2 W/mm for LDMOS). Moreover, their mean time between failures (MTBF) reaches one million hours, demonstrating superior durability and overall performance.

In the higher frequency bands and low power ranges, GaAs currently dominates the market with over 90% of shipments. The main advantages of GaN over GaAs RF devices are its bandgap width and thermal conductivity. The bandgap width of GaN is 3.4 eV, much higher than that of GaAs (1.42 eV), so GaN devices have a higher breakdown voltage and can meet the higher power requirements. Moreover, the thermal conductivity of GaN single crystal substrates is more than twice that of GaAs, which means that the power consumption in the GaN-based device can be more easily transferred to the surrounding environment, and the heat dissipation is better. Compared with devices fabricated on heterogeneous substrates (such as Si, SiC), devices on GaN single crystal substrates have lower dislocation density and require no nucleation or buffer layers. This results in lower interfacial thermal resistance and higher thermal conductivity of the epitaxial layer, which significantly improves device heat dissipation.

In large-scale antenna applications, a large number of array antennas are used in the base station transceiver to achieve higher wireless data through put and connection reliability. This architecture requires a corresponding array of RF transceiver units, so the number of RF devices will greatly increase, making the size of the devices critical. The characteristics of GaN such as small size, high efficiency, and high power density can achieve highly integrated solutions, such as modular RF front-end devices. In addition to the significant increase in the number of RF devices required in the base station RF transceiver unit array, the density and number of base stations will also greatly increase. Therefore, compared to the 3G and 4G era, the number of RF devices in the 5G era will increase by tens or even hundreds of times. In 5G millimeter-wave applications, the high power density characteristics of GaN can effectively reduce the number of transceiver channels and the size of the overall solution under the same coverage conditions and user tracking functions.

As an RF component material, GaN has gradually been adopted in telecommunications infrastructure, national defense, and military applications. National defense is the main application area of GaN RF devices. Due to the demand for high performance and price insensitivity, the national defense market offers substantial growth potential for GaN RF devices. According to Yole statistics, in 2018, the market size of GaN RF devices in the national defense field was $201 million, accounting for 44% of the market share of GaN RF devices, surpassing base stations to become the largest application market. In recent years, the global national defense industry has shown no signs of a slowing down. The market size of GaN RF devices in the national defense field will continue to grow with the increase in penetration rate. The market size is expected to reach $454 million by 2023, with a compound annual growth rate of 18% from 2018 to 2023. Base station is the second largest application market for GaN RF devices. With the rapid development of global 5G communications, the demand for mobile communication power amplifiers will grow explosively. Among them, the terminal side power amplifier will continue to use the GaAs process, while on the base station side, the traditional Si-based LDMOS process will be replaced by the GaN process with higher power handling capacity and efficiency to meet the demand of base station miniaturization. According to Yole Développement, the market for GaN RF devices in base stations was valued at $150 million in 2018, representing 33% of the total GaN RF device market. Following the deployment of 5G networks, this market grew significantly. It is projected to reach $521 million by 2023, achieving a CAGR (Compound Annual Growth Rate) of 28% from 2018 to 2023.

8.2 PN Junction Diode

Figure 8.2 illustrates the operating principle and device structure of the PN junction diode. As shown in Fig. 8.2a, the PN junction is formed by a closely contacted n-type region and a p-type region. The region near the interface of the two semiconductors is the PN junction. Due to the interaction between positive and negative charges, a built-in electric field is formed in the space charge region, which is directed from the positively charged N region to the negatively charged P region. This electric field direction is opposite to the direction of carrier diffusion, thereby inhibiting carrier movement.

The PN junction has two operating states:

(1) When the PN junction is forward biased: the p-type region of the PN junction is connected to the positive pole of the power supply, and the n-type region is connected to the negative pole of the power supply. At this time, the electric field generated by the external voltage on the PN junction is opposite to the direction of the built-in electric field of the PN junction, weakening the built-in electric field of the PN junction, allowing the majority carrier to pass through the PN

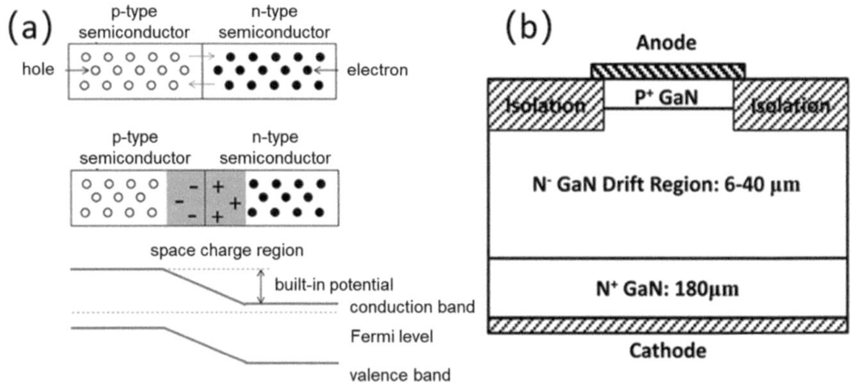

Fig. 8.2 The working principle and device structure of the PN junction diode. **a** The working principle of the PN junction diode, **b** Schematic diagram of a typical device structure

junction to form a forward current, and rapidly increasing with the increase of the external voltage. In other words, the PN junction is in an on-state when a forward voltage is applied.

(2) When the PN junction is reverse biased: the p-type region of the PN junction is connected to the negative pole of the power supply, and the n-type region is connected to the positive pole of the power supply. At this time, the electric field generated by the external voltage on the PN junction is in the same direction as the built-in electric field of the PN junction, which strengthens the built-in electric field of the PN junction, making it difficult for the majority carrier to pass through the PN junction under the action of the electric field force, so that the reverse current is extremely small, that is, the PN junction is in an off-state when a reverse voltage is applied.

The schematic diagram of a typical vertical structure PN junction diode is shown in Fig. 8.2b. The vertical PN junction diodes have superior performance in achieving high breakdown voltage and low conduction resistance compared to planar structure devices. PN junction diodes often operate under high voltage and high power conditions, so improving the breakdown voltage of the device is critical. The breakdown voltage is primarily influenced by the following two factors:

(1) The defect density and doping concentration of the epitaxial layer material.

Based on high-quality GaN single crystal substrates, the dislocation density in the epitaxial layer can be effectively reduced, and the breakdown voltage of the device can be improved [2]. In addition, the doping concentration also has a significant influence on the breakdown voltage of the device, as illustrated in Fig. 8.3. As the carrier concentration inside the GaN material decreases, the thickness of the device drift region gradually increases, and the device breakdown voltage also increases significantly [3]. Therefore, the further development of high-voltage GaN power electronic devices require high-purity GaN single crystal substrates with low impurity concentrations.

(2) The structure of the device

A peak electric field occurs at the anode edge of the PN junction diode, which reduces the device breakdown voltage and prevents the full advantages of GaN material from being realized. The ideal parallel plane depletion region edge is uniform and flat, while the actual PN junction is usually formed within a certain size of the mask window, so the depletion region edge junction has a larger curvature, resulting in a peak in the electric field intensity at the edge of the actual PN junction and the device breaks down prematurely.

At present, the internal electric field distribution can be optimized to improve the breakdown voltage of the device through field plate structure, junction terminal structure and other edge terminal technologies, and drift region gradient doping technology.

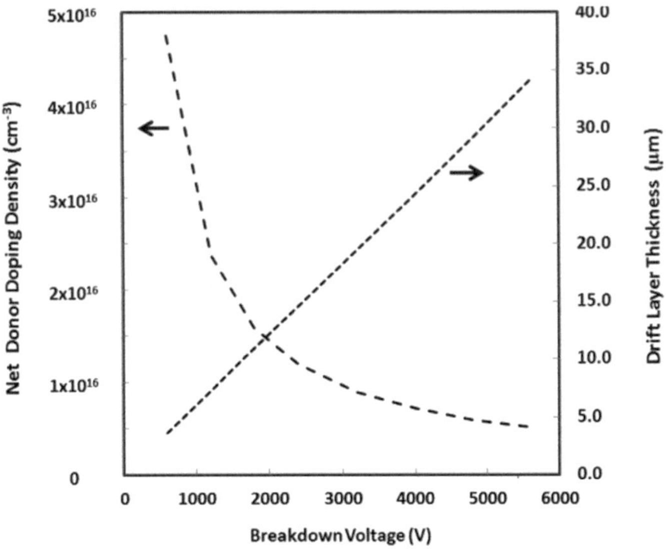

Fig. 8.3 The effect of doping concentration on the drift region thickness and breakdown voltage of a vertical structure GaN-based PN junction diode

(1) Field plate structure. When the device operates in the off state, the curvature of the edge of the PN junction depletion region is large, and the electric field strength is high. By adding a field plate, the entire P region and its edge are covered. As the reverse bias voltage increases, the P region and the field plate simultaneously deplete the N region. The entire depletion region extends along the junction surface to the edge of the field plate, the number of electric field peaks increases from 1 to 2, the extra peak is located at the edge of the field plate, effectively reducing the electric field peak at the edge of the original PN junction depletion region and improve the breakdown voltage of the device.
(2) Junction terminal structure. In addition to using the field plate structure to improve the voltage resistance level of the vertical structure diode, a junction terminal structure can also be formed at the edge of the device by ion implantation. By optimizing and controlling the energy and dose of ion implantation, a uniform vacancy can be formed inside the semiconductor to compensate for the free carriers inside the semiconductor, and finally a high resistance region is formed in the ion-implanted region to suppressing electric field crowding at the junction edge.
(3) Drift region gradient doping. The voltage resistance level of the device is also related to its internal electric field distribution. If the electric field distribution is non-uniform, it will be difficult to effectively improve the voltage resistance of the device. When the device is in the reverse working state, the electric field reaches its peak at the junction surface of the PN junction, and it is relatively low inside the device. The drift region gradient doping technology can make

the doping concentration gradually decrease from the junction surface of the PN junction to the inside of the device, which can increase the width of the depletion region and weaken the electric field peak at the junction surface. Therefore, optimizing the internal electric field distribution by the drift region gradient doping technology can effectively improve the voltage resistance level of the device.

At Cornell University, K. Nomoto et al. [4] designed and fabricated a GaN-based vertical structure PN diode with a field plate structure and a drift layer thickness of 10 μm on a GaN single crystal substrate. By applying a bias voltage to the field plate, the radius of curvature of the PN junction depletion region can be effectively increased, thereby alleviating the electric field concentration effect at the edge. The breakdown voltage in the device without the field plate structure is 806 V, while it increases significantly to 1700 V after the addition of the field plate structure, indicating that the field plate structure can effectively suppress the electric field crowding effect at the junction edge. The average breakdown field strength in the device with the field plate is 1.7 MV/cm, and the FOM is 5.30 GW/cm^2. H. Ohta et al. [5] fabricated a GaN-based vertical PN junction diode with three drift layers and a field plate structure on a GaN single crystal substrate. By applying multiple lightly doped n-GaN drift layers to the PN junction diode, the peak electric field at the device surface is effectively suppressed, making the internal electric field distribution more uniform. At the same time, the design uses a field plate structure to reduce the reverse leakage current and increase the breakdown voltage. The obtained device has a breakdown voltage of 4.7 kV and a specific on-resistance of 1.7 mΩ·cm^2. In 2015, I. C. Kizilyalli and others [6] from Avogy, a US company, successfully obtained a vertical GaN PN junction diode with low leakage current (less than 10 nA) utilizing a free-standing GaN substrate. Its breakdown voltage exceeds 4 kV, the specific on-resistance is 2.8 mΩ·cm^2, and the calculated $V_{BR}{}^2/R_{on}$ is as high as 5714 MW/cm^2. In 2016, K. Nomoto and others [4] from Cornell University in the United States achieved a PN junction diode with a breakdown voltage of 1.7 kV and a specific on-resistance of 0.55 Ω·cm^2 on a GaN single crystal substrate through a field plate structure.

8.3 Schottky Barrier Diode

The Schottky barrier diode (SBD) is a power semiconductor device that achieves rectification by generating a Schottky barrier through metal–semiconductor contact. Its basic structure is shown in Fig. 8.4. The Schottky contact is fabricated on the undoped GaN layer, while the ohmic contact is fabricated on the heavily-doped GaN layer. Under forward bias, the barrier height decreases, and electrons can easily flow from the semiconductor to the metal, resulting in current conduction; Under reverse bias, the barrier height increases, making it difficult for electrons to pass through

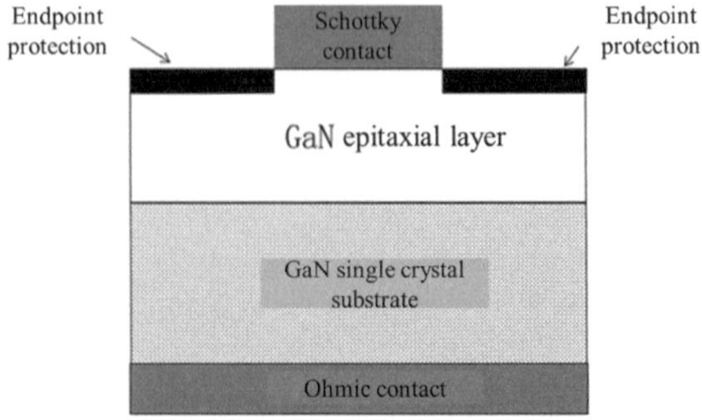

Fig. 8.4 Schematic diagram of the structure of a vertical GaN Schottky diode

the high barrier. This results in reverse blocking and demonstrates the device's unidirectional conductivity.

The Schottky barrier diode is a majority carrier device. Compared with the PN junction diode, it has the advantages of high switching speed, low forward voltage, short reverse recovery time, and low switching loss. It can simultaneously achieve high forward current and high reverse-blocking voltage, making it suitable for high-frequency, high-speed circuits. Under the same withstand voltage, the on-resistance of GaN power devices is two orders of magnitude lower than that of Si-based power devices and one order of magnitude lower than that of SiC power devices. Thus, the conduction loss generated by GaN devices during the switching process could be significantly reduced.

The ideal Schottky diode should exhibit high breakdown voltage, low leakage current, low forward voltage drop, low on-resistance, and fast recovery characteristics. Currently, the performance parameters of GaN-based vertical Schottky barrier diodes are still far from their theoretical values. Whether in terms of material growth or device technology, there are still many challenges. From top to bottom, the key technologies for manufacturing high-performance GaN vertical Schottky barrier diodes mainly include the following five points: (1) high-quality Schottky contacts; (2) terminal protection structures; (3) high-quality lightly-doped homoepitaxial thick films; (4) high-quality bulk GaN single crystal substrates; (5) high-reliability, low-resistivity ohmic contacts.

(1) For Schottky contacts, the commonly used electrode metals are Ti and Pt. But due to interface and surface state problems, different process conditions, such as surface treatment and annealing conditions, cause great differences in Schottky barrier height, ideality factor, and other device performance indicators. Moreover, Schottky diodes for different applications have different requirements for the Schottky barrier. As a fast switch, the Schottky barrier should be as low as possible, while as a rectifier under high voltage, the Schottky barrier should be high to reduce reverse leakage current. Therefore, the study on surface states and interface states is the key to improve the quality of Schottky contacts.
(2) In the design of terminal protection structures, the limited device diminsions cause current crowding and electric field concentration at the edge of the

Schottky electrode, leading to a premature breakdown under reverse bias. Therefore, it is necessary to reasonably design the edge terminal to effectively reduce the edge electric field concentration. Figure 8.5 shows three typical terminal protection structures, mainly including ion implantation termination, trench termination, and field plate termination.

(a) Ion implantation termination. The ions used for ion implantation are mainly Ar^+, Mg^+, F^+, etc., as shown in Fig. 8.5a. These ions will introduce a large number of defects or fixed negative charges into GaN after implantation, which will increase the resistivity of GaN and thus inhibit the edge leakage current.

(b) Trench type termination. Since ion implantation involves accelerating and bombarding high-energy ions at the sample, some ions need to be annealed after implantation to achieve the effect of eliminating damage. In 2016, S. Kone et al. [7] prepared two quasi-vertical structure GaN SBDs with and without a protection ring. C-V and I-V test results show that the protection ring can reduce the reverse leakage current of the device, and the reverse breakdown voltage of the device increases from below 300 V to about 400 V. They used the deep-level transient spectrum to study GaN. The defects introduced by the high-resistance edge terminal protection ring at the edge of the SBD were analyzed. Following the ion implantation to form the terminal structure, the existing defect levels are enhanced and utilized with a combination of dry etching and wet etching techniques. This technique retains the anisotropy of dry etching and uses wet etching for auxiliary repair, thereby eliminating the damage layer produced by dry etching. Currently, this is generally achieved by combining inductively coupled plasma etching technology with tetramethylammonium hydroxide (TMAH). The method of etching to form trenches suppresses the electric field at the boundary, suppresses reverse leakage current, and thus improves the reverse characteristics of the vertical structure GaN-based Schottky diode. This technology is also currently used to fabricate periodic trench structures, the width of these trenches is generally in the size of a few micrometers and combined with the deposition of dielectric layers to passivate the surface of GaN, as shown in Fig. 8.5b. The periodic trench design redistributes and weakens the peripheral electric field, thereby increasing the reverse breakdown voltage and further suppressing leakage current.

(c) Field plate termination. The field plate structure utilizes electrodes and dielectric layers to design different field plate structures to solve the boundary electric field problem, as shown in Fig. 8.5c. This structure utilizes the characteristics of high withstand voltage and passivation of the dangling bonds at the interface of the dielectric layer. It confines the edge electric field within the dielectric layer and simultaneously uses the dielectric layer to passivate the dangling bonds of GaN, thereby achieving the suppression of the reverse leakage current and an increase in the breakdown voltage.

(3) For the GaN epitaxial layer, the carrier concentration and thickness of the epitaxial layer are the key to determine the theoretical values of forward on-resistance and reverse breakdown voltage. As the carrier concentration in GaN decreases, the thickness of the device's drift region gradually increases, and the breakdown voltage would increase significantly [3]. Therefore, to improve the breakdown voltage of the device, it is crucial to precisely control the carrier concentration and thickness of the epitaxial layer. However, it is currently difficult to obtain a drift layer with extremely low carrier concentration (below 10^{15} cm^{-3}) by HVPE, MOCVD, and MBE methods.
(4) The primary challenge in the growth of GaN substrates is the production of large-size, high-quality bulk single crystals. The defect density of the substrate directly affects the quality of the homoepitaxial layer. The majority of the current GaN-based devices are grown on hetero-substrates, resulting in a significant lattice mismatch between the GaN layer and the foreign substrate. Consequently, the defect density of the GaN epitaxial layer is considerably higher than that of Si and GaAs. Experimental evidence indicates that screw dislocations in GaN are significant contributors to current leakage. Consequently, the presence of high-density defects diminishes the quality of the epitaxial wafer, increases the leakage current of the device, and decreases the performance of the device. Currently, the HVPE method for the growth of high-quality bulk GaN single crystals has gradually matured and is being employed in large-scale production, while other methods, such as the ammonothermal method and the Na-flux method, have also made notable advancements.
(5) The fabrication of ohmic contacts on the N-face of GaN presents two main challenges: low resistivity and high reliability. The ohmic contact electrode must be fabricated on the N-face of GaN for vertical devices. However, GaN is a polar compound, and the physicochemical properties of the Ga-face and N-face are markedly disparate. Consequently, the contact process of Ga-face ohmic contact cannot be directly applied on N-face. The N-face ohmic contact typically employs a Ti/Pt/Au (50/100/50 nm) multi-layer as the electrode. The two principal challenges confronting the N-face ohmic contact are as follows: (a) the contact resistivity is high, (b) the thermal stability is poor, especially after annealing at 300 °C. The poor performance of the N-face ohmic contacts is primarily attributed to two factors: (a) the susceptibility of the N-face GaN surface to oxidation, and (b) the predominant diffusion of Ga atoms (with minimal N diffusion) on the surface, which generates acceptor-type Ga vacancies.

Although SBDs exhibit a low turn-on voltage, the Schottky barrier lowering effect becomes pronounced under high reverse bias. This leads to excessive leakage current and premature breakdown, severely limiting the device's breakdown voltage. In contrast, the PN junction diode has a larger breakdown voltage and a lower leakage current, while its turn-on voltage is about 3 V. This higher threshold degrades switching speed and increases power consumption during switching operations. Both structures have their shortcomings. The PN junction's higher turn-on voltage leads to greater forward conduction loss, and the SBD's barrier lowering effect restricts its breakdown voltage. To overcome these limitations, it is of great significance to develop a GaN-based rectifier that combines both the low turn-on voltage of a SBD with the high breakdown voltage of a PN junction. The Junction Barrier Schottky (JBS) diode [8] combines the advantages of both PN junction and SBD structures. This is achieved by fabricating multiple PN junctions within the SBD's epitaxial layer, resulting in a characteristic comb-like structure. Therefore, the JBS diode has both a PN junction and a Schottky contact. During forward conduction, the Schottky contact conducts preferentially, so the forward characteristics of the JBS diode are similar to those of an SBD; During reverse bias, the PN junction depletion region expands to pinch off the current channel and shield the Schottky barrier, thereby suppressing the barrier lowering effect. As a result, its breakdown voltage is higher than that of the SBD structure but slightly lower than that of a PN junction. Therefore, the JBS combines the advantages of SBD and PN junction diodes, resulting in a low forward conduction voltage, a small reverse leakage current, and a high breakdown voltage. The junction barrier diodes with their ability to support large currents, withstand high voltages, and operate at high switching speeds are suitable for automotive electronics, aerospace, medical electronics, and other fields. Many companies and research institutes have conducted quite mature research on the structure and working principle of Si-based and SiC-based JBS, and have successfully launched products. However, there is relatively little research on GaN-based JBS, primarily confined to theoretical and simulation studies.

In 2007, S. Hashimoto et al. [9] from Sumitomo Electric Industries in Japan obtained a GaN SBD device with a breakdown voltage of 580 V by using a free-standing GaN substrate. By further reducing the dislocation density and impurity concentration in the GaN material, the specific on-resistance of the device is only 1.3 mΩ·cm^2. Recently, with the introduction of high-temperature hydride vapor phase epitaxy grown free-standing GaN substrates, the production of 600 V–1.2 kV GaN SBDs will be greatly improved, and it is expected to compete with the current SiC SBDs. Research into GaN Junction Barrier Schottky (JBS) diodes is ongoing, promising significant performance enhancements for GaN power rectifiers in the 600 V to 3.3 kV range. Despite this potential, a key challenge remains: improving the high contact resistance associated with ohmic contacts to p-type GaN [10].

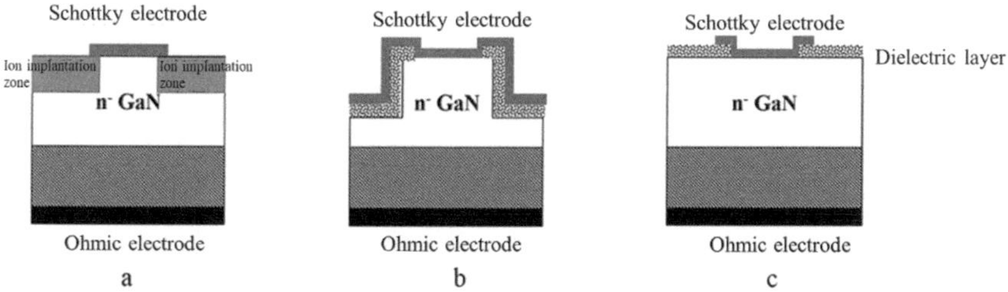

Fig. 8.5 Three typical terminal protection structures. a Ion implantation structure, b Trench structure, c Field plate structure

8.4 GaN MOSFET Devices

Figure 8.6 shows two typical GaN MOSFET structures. In 2008, H. Otake et al. [11] first proposed the vertical trench MOSFET and realized the enhancement mode of operation, as shown in Fig. 8.6a. Compared with the HEMTs, the vertical trench MOSFET does not have an AlGaN/GaN heterostructure, which makes material growth relatively simple. It uses an insulating gate structure, which significantly reduces the gate leakage current, making it more suitable for high-voltage power conversion applications. The enhancement mode GaN-based MOSFET has the advantages of normal off-state and large conduction band offset. It is less affected by issues like hot electron injection, surface state, and current collapse, thus achieving higher reliability and effectively compensating for SiC-based MOSFET's disadvantages in this regard [12].

For the enhancement mode MOSFET device, when the absolute value of the gate voltage is greater than that of the threshold voltage, an inversion layer is formed under the gate, creating a conductive channel between the source and drain, thus turning the device on. When a voltage is applied between the source and drain electrodes, carriers will drift from the source to the drain, forming a drain current. When a strong inversion forms a conductive channel on the semiconductor surface, the channel exhibits resistive characteristics. When the drain current passes through the channel resistance, a voltage drop occurs across the channel. Depending on the different source-drain voltage conditions, the MOSFET device is divided into two operating regions: the linear region and the saturation region. When the source-drain voltage is low, the potential change along the channel direction is small, and the drain current changes linearly with the source-drain voltage, which is called the linear region. At elevated drain–source voltage, the channel is pinched off near the drain, creating a high-resistance region across which the majority of the voltage is dropped. The source-drain voltage drops mainly in the high resistance region, so the electric field between the pinch-off point and the drain is very strong. The carriers in the conductive channel continuously drift from the source to the drain under

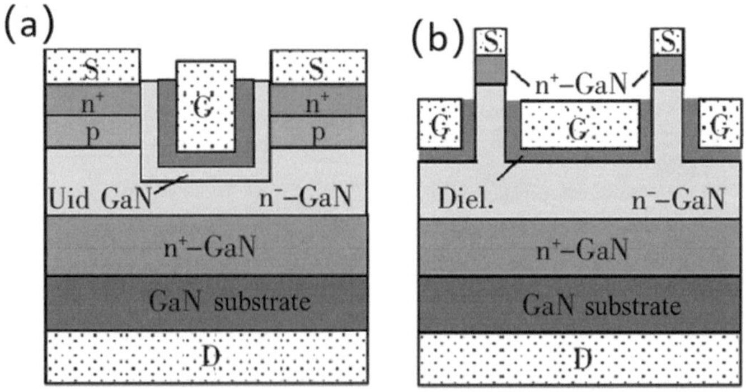

Fig. 8.6 Two typical GaN MOSFET diagrams. **a** Vertical trench MOSFET, **b** Vertical fin MOSFET

the source-drain voltage. When these carriers drift to the pinch-off point, they are immediately swept into the drain region by the strong electric field in the pinch-off region, forming a drain current. As the source-drain voltage increases, the increased voltage drops mainly in the high resistance region, and the drain current remains essentially constant with the increase of the source-drain voltage, which is called the saturation region.

One disadvantage of the vertical trench MOSFET is that the electron mobility in the p-GaN region is relatively low, which makes it challenging for the source electrode to form a good ohmic contact on the buried p-GaN layer. This can result in a negative shift in the threshold voltage under high drain bias. The fin MOSFET does not require a p-GaN layer, fundamentally avoiding the above problems. Additionally, it can achieve normally-off operation, but the process is complex and difficult to implement. M. Sun and others [13] reported a GaN vertical fin MOSFET structure, as shown in Fig. 8.6b. The device channel is completely wrapped by the gate metal, enabling normally-off operation. At $V_{GS} = 0$, due to the work function difference between the gate metal and GaN, the electrons in the channel are depleted. As V_{GS} increases, the depletion region width decreases and a conductive channel forms. The manufactured transistor does not require a p-GaN layer, exhibits a threshold voltage of 1 V, and a specific on-resistance of 0.36 $m\Omega \cdot cm^2$. The breakdown voltage can be further improved by improving the quality of the gate dielectric and adding the edge terminal structure. This approach has led to a breakthrough, resulting in a breakdown voltage of 800 V at a gate bias of 0 V.

In 2014, T. Oka and colleagues [14] from the Japanese Toyoda Gosei Corporation employed field plate technology on an isolated region of a free-standing GaN substrate to reduce the aggregation of the electric field. The device's performance was significantly improved, with its breakdown voltage increased to 1605 V from 775 V (field-free structure) and a specific on-resistance of 12.1 $m\Omega \cdot cm^2$. At the same time, this structure also achieved the normally-off operation of the device, with a threshold voltage of 7 V. However, due to the large device area, its conduction resistance is considerable. In 2020, the Japan's Fuji Electric Company achieved a breakdown voltage of 1200 V and a specific on-resistance of 2.78 $m\Omega \cdot cm^2$ for a MOSFET device on a GaN single crystal substrate through ion implantation. This method does not require the growth of p-type GaN epitaxial layer.

GaN-based HEMT devices have considerable gate leakage current and high power consumption. Furthermore, the channel electron mobility of GaN-based MOSFET devices is low due to the influence of interface states, roughness, and scattering phenomena, resulting in large conduction resistance. In order to further improve the performance of GaN devices, researchers have combined HEMT and MOSFET structures to successfully prepare GaN MOS-HEMT devices, as shown in Fig. 8.7.

GaN MOS-HEMT devices have both the advantages of low gate leakage current of MOS devices and high electron mobility of HEMT devices. These properties suggest a promising future for their application. In 2012, H. S. Lee and others [15] from the Massachusetts Institute of Technology employed the AlGaN back barrier structure to reduce the buffer layer leakage current, and utilized SiO_2 dielectric to reduce the gate leakage current, thereby improving the breakdown voltage of the device.

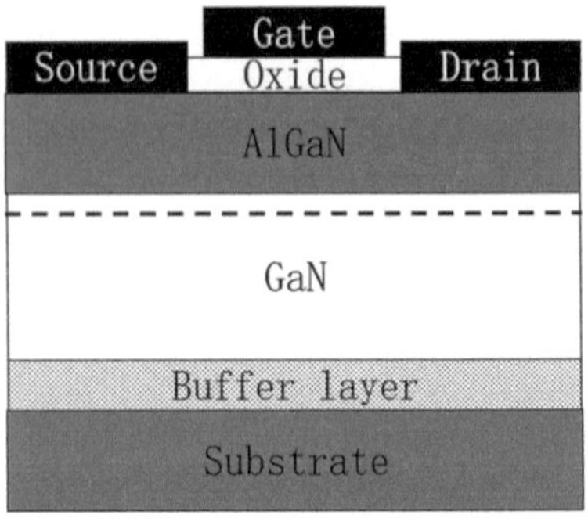

Fig. 8.7 Schematic diagram of the structure of GaN MOS-HEMT device

The obtained InAlN/GaN MOS-HEMT device exhibited a breakdown voltage of up to 3000 V, a specific on-resistance of 4.25 mΩ·cm^2, and a calculated V_{BR}^2/R_{on} of 2117 MW/cm^2.

8.5 GaN-Based HEMT Devices

In the heterojunction formed between GaN and wide bandgap AlGaN (or AlInN), the polarization electric field can significantly modulate the energy band structure and charge distribution. Even if the entire heterostructure is not doped, a high-mobility two-dimensional electron gas (2DEG) can be formed at the AlGaN/GaN interface with a sheet carrier density of up to $1 \times 10^{13} \sim 2 \times 10^{13}$ cm^{-2}. This device is called a high electron mobility transistor (HEMT). The device structure and band structure are shown in Fig. 8.8. The source and drain electrodes form ohmic contact with the 2DEG channel, while the gate electrode forms a Schottky contact with the AlGaN layer. By controlling the relationship between the gate voltage (V_g) and the threshold voltage (V_{TH}), the on-state or off-state of the 2DEG channel under the gate electrode can be realized. It should be noted that the electric field at the edge of the gate electrode is concentrated, which may cause the device to break down prematurely. The working mechanism of AlGaN/GaN HEMT is as follows:

(1) When $V_g > V_{TH}$, the drain electrode is positively biased and the source electrode is grounded. This configuration results in the conduction band of the AlGaN/GaN heterojunction under the gate electrode being below the Fermi level. This implies the presence of a high concentration of 2DEG under the gate electrode, and the source and drain electrodes can be electrically connected through the 2DEG channel, thus establishing that the device is in the on state.
(2) When $V_g < V_{TH}$, the drain electrode is positively biased and the source electrode is grounded. The conduction band of the AlGaN/GaN heterojunction under the gate electrode is above the Fermi level. This indicates that the 2DEG at the AlGaN/GaN heterojunction interface under the gate electrode is depleted, and the device is in the off state.

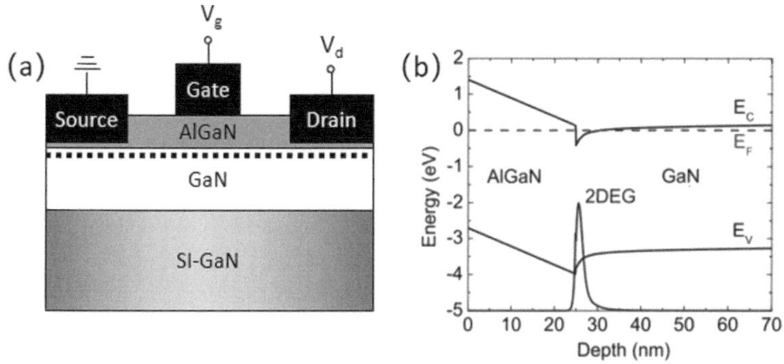

Fig. 8.8 Schematic diagrams of the device structure and band structure of the GaN HEMT device. **a** Schematic diagram of the GaN HEMT device structure, **b** Band diagram and charge distribution of AlGaN/GaN heterojuction

At present, there are still many problems to be solved in various areas of nitride electronic devices, including the basic theory, material defects, device performance and reliability, and device structure.

(1) With regard to fundamental theoretical issues, further research is required in the formation mechanisms and control methods of material defects, the transport properties of two-dimensional electron gas, the surface states/interface states and leakage mechanisms of devices, and device physics. An important issue is the limited research on extending the power output frequency band of GaN-based HEMT to sub-millimeter waves due to the nonlinearity of short-channel devices. This nonlinearity manifests as a rapid decline in the cutoff frequency (f_T) (or transconductance), which drops rapidly with the increase of bias under high drain voltage (high gate voltage). This limits the high-speed operating characteristics of the device under high voltage and results in a maximum current density that is significantly lower than the theoretical prediction. The primary cause of the nonlinearity of drain voltage is the lateral extension of the depletion region under the gate towards the drain electrode under high drain voltage, which increases the effective gate length and gate capacitance, resulting in the phenomenon known as drain delay. The deterioration of gate control ability (transconductance) with drain voltage is more significant in deep sub-micron short-channel devices. To solve this problem, it is essential to begin by optimizing the electric field distribution of the depletion region, ensuring device reliability under proportional longitudinal size reduction. The mechanism of gate voltage nonlinearity of f_T (or transconductance) of GaN-based HEMT is still controversial. It is commonly accepted that this nonlinearity originates from the reduction of the effective speed of electrons under high gate voltage. This reduction can be attributed to a number of factors, including the enhancement of interface scattering by longitudinal electric field, the degradation of electron transport due to self-heating under large current, the emission of optical phonons

under large current, and the increase in gate-to-source channel resistance under large current, etc.

(2) The defect density of nitride materials still requires significant reduction. The current threading dislocation density of nitride heteroepitaxial materials is still on the order of 10^8–10^9 cm^{-2}, which is considerably higher than that of Si (less than 100 cm^{-2}) and GaAs (10^4–10^5 cm^{-2}) semiconductor materials. Defects in nitride materials limit their thermal conductivity, electron saturation velocity, and other properties, and further deteriorate the performance, reliability, and yield of GaN electronic devices. A prominent issue is that threading screw dislocations (which account for approximately 10% of the total dislocation density) can conduct electricity, and thus serve as leakage paths in devices. In particular, the screw dislocations in the high field region near the gate and drain of the device can result in a significant gate leakage current even in the initial state. The performance degradation of microwave power devices is closely related to the magnitude of the initial gate leakage. Therefore, it is crucial for power electronic devices to strictly control the leakage problem. In recent years, the technology of preparing GaN substrates by ammonothermal method and hydride vapor phase epitaxy (HVPE) method has developed rapidly. It is now possible to prepare GaN substrates with a diameter of 2 to 4 inches and a dislocation density on the order of 10^4 cm^{-2} to 10^5 cm^{-2}. In light of these developments, the pursuit of homoepitaxy technology for the preparation of GaN materials with an extremely low dislocation density has emerged as a novel research direction.

(3) Thermal dissipation remains a significant challenge limiting the performance of GaN microwave power devices. The high power density advantage of GaN is difficult to maintain in high power devices, manifesting as a rapid decrease in output power density with increasing gate width. This is because the gate of high power devices is a multi-finger gate structure, and the dense channel exhibits a significant self-heating effect during operation, resulting in a rapid rise in junction temperature. Even SiC substrates present significant challenges in meeting the heat dissipation requirements, severely limiting the output power density of the device. As the operating frequency of the device increases, its size is proportionally reduced, resulting in lower breakdown voltage and efficiency, while the high junction temperature causes significant degradation in carrier transport characteristics. This ultimately results in the continuous loss of the high power density advantage of GaN. Improving the heat dissipation problem can start from improving the thermal conductivity of the substrate. Currently, there is a notable technical trend towards the utilisation of diamond substrates to replace SiC substrates. The thermal conductivity of single crystal diamond is as high as 2200 W/(m K), and the thermal conductivity of polycrystalline diamond deposited by CVD method can also reach 1200–1500 W/(m K), which can greatly improve the heat dissipation ability of the substrate (the thermal conductivity of 4H-SiC is 490 W/(m K)). The mainstream approach involves bonding GaN heterojunction epitaxial wafers onto CVD diamond substrates using chip bonding techniques. The device prepared by this technology has achieved a wafer size of 4 inches and RF power characteristics of more than

7 W/mm@10 GHz, with the power per unit area reaching more than three times that of devices on SiC substrates. Another measure to address the heat dissipation problem is to improve heat removal from the device surface. One potential solution is to replace the SiN passivation layer with an AlN layer that has higher thermal conductivity.

(4) The reliability of GaN devices is currently insufficient due to various factors. The two-dimensional electron gas in GaN heterojunctions is highly sensitive to the piezoelectric polarization strength of the material (which is determined by the strain state) and surface conditions (including cleanliness, oxidation degree, traps, and surface states, among other factors). Therefore, the instability of strain and surface conditions represents an inherent factor limiting the reliability of GaN devices. Secondly, the deterioration of GaN microwave power devices under operational conditions is primarily attributable to two significant external factors: electric field and junction temperature. In the presence of a strong electric field, nitride materials exhibit an inverse piezoelectric effect, manifested as additional tensile strain in AlGaN caused by the electric field, which superposes with the original tensile strain of AlGaN itself. Thus, when the voltage reaches a specific critical voltage value, cracks and other defects may emerge in the high field region near the drain side of the gate. In the presence of a strong electric field, a considerable number of hot electrons can undergo real-space transfer, resulting in a reduction in the channel electron density and a decline in the device saturation current and transconductance. Additionally, hot electrons can also collide with the lattice, leading to the generation of new defects, which further exacerbates the device's degradation. In high power working conditions, the junction temperature in GaN-based HEMT devices (especially those with a large gate width devices) rises, severely limiting the output power density and reliability of the device. The severity of this problem lies in the fact that high junction temperatures reduce the thermal conductivity of the heated substrate, further increasing the junction temperature and severely deteriorating the reliability of the device. Furthermore, the presence of high-density dislocation defects and large residual stress in the material are parasitic factors that affect the reliability of GaN devices. Improving GaN device reliability is a significant challenge, requiring a comprehensive approach including reducing material defect density, enhancing process stability and repeatability, optimizing electric field distribution, and improving heat dissipation.

(5) Current research on nitride electronic devices, encompassing both microwave power and power electronic devices, is primarily focused on lateral high-electron-mobility transistor (HEMT) structures due to their relative simplicity. In contrast, alternative architectures have received considerably less attention. This research bias stems from the prevailing heteroepitaxial growth methods used for nitride semiconductors. This approach offers two key advantages: first, it facilitates the epitaxial growth of multilayer heterostructures essential for realizing the field effect principle; second, it circumvents the detrimental impact of high threading dislocation density (TDD) inherent in heteroepitaxial materials on device performance. In vertical structures, these dislocations would severely

degrade critical characteristics such as breakdown voltage. However, with the rapid advancement of GaN homoepitaxy technology, which enables the production of low-dislocation-density native substrates, vertical device architectures are emerging as a promising and transformative new research direction. Which are relatively simple. In contrast, other structures have been less studied. This is because nitride materials are predominantly prepared by heteroepitaxy methods. On the one hand, it is easier to prepare multilayer heterojunctions to realize the field effect principle. On the other hand, the threading dislocation density of the material is high, the preparation of vertical devices will significantly affect the breakdown and other characteristics of the device. With the rapid progress of homoepitaxy technology today, vertical structure devices are set to become a new research direction.

(6) Material epitaxy and subsequent cooling cause wafer warpage, which severely limits the yield of large-diameter wafers. In order to mitigate the warpage of 4-inch nitride epitaxial wafers on Si substrate and thus achieve a higher yield, it is necessary to increase the substrate thickness to more than 1 mm, which significantly increases the cost of material. The epitaxial growth of high quality single crystal GaN on 2D material graphene was realized by the IBM T. J. Watson Research Center in the United States, which can conveniently transfer the GaN thin film to any substrates, potentially bypassing the warpage issue altogether and presenting a vital technology for lowering the production costs of nitride electronic devices [16].

8.6 GaN-Based CAVET Devices

By combining the high-concentration 2DEG of lateral structure AlGaN/GaN HEMTs with the advantages of vertical GaN-based SBD and PN structures, a novel vertical structure device is obtained, namely current aperture vertical electron transistors (CAVET). The typical device structure is shown in Fig. 8.9. CAVET was first proposed by Y. I. Ben et al. in the United States [17]. They used Mg-doped p-GaN as the current blocking layer to make a vertical structure AlGaN/GaN CAVET, transferring the high surface electric field to the internal of the device to avoid current collapse.

The conventional CAVET device structure consists mainly of a light-doped n-GaN buffer layer, a current aperture, and a current blocking layer (CBL), as shown in Fig. 8.9a [18]. When the device is in the operation state, electrons flow from the source at the top of the device, through the two-dimensional electron gas channel to the current aperture, and into the buffer layer, where the CBL prevents the electrons from passing through. When the device is in the off state, its operation is similar to that of a lateral AlGaN/GaN HEMTs. When the gate voltage is applied to deplete the 2DEG, the n-GaN buffer layer acts as a voltage-resisting layer. Compared with the lateral structure device, the drain of the vertical structure device is on the substrate, and the electric field is in the vertical direction instead of the horizontal direction. This not

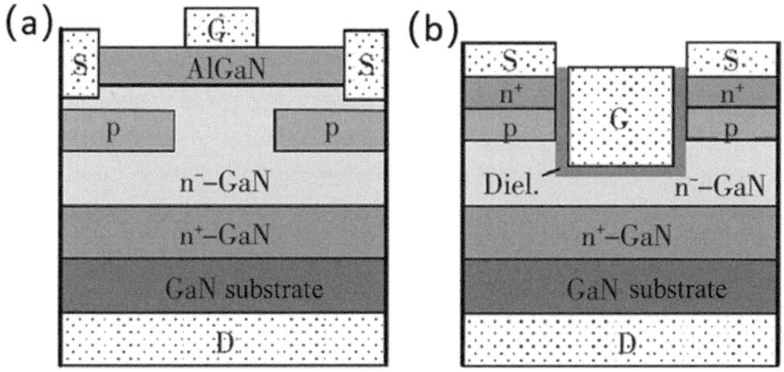

Fig. 8.9 Schematic diagram of two typical GaN CAVET device structures. **a** conventional CAVET, **b** trench CAVET

only passivates surface states to suppress current collapse but also permits a thicker lightly-doped GaN drift region to realize high breakdown voltage. For lateral devices, high voltage resistance is generally achieved by increasing the lateral size, which greatly increases the chip area and contradicts the requirement for miniaturization.

There are three key issues with GaN CAVET devices:

(1) Preparation of high-quality CBL. The CBL in CAVET is crucial for achieving high blocking voltage, so how to fabricate high-quality CBL has always been the focus of device development. CBL can be made of materials with good insulation properties such as SiN_x, SiO_2, but it requires secondary epitaxial growth, complicates the process, and may result in poor device off-state characteristics and increased leakage current. Typically, a p-type GaN layer obtained by epitaxial growth or ion implantation is used as the CBL. Mg is commonly used to achieve p-type doping of GaN, but the low activation efficiency and memory effect of Mg remain challenging. In addition, in this case, the aperture region must be dry etched to remove the p-GaN layer and grow the n-GaN channel layer. It has been reported [17] that the etching of the aperture region exposes non-c-plane GaN, and epitaxial growth on these facets can lead to non-planar morphology, resulting in severe gate leakage current. In addition, high concentrations of n-type impurities tend to penetrate the aperture region. This leads to an increase in the peak electric field in the channel and a decrease in the breakdown voltage in the device.
(2) Negative threshold voltage. D. Shibata and others [19] proposed a novel vertical GaN transistor based on the trench CAVET structure, which is characterized by the regrowth of the p-type gate/AlGaN/GaN triple layer on the V-groove, as shown in Fig. 8.9b. The results show that the polarization charge at the AlGaN/GaN heterojunction decreases with the increase of the angle between the heterojunction and c-plane. By implementing the trench CAVET structure, a positive high threshold voltage can be achieved.
(3) Current leakage. In conventional CAVET and trench CAVET structures, the punch-through current caused by current leakage in the CBL limits the high

voltage operation of the device, which is mainly due to the low activation efficiency of Mg doping in the p-type layer. By depositing a thin C-doped GaN layer over the p-GaN, the leakage current in the p-GaN layer can be reduced [20]. The C-GaN/p-GaN layer as a hybrid blocking layer (HBL), which can suppress the punch-through effect and increase the device breakdown voltage from 580 V to 1.7 kV.

In 2015, R. Yeluri and others [20] have fabricated vertical GaN current aperture transistors on GaN substrates. Compared with the traditional p-GaN current blocking layer formed by Mg implantation, they used MOCVD to grow Mg-doped p-GaN as the CBL, and then used the ICP method to etch a narrow trench. By connecting the source electrode with the p-GaN current blocking layer, the source electrode and the p-GaN layer can be annealed simultaneously, which not only activates the Mg impurities, but also reduces the number of annealing times and avoids the high-temperature redistribution problem of Mg ions in the ion implantation process. By designing the net doping concentration of the drift layer to be 1×10^{16} cm^{-3} and the thickness of the drift layer to be 15 μm, the breakdown voltage of the device is improved. The thickness and doping concentration of the p-GaN layer under the gate are adjusted to achieve normally-off operation state of the device. The terminal structure can effectively suppress electric field crowding at the junction edge, and its fabrication method is similar to that of power diodes. Based on this structure, a device with a breakdown voltage of 1.5 kV and a specific on-resistance of 2.2 mΩ·cm^2 was obtained.

8.7 Development Trends

Nitride electronic devices have made great progress in performance in the last 20 years. Internationally, GaN microwave power devices and power electronic devices have begun to enter the application field. In China, GaN microwave power devices have also begun to enter the application stage, and the pace has significantly accelerated in recent years. However, the problems in this field are still prominent, such as material defects, reliability and yield rate. The related mechanisms of GaN electronic devices and materials still need to be studied in depth.

(1) Improve the breakdown voltage. Theoretically, under the same breakdown voltage, the on-resistance of GaN power devices is lower than that of Si and SiC power devices, but its performance is far from the theoretical value. Research has found that the main reason is source-drain vertical punch-through in the GaN buffer layer, resulting in a leakage along the interface between the Si substrate and the GaN buffer layer. Therefore, the current plan to improve the device breakdown voltage mainly focuses on the following three directions: (1) optimizing the substrate structure; (2) enhancing the buffer layer design; (3) improving the device architecture.

(2) Realize enhancement mode (normally-off) devices. AlGaN/GaN-based devices are inherently depletion mode (normally-on) devices, while enhancement mode (normally-off) power devices with positive threshold voltage can ensure system safety and reduce system cost and complexity. The enhancement mode power devices are the preferred devices in power systems. Therefore, for GaN power devices, the realization of enhancement mode devices is also a problem of great concern to researchers. At present, methods such as grooved-gate, p-GaN gate, and fluorine ion implantation are used to directly realize enhancement mode. In addition, cascode technology is also used to indirectly realize normally-off devices.
(3) Suppress the current collapse effect. The methods for suppressing current collapse are mainly as follows: (a) Surface passivation. The problem concern with surface passivation is that the passivation process is relatively complex with limited repeatability, and thus it cannot completely eliminate the current collapse effect. It affects the gate leakage current and cut-off frequency of the device, and increases the heat dissipation problem of the device. (b) Field plate. In 2011, HRL in the United States used a three-field plate structure (one gate field plate and two source field plates) combined with SiN passivation to achieve a high-voltage resistant, low dynamic resistance Si-based GaN transistor. The ratio of dynamic to static R_{on} ratio of the device was 1.2 at 350 V and 1.6 at 600 V [21]. (c) Growing a cap layer, such as a p-type GaN cap layer, to isolate the host impurities to form a negative space charge layer, which shields the channel from surface potential fluctuations. This method is relatively simple and easy to control in the material growth process, but it increases the process difficulty, such as the gate fabrication process. (d) Barrier layer doping. This method increases the channel electron concentration, or reduces the surface state density of the barrier layer. Generally, this type of device often incorporate a thin layer of undoped GaN or AlGaN cap layer.
(4) Manufacturing process. The manufacturing process for GaN power device is compatible with the existing Si process, which is an important factor driving their industrialization and widespread adoption. The key to developing a GaN power device manufacturing process that is compatible with the existing Si process is to develop a gold-free process. In 2012, at the ISPSD annual conference, IMEC reported the enhancement-type GaN power transistor fabricated on an 8-inch GaN-on-Si wafer using a CMOS-compatible gold-free process combined with a grooved-gate process. IMB-CNM-CSIC reported the MIS-HEMT and i-HEMT fabricated on a 4-inch Si wafer using the CMOS-compatible gold-free process. The development of a gold-free process has received great attention from academia and industry in recent years, and it is an important way to reduce costs for mass production and large-scale commercial applications.
(5) Power integration technology represents a pivotal development direction for GaN power devices. It involves the creation of independent, fully-functional modules tailored for direct end-use applications. These modules integrate GaN power core devices, drivers, protection circuits, and peripheral passive components into a single system. This high level of integration, often termed "smart

power" technology, enables performance benchmarks—such as high efficiency, operational safety, switching speed, and high-temperature resilience—that are unattainable with traditional silicon-based power ICs. Driven by advancements in core GaN device technology, power integration has emerged as a major research focus within the wider GaN ecosystem in recent years.

In 2021, the Korea Institute of Energy Technology reported a 32 × 32 pixel high-power flip-chip blue Micro-LED-on-HEMT device for submarine optical communication [22]. In 2022, Prof. Mao Junfa's group at Shanghai Jiao Tong University has developed a low-loss, heterogeneously integrated W-band millimeter-wave radar system [23]. This system employs a novel heterostructure architecture to three-dimensionally integrate diverse components, including a Si-based phase-locked loop (PLL) chip, a SiGe-based transceiver chip, a GaN power amplifier (PA), packaged antennas, and capacitors. The reported radar demonstrates outstanding performance: it achieves a detection range exceeding 800 meters, a range resolution finer than 0.08 meters, and a total weight of merely 78 grams. In addition, Professor Chen Jing's group at the Hong Kong University of Science and Technology demonstrated monolithic GaN-based complementary logic integrated circuits by co-integrating enhancement-mode n-channel and p-channel field-effect transistors [24].

(6) Reliability. With the continuous progress of various device technologies, GaN devices have gradually moved from the laboratory to the industry. Consequently, reliability has become a critical concern. Compared with Si power device technology, the research on the reliability and stability of GaN power devices remains relatively underdeveloped. Although studies have been conducted on degradation mechanisms, failure modes, and strategies for reliability enhancement, these efforts are still insufficient to support the stringent requirements for large-scale practical deployment.

The factors affecting the reliability of GaN power devices are quite complex, including material quality, device structure, and device process and others. Based on the operating mode characteristics and operating environment of power devices, the main focus of GaN power device reliability research includes the following points: (a) gate leakage current and surface state; (b) gate metal degradation; (c) thermoelectric/thermoacoustic effects under high electric field and high temperature; (d) material quality.

(7) A succession of mature products is being launched. With the reduction in the cost of GaN power devices, improvement in their electrical characteristics, and the expansion of the peripheral technology of GaN power devices, the environment for the use of GaN power devices is rapidly taking shape. Since the latter half of 2011, numerous companies have initiated the development of comprehensive GaN product series and commenced volume supply. In parallel, the peripheral technologies essential for utilizing these devices have become increasingly sophisticated, further accelerating market penetration.

Although there is still a discrepancy between the actual performance and theoretical performance of GaN power devices, the test results of current devices and their functional circuits indicate that they already exhibit an obvious performance advantage compared to traditional Si technology. As the material quality, device technology, power integration technology, and reliability of GaN power devices continue to improve, it is likely that GaN power devices will replace Si power devices and become the preferred technology solution in power electronics applications.

References

1. Xu K, Wang JF, Ren GQ (2015) Progress in bulk GaN growth. Chin Phys B 24(6):066105
2. Fu HQ, Huang X Q, Chen H et al (2017) Effect of buffer layer design on vertical GaN-on-GaN p-n and schottky power diodes. IEEE Electron Device Lett 38(6):763–766
3. Kizilyalli IC, Edwards AP, Aktas O et al (2015) Vertical power p-n diodes based on bulk GaN. IEEE Trans Electron Devices 62(2):414–422
4. Nomoto K, Song B, Hu ZY et al (2016) 1.7-kV and 0.55-m Omega. cm(2) GaN p-n Diodes on Bulk GaN Substrates with Avalanche Capability. IEEE Electron Device Lett 37(2):161–164
5. Ohta H, Nakamura T, Mishima T (2016) High quality free-standing GaN substrates and their application to high breakdown voltage GaN p-n diodes. IEEE International Meeting for Future of Electron Devices, Kansai (IMFEDK), 2016 Jun 23–24, Ryukoku Univ, Kyoto, Japan, 2016:90–91
6. Kizilyalli I C, Prunty T, Aktas O (2015) 4-kV and 2.8-m Omega-cm(2) Vertical GaN p-n Diodes with Low Leakage Currents. IEEE Electron Device Lett 36(10):1073–1075
7. Kone S, Cayrel F, Yvon A et al (2016) DLTS analysis of high resistive edge termination technique-induced defects in GaN-based schottky barrier diodes. Physica Status Solidi a-Applications and Materials Science, 213(9):2364–2370
8. Radhakrishnan R, Zhao JH (2011) A 2-dimensional fully analytical model for design of high voltage junction barrier Schottky (JBS) diodes. Solid-State Electronics, 63(1):167–176
9. Hashimoto S, Yoshizumi Y, Tanabe T et al (2007) High-purity GaN epitaxial layers for power devices on low-dislocationdensity GaN substrates. J Cryst Growth 298:871–874
10. Placidi M, Perez-Tomas A, Constant A et al (2009) Effects of cap layer on ohmic Ti/Al contacts to Si+ implanted GaN. Appl Surf Sci 255(12):6057–6060
11. Otake H, Chikamatsu K, Yamaguchi A et al (2008) Vertical GaN-based trench gate metal oxide semiconductor fieldeffect transistors on GaN bulk substrates. Applied Physics Express 1(1):011105
12. Huang W, Chow TP, Niiyama Y et al (2008) Lateral implanted RESURF GaN MOSFETs with BV up to 2.5 kV. 2008 20th International Symposium on Power Semiconductor Devices and ICs, 2008 May 18–22, Orlando, FL, 2008: 291
13. Sun M, Zhang Y, Gao X et al (2017) High-performance GaN vertical fin power transistors on bulk GaN substrates. IEEE Electron Device Lett 38(4):509–512
14. Oka T, Ueno Y, Ina T et al (2014) Vertical GaN-based trench metal oxide semiconductor field-effect transistors on a free-standing GaN substrate with blocking voltage of 1.6 kV. Applied Physics Express 7(2):021002
15. Lee HS, Piedra D, Sun M et al (2012) 3000-V 4.3-Omega.cm(2) InAlN/GaN MOSHEMTs With AlGaN Back Barrier. IEEE Electron Device Lett 33(7):982–984
16. Kim J, Bayram C, Park H et al (2014) Principle of direct van der Waals epitaxy of single-crystalline films on epitaxial graphene. Nature Communications, 5:4836
17. Ben YI, Seck YK, Mishra UK et al (2004) AlGaN/GaN current aperture vertical electron transistors with regrown channels. J Appl Phys 95(4):2073–2078

18. Chowdhury S, Swenson BL, Mishra UK (2008) Enhancement and depletion mode AlGaN/GaN CAVET with Mg-ionimplanted GaN as current blocking layer. IEEE Electron Device Lett 29(6):543–545
19. Shibata D, Kajitani R, Ogawa M et al (2016) 1.7 kV/1.0 m Omega cm(2) normally-off vertical GaN transistor on GaN substrate with regrown p-GaN/AlGaN/GaN semipolar gate structure. 62nd Annual IEEE International Electron Devices Meeting (IEDM), IEEE International Electron Devices Meeting, 2016 Dec 03–07, San Francisco, CA, 248–251
20. Yeluri R, Lu J, Hurni CA et al (2015) Design, fabrication, and performance analysis of GaN vertical electron transistors with a buried p/n junction. Appl Phys Lett 106(18):183502
21. Chu RM, Corrion A, Chen M et al (2011) 1200-V Normally Off GaN-on-Si Field-Effect Transistors With Low Dynamic ON-Resistance. IEEE Electron Device Lett 32(5):632–634
22. Kim TK, Islam A MH, Cha YJ et al (2021) 32 × 32 Pixelated High-Power Flip-Chip Blue Micro-LED-on-HFET Arrays for Submarine Optical Communication. Nanomaterials 11(11):3045
23. Yang X, Huang YS, Zhou L et al (2022) Low-loss heterogeneous integrations with high output power radar applications at W-band. IEEE J Solid-State Circuits 57(6):1563–1577
24. Zheng ZY, Zhang L, Song WJ et al (2021) Gallium nitride-based complementary logic integrated circuits. Nature Electronics 4(8):595–603

MIX
Papier aus verantwortungsvollen Quellen
Paper from responsible sources
FSC® C105338

If you have any concerns about our products,
you can contact us on
ProductSafety@springernature.com

In case Publisher is established outside the EU,
the EU authorized representative is:
Springer Nature Customer Service Center GmbH
Europaplatz 3, 69115 Heidelberg, Germany

Printed by Libri Plureos GmbH
in Hamburg, Germany